HERMANN HISTORICA OHG

International Auctioneers
of Antique Arms and Armour, Orders, Militaria and Historical Collectibles

Imperial German Garde du Corps troopers helmet, M 1894.

For further information or to order a catalogue please contact

HERMANN HISTORICA OHG Sandstr.33 D-80335 Munich
Phone +49-89-5237296   Fax +49-89-5237103

Historex Agents

Verlinden Productions
Elite Miniatures Spain
Fonderie Miniatures
Soldiers Lucchetti
Andrea Miniatures
Fort Royal Review
G & M Miniatures
Mini Art Studios
Windrow & Greene
Poste Militaire
Friulmodelismo
Wolf Miniatures
New Connection
Metal Models
Hecker & Goros
Mascot Models
Trophy Models
Quardi Concept
Show Modelling
Tiny Troopers
White Models
Fort Dusquesne
Hornet Models
Model Cellar
Nuts & Bolts
Shenandoah
Grandt Line
Wild West
Le Cimier
Sovereign
Historex
Puchala
Almond
Nemrod
Ironside
Mil–Art
Pegaso
Pili Pili
Starlight
Tomker
Beneito
Azimut
Decima
Glory
Aber

HISTOREX AGENTS

Wellington House, 157 Snargate Street, Dover, Kent CT17 9BZ Tel: 01304 206720 Fax: 01304 204528

BEWARE OF MEN IN DRESSES

THERE COMES A TIME IN A MAN'S LIFE WHEN HE NEEDS TO WEAR A DRESS. THIS, FOR REGIA MEMBERS OCCURS AT LEAST ONCE A MONTH THROUGHOUT THE YEAR. THEY WEAR THEIR KIT PROUDLY INTO BATTLE; ON HORSEBACK; IN THEIR SHIPS; BY THE GLIMMERING CAMPFIRE AND MOST IMPORTANTLY IN FRONT OF THE PUBLIC.

REGIA ANGLORUM RECREATES LIFE IN THE 100 YEARS PRIOR TO THE NORMAN CONQUEST, WHEN SAXONS MINGLED WITH MANY DIFFERENT PEOPLES IN NORTHERN EUROPE. WE STRIVE TO DO THIS AS AUTHENTICALLY AS IS POSSIBLE, WHICH MEANS WE SEW OUR KIRTLES - AS WE CALL OUR 'DRESSES' - TOGETHER BY HAND.

WHETHER YOU ARE THE KIND OF PERSON WHO ENJOYS JUST WEARING ARMOUR AND BEING IN THE THICK OF THE BATTLE; OR SAILING IN OUR HISTORICAL VESSELS; RIDING ON HORSEBACK, OR SIMPLY REDISCOVERING FORGOTTEN CRAFTS, THEN REGIA IS THE PLACE FOR YOU. IF YOU WANT TO JOIN US, OR HIRE US THEN GIVE KIM A CALL ON 01179 646 818. AND INCIDENTALLY, WOMEN ARE ALLOWED TO WEAR DRESSES TOO.

Regia Anglorum

Head Office, 9 Durleigh Close, Headley Park, Bristol, BS13 7NQ. Tel. 01179 646 818

E-Mail us on 101364.35@compuserve.com or see our Web Site at http://www.ftech.net/~regia

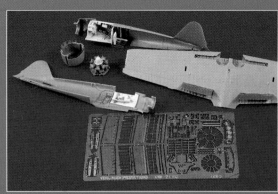

1289 A6M2b Zero Type 21 Detail Set. 1:48

**1292
"Spoils of War" WWII GI's Ardennes 1944
Vignette 1:35**

NOVEMBER
1997
RELEASES

1294   Panzer IV Engine & Compartment (for Tamiya Pz.IV's)   1:35

**1293
USAAF Fighter Pilot Europe WWII
200mm Bust**

1296 WWII Salvage Diver   120mm

**1290
Cuirassier General
120mm**

1295 US Navy WWII Flight Deck Crew 1:48

**1291
German WWII Mounted Food Carrier 1:35**

**UK Distributor
HISTOREX AGENTS**
Wellington House
157 Snargate Street
Dover, Kent, CT17 9BZ
Tel.: 01304/206720
Fax.: 01304/204528

**US Distributor
The V.L.S. Corp.**
Lone Star Industrial Park
811 Lone Star Drive
O'Fallon, MO 63366
Tel.: 314/281 5700
Fax.: 314/281 5750

Windrow & Greene's
MILITARIA DIRECTORY
AND
SOURCEBOOK

Windrow
&
Greene PUBLISHING

© 1997 Windrow & Greene Ltd

This edition published in Great Britain 1997 by
Windrow & Greene Ltd
5 Gerrard Street
London W1V 7LJ

A CIP record for this book is available from the British Library

ISBN 1 85915 068 3

Printed and bound in Great Britain by
Hillman Printers through Amon-Re Ltd

The descriptive text entries in this Directory have been supplied by the
companies and individuals listed. Whilst we try to ensure that the factual
details are correct at the time of going to press, we apologise in advance for
any changes which may occur during the life on this edition. Suggestions,
amendments or applications for inclusion in future editions are always
welcome.

Advertising agents for this edition:
Boland Advertising
1 Middle Row, High Street, Ashford, Kent TN24 8SQ
Tel: 01233 635216 Fax: 01233 635199

Designed and produced by:
Hodgson Williams Associates
71 Bridge Street, Cambridge CB2 1UR

CONTENTS

INTRODUCTION TO 1998 EDITION

by Ken Jones, Editor of *Military Modelling* magazine

If I had a penny for every time a telephone caller to the *Military Modelling* editorial office has asked, "I wonder if you could help me with an address for...?", then I wouldn't be writing this introduction – I'ld be lying on a beach somewhere ... well, you know the rest.

This constant foraging by our readers for information and contacts of all kinds has been an almost daily phenomenon ever since I took over the editorship of *Military Modelling* in 1979. The first edition of *Windrow & Green's Militaria Directory & Sourcebook* appeared as long ago as 1992; but still the calls keep on coming – the directory's up-dating and re-appearance in 1993, 1994 and 1996 has not eased my burden one jot.

It is well known amongst information-seekers that I am normally the last port of call; previous contacts further down the line, eager to alleviate themselves of telephone enquiries requiring a certain amount of response time, take the easy way out by saying "Oh, ring *Mil Mod*, they'll know". How I wish they would say, "Why don't you buy yourself a copy of *Windrow & Greene's Militaria Directory & Sourcebook*"? (In dark moments I even suspect Windrow & Greene themselves of passing on "customers'" to me; then they have the nerve to talk me into writing this introduction! (Guilty – Ed.)

Shortening the Directory's long, if necessarily descriptive title to simply "*Windrow & Greene's*" is a very British trait, and this makes it sound like a name you'ld find on the side of a packet of biscuits. In our office this book is a good deal more useful than a biscuit; it's a bible, which we reach for automatically when answering many of the day-to-day enquiries which pursue us. The Directory is also very easy to use, with its multitude of contact addresses and telephone numbers divided into generally logical chapters and, in recent editions, thoroughly indexed and cross-indexed.

However, it seems that there are still some who fail to appreciate the existence of "*Windrow & Greene's*", which is to the military hobby what Yellow Pages is to the rest of the phone-using public. Why are so many people still ignorant of this splendid book? It has been advertised in the model press and elsewhere, and is readily available on open sale. Our well-thumbed office copy is constantly in use in the editorial and advertising departments alike, and I for one am very grateful to the publishers for producing it in the first place.

A little thought on the practicalities of compiling and republishing this book at regular intervals will soon suggest that it is no mean task. Granted, once you have the data base up and running it may seem to be plain sailing; but ask anyone who has to keep even a small, simple list of anything up to date how easy it is – and then multiply by a thousand. At the best of times it is difficult to keep on top of it; even without the occasional treachery of computers (which can sadistically decide to reduce hundreds of thousands of keystrokes to alphabet soup before your appalled eyes), such lists soon become a full-time job and, by their very nature, they become out of date almost as soon as they are published.

On the positive side, this Directory has to get bigger as increasing numbers of entries from overseas countries are included. (To make a factual comparison, about a third of *Military Modelling* magazine's monthly circulation goes overseas in subscriptions and news stand sales; and this latest edition of the Directory now includes 22 per cent of non-British entries, from countries all over Europe and, in a few cases, from the USA, Canada and Australasia.) It should not be too long, it seems, before Windrow & Greene can legitimately add "International" to the title.

*   *   *

Naturally, I have seen the military modelling hobby change over the past 40 or so years that I have been involved with it. The changes include not only the expected temporal shifts induced by trends or transformation of interest; as these come and go the standards demanded and achieved have also changed for the better. The thirst for knowledge has also increased a hundredfold – though I sometimes think that this has not always been for the better. What once was "near enough is good enough, and didn't we have a lot of fun doing it"? has sometimes given way to obsessive "rivet-counting". As the editor of the major magazine serving the hobby, with a lively correspondence column, I have watched with some bemusement the birth, relentless progress, and eventual exhausted collapse of various ill-founded "crusades", which ultimately become bogged down from attempts to prove singular points so insignificant that they are virtually pointless. Hobbies should be fun – a fact, it appears, that is now too easily forgotten.

The chapters now found in the Directory have come to reflect the make-up of the whole broad spectrum of a hobby whose core interest was once based quite narrowly upon the modelling of miniature figures and vehicles. Around this core have grown the interests in re-enactment and "living history", full size vehicle restoration, wargaming, toy soldier collecting, uniforms, weapons, insignia – and each aspect has now spun off supporting interests in books, paintings, prints and postcards, museums ... the list is, so far, endless.

Editing *Military Modelling* magazine for nearly twenty years has put me in a unique position to notice and react to these trends as they rise and fall. At the time of writing the miniature armoured vehicle is the supreme modelling interest, reborn after a lull following its popular establishment in the 1970s. The major kit manufacturers have been quick to redress an imbalance during which the "cottage industries" took the leading innovative role; now the big battalions are making new military vehicle kits once again – thus flooding the market, resulting in the present decline in sales. There are simply too many kits on the market for all to sell in equally large numbers.

Interest in miniature figure modelling has declined for the time being in favour of vehicles; but like all other trends this will probably prove temporary. The hard core of dedicated figure modelling aficionados will inevitably see, once again, an increase in numbers. Although a trip to the annual military modelling exhibition and competition – Euro Militaire, at Folkestone late each September – may seem to cast doubt on this statement, trends abound in this hobby; there is always a search for something different, even though it may not be apparent in the range of work to be seen in a single year. In particular, the interest of those modellers who combine vehicles and figures – the diorama builders – never flags, and competition in the creation of scenic settings for models flourishes.

Consider also the crossover interests: to generalise, many wargamers make models, and also dress as re-enactors, as do military vehicle restorers. The collectors of prints and postcards paint miniature military figures – as do uniform collectors. Those following narrowly specific interests really are in the minority. You can't paint a model figure properly without consulting published works – books and magazines – and studying available illustrations. Neither can you "re-enact" without the correct clothing and weapons – and where do you find out about it all? You could be one of the multitude who ring me; or you could buy this Directory! It's all here; and, especially if you're professionally involved, the book will soon pay for itself many times over.

I am pleased that Windrow & Greene have revised and updated the Directory and that it has made it to the bookstands once again – not least, because our office copy is falling to bits! Not because of ill use, I hasten to add, but – like a pair of comfortable old slippers – simply due to the wear and tear from constant use. I commend it to you.

Ken Jones
Military Modelling
Hemel Hempstead, 1997

AUCTION HOUSES

Berliner Auktionhaus fur Geschichte
Motzstrasse 22, 10777 Berlin, Germany
Tel/fax: (0049) 302119538

Bonhams
Montpelier Street, Knightsbridge, London SW7 1HH
Contact: Angus Barnes
Tel: 0171 393 3947
Fax: 0171 393 3905

Bonhams auctioneers hold regular sales of militaria, medals, antique guns, arms and armour. Free valuations without obligation. Collection easily arranged, confidentiality assured.

Bosley's Military Auctioneers
42 West Street, Marlow, Buckinghamshire SL7 2NB
Contact: Bernard Pass
Tel: 01628 488188
Fax: 01628 488111

6 regular sales per annum. Each over 1000 lots of original badges, medals, books, headress, uniforms, swords, avaiation etc. Annual subscription £50 (£60 airmail).

Buckland Dix & Wood
1 Old Bond Street, London W1X 3TD
Tel: 0171 499 5022
Fax: 0171 499 5023

Specialist auctions of orders, decorations and war medals.

Christie's
85 Old Brompton Road, London SW7 3LD
Tel/fax: 0171 581 7611

Head Office: 8 King Street St James's London SW1 Tel: 0171 839 9060
City Office: 50-60 Gresham Street London EC2 Tel: 0171 588 4424

D Delavenne / D Lafarge
26 Rue Bergere, 75009 Paris, France
Tel: (0033) 147704596
Fax: (0033) 145231892

Regular auctions of historic militaria, toy soldiers, etc.

Danyela Petitot
54 bis Avenue de la Motte Picquet, Paris 75015, France
Tel/fax: (0033) 145674235

Auction adviser; we act as counsellor for our customers. Sale, purchase, research on old and ancient books covering history, military art, costumes and uniforms.

Forman International Auctions
PO Box 25, Minehead, Somerset TA24 8YX
Contact: Adrian Forman
Tel/fax: 01643 862511

For original military and historical collectors' pieces. Forman guarantee of originality and no buyer's commission. Colour illustrated (4) catalogues £12.50 or US$20. Send for free newsletter and preview of forthcoming sale. Single copy £3 or US$5 plus free illustrated book list. Third Reich awards and militaria a speciality.

Glendining's
101 New Bond Street, London W1Y 9LG
Contact: Andrew Litherland; Chris Allen
Tel: 0171 493 2445
Fax: 0171 491 9181

Auctioneers and valuers of all Orders, Decorations and Medals, as well as Arms, Armour and Militaria. Three sales a year. Valuations for insurance, probate, etc. Catalogues by subscription. Open Monday to Friday 8.30am to 5.00pm.

Hermann Historica OHG
Maximilianstrasse 32, 80539 Munchen, Germany
Tel/fax: (0049) 89 296391

Hermann Historica OHG
Sandstrasse 33, Munich 80335, Germany
Tel: (0049) 89 5237296
Fax: (0049) 89 5237103

Specialist international auctioneers of arms, armour, historic militaria, orders, medals, uniforms, etc.

Jan K Kube
Altes Schloss, 91484 Sugenheim, Germany
Tel/fax: (0049) 91 65650

Onslow's
The Old Depot, 2 Michael Road, London SW6 2AD
Contact: John Jenkins
Tel: 0171 371 0505
Fax: 0171 384 2682

Specialist collectors' auctions. Travel, advertising and war posters, Printed ephemera, aeronautical, maritime, motoring and railways. Photograph library. Free calendar of sales. Illustrated catalogues £7.00 each.

Phillips
7 Blenheim Street, New Bond Street, London W1Y 0AS
Contact: Christopher Allen
Tel: 0171 629 6602
Fax: 0171 495 3536

Phillips hold regular specialised sales of arms, armour and militaria which include all types of antique edged weapons, antique and modern firearms and uniforms etc.

Phillips Bayswater
10 Salem Road, Bayswater, London W2 4DL
Contact: James Opie
Tel: 0171 229 9090
Fax: 0171 792 9201

Phillips Bayswater continues a twenty-seven year series of specialist toy soldier and figure auctions held four times a year under the guidance of world expert James Opie.

Sotheby's
34-35 New Bond Street, London W1A 2AA
Tel/fax: 0171 493 8080

Sotheby's / Billingshurst
Summers Place, Billingshurst,
Sussex (W.) RH14 9AD

Contact: Gordon Gardiner
Tel: 01403 783933
Fax: 01403 785153

Regular auctions of militaria, antique firearms, air pistols & rifles, edged weapons, aeronautica, medals, orders & decorations. Free auction valuations & advice.

Spink & Son Ltd
5 King Street, St James's, London
SW1 6QS

Contact: David Erskine-Hill
Tel: 0171 930 7888
Fax: 0171 839 4853

Spink auction Orders, Decorations and Medals of the World and militaria, the latter including military uniforms, swords and headdress. Three specialist catalogues per annum, available by subscription. Also three issues of the 'The Medal Circular' a retail list. Open Monday to Friday, 9am to 5.30pm.

Wallis & Wallis
West Street Auction Galleries,
Lewes, Sussex (E.) BN7 2NJ

Contact: Roy Butler
Tel: 01273 480208
Fax: 01273 476562

Specialist auctioneers of arms, armour, militaria, medals. Nine regular two-day sales per year with spring and autumn Connoisseur Collectors' sales. Sales catalogues (including U.K. postage): Arms & armour; Militaria & medals (both £4.50); Connoisseur Collector's – full colour (£12.50). Postal bids welcome. Open Mon-Fri, 9.00am-5.00pm; Sat until 1pm on viewing days.

Weller & Dufty Ltd
141 Bromsgrove Street,
Birmingham, Midlands (W.)

Contact: M R Scott
Tel: 0121 692 1414
Fax: 0121 622 5605

Specialist auctioneers of antique and modern arms, armour, militaria and ammunition. Normally about ten sales annually; catalogues available by annual subscription.

BOOK PUBLISHERS

Acme Publishing Co
The Street, Horton Kirby,
Dartford, Kent DA4 9BY

Tel/fax: 01322 864919

After The Battle
Church House, Church Street,
London E15 3JA

Contact: Winston Ramsey
Tel: 0181 534 8833
Fax: 0181 555 7567

Publisher of books and magazines that specialises in revisiting the battlefields of World War II and presenting stories through 'then and now' photographic comparisons.

Air Data Publications
Southside Manchester Airport,
Wilmslow, Cheshire SK9 4LL

Contact: Barry Richards
Tel: 0161 499 0023
Fax: 0161 499 0298

Publishers of military and aviation books, mostly in hardback. Also of the 'Pilots' Notes' series. Distributors of American Flight Manuals and Flugzeug.

Airlife Publishing
101 Longden Road, Shrewsbury,
Shropshire SY3 9EB

Contact: A D R Simpson
Tel: 01743 235651
Fax: 01743 232944

Airlife is a specialist publisher covering all aspects of aviation both contemporary and historical, military and civil. Proposals for new books welcomed.

Andrew Mollo
10 Rue Jean Juares, 03320
Lurcy-Levis, France

Tel/fax: (0033) 70679183

Military consultant to film and TV production companies; author and collector, specialising in German and Russian subjects; publisher of Historical Research Unit books.

Anglo-Saxon Books
Frithgarth, Thetford Forest Park,
Hockwold-cum-Wilton, Norfolk
IP26 4NQ

Contact: Tony Linsell
Tel/fax: 01842 828430
Email:
anglosaxon@compuserve.com

'The English Warrior from earliest times to 1066', examines strategy, tactics, weapons and the attitudes and rituals of warriors. 'English

Martial Arts' deals mainly with 15th and 16th century fighting techniques. Contact us for a catalogue or see it at http://www.anglo-saxon.demon.co.uk/asbooks/.

Anschluss Publishing
Rivendell, Wathen Way,
Marsham, Aylsham, Norfolk
NR10 5PZ

Argus Books
Eddington Hook Ltd, 406 Vale
Road, Tonbridge, Kent TN9 1XR

Arms and Armour Press
Cassell plc, Wellington House,
125 Strand, London WC2R 0BB

Contact: Roderick Dymott
Tel: 0171 420 5555
Fax: 0171 240 7261

The most active military book imprint, producing over 50 titles each year on military, naval and aviation history, collectables, uniforms etc.

Athena Publishing
34 Imperial Crescent, Town
Moor, Doncaster, Yorkshire (S.)
DN2 5BU

Tel: 01302 322913
Fax: 01302 730531

Publishers of Napoleonic facsimiles and UK distributor for Ryton Publishing and Cannon Books.

B T Batsford
583 Fulham Road, London SW6
5BY

Contact: Alan Ritchie
Tel: 0171 471 1100
Fax: 0171 471 1101
Email: info@batsford.com

English Heritage series: Channel Defences (Saunders) and Roman Britain (Millett). Historic Scotland series: Roman Scotland (Breeze), Fortress Scotland and the Jacobites.

Barett Verlag GmbH
PO Box 110170, 42661 Solingen,
Germany

Contact: Karl Heinz Dissberger
Tel/fax: (0049) 212 267100

Barrie & Jenkins Ltd
Random House, 20 Vauxhall
Bridge Road, London W11 2QA

Tel: 0171 973 9710
Fax: 0171 233 6057

Berg Publishers
150 Cowley Road, Oxford,
Oxfordshire OX4 1JJ

Contact: Kathryn Earle
Tel: 01865 245104
Fax: 01865 791165
Email: enquiry@berg.demon.co.uk

Academic publisher with a strong list in European – especially German – military studies.

Blackwell Publishers
108 Cowley Road, Oxford,
Oxfordshire OX4 1JF

Tel: 01865 791100
Fax: 01865 791347

Academic publisher with some military history titles including Civil War and World War I. Information sheets and current titles available.

Border Press
Unit 27, Monument Industrial
Park, Chalgrove, Oxfordshire
OX44 7RW

Contact: Alan Badger
Tel: 01865 400256
Fax: 01965 400257

Specialist firearms books and manuals; trade enquiries welcome.

Boydell & Brewer Ltd.
PO Box 9, Woodbridge, Suffolk
IP12 3DF

Tel/fax: 01394 411320

Publish medieval history including warfare, arms and armour, tournaments, battles, editions of campaign diaries, and modern military history. Military History catalogue. Mastercard & Visa accepted; direct orders welcome.

Brassey's (UK) Ltd
33 John Street, London
WC1N 2AT

Contact: Warren Prentice
Tel: 0171 753 7777
Fax: 0171 753 7794
Email: brasseys@dial.pipex.com

Publishers of a wide range of highly acclaimed titles on military history and technology, defence and international affairs; free catalogue on request.

Brown Books
Bradley's Close, 74-77 White
Lion Street, London N1 9PF

Contact: Sara Ballard
Tel: 0171 520 7600
Fax: 0171 520 7606

Publishers of military books
ranging from highly illustrated
general histories, personalities,
battles, units and equipment to
fact-filled yearbooks.

Buchan & Enright Publishers
53 Fleet Street, London EC4Y
1BE

Tel/fax: 0171 353 4401

Cambridge University Press
The Edinburgh Building,
Shaftesbury Road, Cambridge,
Cambridgeshire CB2 2RU

Tel/fax: 01223 312393

Centre for Security & Conflict Studies
12a Golden Square, London
W1R 3AF

Tel/fax: 0171 439 7381

Collectors Books Ltd
Bradley Lodge, Kemble,
Cirencester, Gloucestershire
GL7 6AD

Contact: Christian Braun
Tel: 01285 770 239
Fax: 01285 770 896

Publishers of 'The Little Ships of
Dunkirk' and of 'Bluebird' – both
describing the privately owned
pleasure cruisers that took part in
the Dunkirk rescue.

Conway Maritime Press
Brassey's (UK) Ltd, 33 John
Street, London WC1N 2AT

Contact: Warren Prentice
Tel: 0171 753 7777
Fax: 0171 753 7794
Email: brasseys@dial.pipex.com

The world's leading publisher of
authoritative, highly illustrated
reference books on maritime
history. Free catalogue available.

Crecy Books
Unit 2A, Newbridge Trading
Estate, Bristol, Avon BS4 4AX

Tel: 0117 9724248
Fax: 0117 9711056

Crowood Press
The Stable Block, Crowood
Lane, Ramsbury, Marlborough,
Wiltshire SN8 2HR

Tel: 01672 20320
Fax: 01672 20280

D P & G
PO Box 186, Doncaster,
Yorkshire (S.) DN4 0HN

Publications on the Victorian and
Edwardian British Army,
covering: uniforms, badges,
buttons, colours & standards,
sabretaches, side caps. Illustrated
books (colour) by H. Payne, R.
Simkin, Major Seccombe. Also
campaigns of NWF, Egypt,
Malakand 1897, Ladysmith, Zulu
War, uniforms of 1854. Special
section on the Indian Army. Send
4 x 1st class stamps or 4 IRCs for
illustrated catalogue.

Darr Publications
Thorshof, 106 Oakridge Road,
High Wycombe,
Buckinghamshire HP11 2PL

Contact: Thorskegga Thorn
Tel: 01494 451814
Fax: 01494 784271

Large number of historical
booklets, both theoretical studies
and 'how-to-do-it' manuals.
Heavily researched but
inexpensive, practical and easy to
read. SAE for details. Email:
thorskegga@calltoarms.com

Datafile Books
10 White Hart Lane, Wistaston,
Crewe, Cheshire CW2 8EX

Contact: Malcolm Bellis
Tel/fax: 01270 663296

Specialist military sourcebooks,
e.g. WWII & post-war British
Army orders of battle, etc.

David & Charles Publishers
Brunel House, Newton Abbot,
Devon TQ12 4PU

Tel: 01626 61121
Fax: 01626 42904

Publishers of aviation art books
including those of Robert Taylor,
Frank Wooton, Gerald Coulson
and John Young. Mail order
available.

E P A Editions
9/13 Rue du Col Pierre Avia, BP
501, 72725 Paris Cedex 15,
France

Tel/fax: (0033) 146429191

Publisher specialising in French
language transport books (cars,
trucks, fire engines, military
vehicles, aviation); 15 new titles a
year (e.g. books on the US Army
GMC, Dodge, Jeep etc.).

El Dorado Books
27 Hallgate, Cottingham,
Yorkshire (E.) HU16 4DN

Contact: Terry Hooker
Tel/fax: 01482 847068
Email: scamhs@aol.com

South and Central American
Military Society booklets on single
subjects such as 'Notes on the
Mexican Army 1900-20'. All relate
to Latin American military history
subjects.

Ermanno Albertelli Editore
Via S Sonnino 34, 43100 Parma,
Italy

Contact: Ermanno Albertelli
Tel: (0039) 521 292733
Fax: (0039) 521 290387

Excelsior Publications
1 Rue du Col Pierre Avia, 75015
Paris, France

Fieldbooks
22 Callendar Close, St Nicolas
Park, Nuneaton, Warwickshire
CV11 6LU

Contact: Paddy Griffith
Tel/fax: 01203 350763

Also incorporating Paddy Griffith
Associates, the company
publishes a range of military
books.

Firebird Books Ltd
PO Box 327, Poole, Dorset
BH15 2RG

Tel: 01202 715349
Fax: 01202 736191

Publishers of illustrated military
history books including the
'Heroes and Warriors' series. Free
illustrated catalogue on request.
Mail order.

Flicks Books
29 Bradford Road, Trowbridge,
Wiltshire BA14 9AN

Tel: 01225 767728
Fax: 01225 760418

Publishers of books on film and
cinema, e.g. The Imperial War
Museum Film Catalogue, The
First World War Archive; The
Biograph in Battle (Boer war
filming).

Goodall Publications Ltd
Southside, Manchester Airport,
Wilmslow, Cheshire SK9 4LL

Contact: Barry Richards
Tel: 0161 499 0024
Fax: 0161 499 0298

Publishers of military and aviation
books in paperback, eg 'Enemy
Coast Ahead', 'Wing Leader',
'Lancaster Target' and 'Beyond
the Dams to the Tirpitz'.

Gosling Press
35 Cross Street, Upton,
Pontefract, Yorkshire (W.) WF9
1EU

Greenhill Books
Park House, 1 Russell Gardens,
London NW11 9NN

Tel: 0181 458 6314
Fax: 0181 905 5245
*Email: lionelleventhal
@compuserve.com*

Publishers of books on military
history. Their Napoleonic Library
now has thirty-two volumes, they
have reprinted classic books by Sir
Charles Oman, etc.

H B T
67c Port Street, Stirling,
Grampian FK8 2ER, Scotland

Tel/fax: 01786 78979

H M S O Books
St Crispins, Duke Street,
Norwich, Norfolk NR3 1PD

Tel: 01603 622211
Fax: 01603 7582

The Hallamshire Press
Broom Hall, 8-10 Broomhall
Road, Sheffield, Yorkshire (S.)
S10 2DR

Contact: Andrew Fyfe
Tel: 0114 266 3789
Fax: 0114 267 9403

Publishers of Sheffield books.
Recent titles include 'The Sheffield
Knife Book' which features many
knives from the WWI & WWII
period. Call for a free catalogue.

Harper Collins
77-85 Fulham Palace Road,
London W6 8JB

Contact: Ian Drury
Tel: 0181 307 4053
Fax: 0181 307 4198
*Email: ian.drury
@harpercollins.co.uk*

20th century military reference
works.

Helion & Company
26 Willow Road, Solihull,
Midlands (W.) B91 1UE

Contact: Duncan Rogers
Tel: 0121 705 3393
Fax: 0121 711 1315

Enormous selection of in-print
military books from around the
world – over 12,000 titles listed in
7 catalogues. Free catalogues
available – Warfare to 1700;
1700–1914; WWI; WWII;
Post-1945; Maritime; Aviation.
Expert booksearch service
available.

Histoire & Collections (UK)
463 Ashley Road, Parkstone, Poole, Dorset BH14 0AX

Contact: Chris Lloyd
Tel: 01202 715349
Fax: 01202 736191

Paris-based publisher of military books, now publishing in English. Full colour illustrated books on historical and contemporary militaria, and World War II.

Holyboy Publications & Distributors
7 Grammar School Road, North Walsham, Norfolk NR28 9JH

Contact: Neil Storey

We supply a varied selection of military books, reproduction photographs and research, specialising in East Anglian military history. Mail order only.

I B Tauris & Co Ltd
Victoria House, Bloomsbury Square, London WC1B 4DZ

Contact: Jonathan McDonnell
Tel: 0171 916 1069
Fax: 0171 916 1068
Email: jmcdonnell@ibtauris.com

Publishers of academic and specialist books on military history worldwide, and on firearms from the Islamic world.

I S O Publications
137 Westminster Bridge Road, London SE1 7HR

Tel: 0171 261 9588
Fax: 0171 261 9179

Ian Allan Ltd
Coombelands House, Coombelands Lane, Addlestone, Surrey KT15 1HY

Contact: W Y Myers
Tel: 01932 855909
Fax: 01932 854750

Publishers of military, aviation and transport subjects, including the 'Collectors' Guides' and 'ABC' series. Specialist bookshops near Waterloo, Manchester Piccadilly and Birmingham New Street stations.

Jane's Information Group Ltd
Sentinel House, 163 Brighton Road, Coulsdon, Surrey CR5 2NH

Contact: Claire Brunavs
Tel: 0181 700 3700
Fax: 0181 763 1006
Email: info@janes.co.uk

In 1997 into 1998 Jane's is celebrating 100 years of being the world's leading provider of defence, aerospace, geopolitical and transportation related information solutions.

Kipling & King
3 Saxon Croft, Franham, Surrey GU9 7QB

Contact: H L & P King
Tel/fax: 01252 716303

Head-dress badges of the British Army. The definitive work on the subject: Volume One 1800–1920, £65.00; Volume Two 1920–1995, £42.50. Totalling 3,000 illustrations on 727 pages.

Manchester University Press
Oxford Road, Manchester, Manchester, Gt. M13 9PL

Contact: Catherine Whelan
Tel: 0161 273 5539
Fax: 0161 274 3346

Publisher with thriving list in military studies including 'War, Armed Forces and Society' series, and their new series 'Manchester History of the British Army'. A listing of all titles is available on request.

The Medals Yearbook
3rd Floor, 1 Old Bond Street, London W1X 3TD

Tel: 0171 499 5026
Fax: 0171 499 5023

The only up-to-date descriptive price guide to all British medals. Fully illustrated.

Midland Publishing Ltd
24 The Hollow, Earl Shilton, Leicester, Leicestershire LE9 7NA

Tel: 01455 847256
Fax: 01455 841805
Email: midlandbooks@compuserve.com

Aviation, railway and military publishers – associated with Midland Counties Publications, see entry in Section 3 (Booksellers).

Montvert Publications
PO Box 25, Stockport, Cheshire SK5 6RU

Book publishers with an expanding range specialising in well-written works dealing with the history, organisation, costume and equipment of ancient and medieval armies.

New Cavendish Books
3 Denbigh Road, London W11 2SJ

Tel: 0171 229 6765
Fax: 0171 792 0027

Specialist publishers of quality illustrated books on toys and other collectables.

Norman D Landing Publishers
216 Lightwoods Hill, Warley Woods, Warley, Midlands (W.) B67 5EH

Tel/fax: 0121 434 4580

Publishers of the US Army uniform reference book 'Doughboy to GI: US Army Uniforms and Equipment 1900-1945'. Trade enquiries welcomed; 24 hour fax.

Osprey Publishing Ltd
2nd Floor, Unit 6, Citadel Place, Spring Gdns, Tinworth Street, London SE11

Contact: Lee Johnson (Editorial)
Tel/fax: 0171 225 9861

Publishers of the Men At Arms, Elite, Vanguard, Campaign, and Warrior series of illustrated military history reference books.

Oxford University Press
Walton Street, Oxford, Oxfordshire OX2 6DP

Tel: 01865 56767
Fax: 01865 56646

International publishing house with titles on strategic studies, war studies, peace studies, current affairs and political economy. Also distribute titles on behalf of the Stockholm International Peace Research Institute – including SPRI Yearbook, reviewing developments in nuclear weapons, world military expenditure and international arms trade.

Picton Publishing (Chippenham) Ltd
Queensbridge Cottages, Patterdown, Chippenham, Wiltshire SN15 2NS

Contact: David Picton-Phillips
Tel/fax: 01249 443430

Publishers of regimental histories, reprints of classic military histories and allied subjects, military biographies and autobiographies. Booklist and sales leaflets available on request. No callers, please.

Pitkin Guides Ltd
Healey House, Dene Road, Andover, Hampshire SP10 2AA

Contact: Ian Corsie
Tel: 01264 334303
Fax: 01264 334110
Email: guides@pitkin.u-net.com

Highly-illustrated souvenir guides for tourists. List includes subjects of military interest – battlefields, campaigns, etc.

The Pompadour Gallery
PO Box 11, Romford, Essex RM7 7HY

Contact: George Newark
Tel/fax: 01375 384020
Email: christopher.newark @virgin.net

Publishers and distributors of military books, postcards and reproduction cigarette cards. Books: 'Kipling's Soldiers', 'Uniforms of the Foot Guards' and our latest book 'Uniforms of the Royal Marines 1664 to the Present Day'. Send SAE for full details of all our publications.

R I G O / Le Plumet
Louannec, 22700 Perros-Guirrec, France

R J Leach & Co.
73 Priory Grove, Ditton, Aylesford, Kent ME20 6BB

Contact: Richard Leach
Tel/fax: 01622 791243

Publishers on all aspects of military history producing approximately six titles yearly. Reprints and new titles. Mail order. List of titles supplied on request.

Robert Hale Ltd
Clerkenwell House, Clerkenwell Green, London EC1R 0HT

Tel: 0171 251 2661
Fax: 0171 490 4958

Roundhouse Publishing Group
PO Box 140, Oxford, Oxfordshire OX2 7FF

Contact: Alan Goodworth
Tel: 01865 512682
Fax: 01865 559594
Email: 100637.3571 @compuserve.com

Distributors of a range of North American publishers' titles on Civil War and other conflicts.

Routledge
11 New Fetter Lane, London EC4P 4EE

Tel/fax: 0171 583 9855

Salamander Books Ltd
129-137 York Way, London N7 9LG

Contact: Colin Gower

Publishers of high quality illustrated books dealing with aircraft, weapons, naval and military history.

Selous Books Ltd
40 Station Road, Aldershot,
Hampshire GU11 1HT

Contact: Trevor Hudson
Tel: 01252 333611
Fax: 01252 342337
Email: 100307,1735@compuserve

Specialist book dealers in military history, historic travel and ethnographic books. Subjects covered include military histories, regimental histories, uniforms, insignia, medal rolls, firearms and equipment.

Severn House Publishers Ltd
9-15 High Street, Sutton, Surrey
SM1 1DF

Tel/fax: 0181 770 3930

Shire Publications Ltd
Cromwell House, Church Street,
Princes Risborough, Aylesbury,
Buckinghamshire HP27 9AA

Contact: J Rotheroe
Tel: 01844 344301
Fax: 01844 347080

Paperbacks on aspects of military history.

Sidgwick & Jackson Ltd
18-21 Cavaye Place, London
SW10 9PG

Tel: 0171 373 6070
Fax: 0171 370 0746

We publish mainly on major commanders, battles and campaigns but also specialist publishing in the areas of Elite Forces.

Souvenir Press Ltd
43 Great Russell Street, London
WC1B 3PA

Contact: Ernest Hecht
Tel: 0171 580 9307
Fax: 0171 580 5064

Spa Books & Strong Oak Press
PO Box 47, Stevenage,
Hertfordshire SG2 8UH

Contact: Steven Apps
Tel/fax: 01438 816896

Publishers of works relating to military history: WWI, WWII, Napoleonic, etc. Annual/seasonal catalogues issued.

Spellmount Ltd, Publishers
The Old Rectory, Staplehurst,
Kent TN12 0AZ

Contact: Jamie Wilson
Tel: 01580 893730
Fax: 01580 893731

History and military history publishers specialising in Napoleonic, Victorian, World War I and World War II periods, plus general military topics. Send for free catalogue.

Stratagem
18 Lovers Lane, Newark,
Nottinghamshire NG24 1HZ

Contact: Duncan McFarlane
Tel/fax: 01636 71973

Stuart Press
117 Farleigh Road, Backwell,
Bristol, Avon BS19 3PG

Contact: Stuart Peachey
Tel/fax: 01275 463041

Publish and sell books on the English Civil War and civilian topics of the period 1580-1660. Send SAE for catalogue. Also banquets and lectures.

Sutton Publishing Ltd
Phoenix Mill, Far Thrupp, Stroud,
Gloucestershire GL5 2BU

Contact: R Schinner
Tel: 01453 731114
Fax: 01453 731117
Email: spluk1

Sutton Publishing are leading book publishers specialising in: archaeology, aviation, classic fiction, history (national, regional, social), military (Roman, Medieval, 16-19 Cent, WWI & WWII), transport.

T L O Publications
Longclose House, Common Road, Eton Wick, nr Windsor,
Berkshire SL4 6QY

Contact: Tony Oliver
Tel: 01753 862637
Fax: 01753 841998

Reference books on German subjects for collectors, eg. 'German Order, Decoration and Medal Citations', and DDR subjects.

Token Publishing
Orchard House, Duchy Road,
Heathpark, Honiton, Devon
EX14 8YD

Contact: Philip Mussell
Tel: 01404 831878
Fax: 01404 831895

Publishers of Medal News, the world's only magazine devoted to medals and battles. Also The Medal Yearbook, the definitive guide to British orders and medals.

University of Nebraska Press
1 Gower Street, London WC1
6HA

Tel: 0171 580 3994
Fax: 0171 580 3995

V S-Books
PO Box 20 05 40, D 44635
Herne, Germany

Contact: Torsten Verhülsdonk
Tel: (0049) 2325 73818
Fax: (0049) 201 421396

Publishers of books on military history, uniforms and weapons from ancient times to the present day. Potential authors may contact us with details of manuscripts and illustrations.

Viking
27 Wright's Lane, London W8
5TZ

Tel: 0171 938 2200
Fax: 0171 937 8704

Vilmor Publications
Morton Villa Farm, Misson
Springs, Doncaster, Yorkshire
(S.) DN10 6ES

Contact: Philip Baker
Tel/fax: 01302 770295

Publisher and stockholder of 'Youth Led By Youth: Some Aspects of the Hitlerjugend'. Volume II is now in preparation.

W S Curtis (Publishers) Ltd
PO Box 493, Rhyl, Denbighshire
LL18 5XG, Wales

Contact: W S Curtis
Tel/fax: 01745 584 981

Specialising in the re-publication, with new researched Introductions, of rare 19th century books on shooting, firearms and artillery for the modern shooter and collector.

Wargames Research Group
The Keep, Le Marchant
Barracks, London Road, Devizes,
Wiltshire SN10 2ER

Tel/fax: 01380 724558

Publisher of wargame rules and reference books; also wholesaler of some allied products, publications, and imported books and rules.

Windrow & Greene Ltd
5 Gerrard Street, London
W1V 7LJ

Contact: Alan Greene
Tel: 0171 287 4570
Fax: 0171 494 0583

The country's leading publisher of high quality colour-illustrated books on military history, costume etc; also automotive and and general titles. The Europa Militaria series provides superb full colour photographic reference for modellers and has featured the leading re-enactment groups in recent Europa Militaria Specials, including 'Napoleon's Imperial Guard' and 'Napoleon's Line Cavalry'. The 'Masterclass' series for modellers includes Tony Greenland's much acclaimed 'Panzer Modelling Masterclass' and a new title from Roy Porter on Model Buildings. The new series 'Live Firing Classic Military Weapons in Colour Photographs' gives a unique insight into the use of 'German Automatic Weapons of WWII', 'Machine Guns of WWI' and 'Powder and Ball Small Arms'. Contact Alan Greene for catlogues and prices.

BOOKSELLERS

Adrian Forman Books
PO Box 25, Minehead, Somerset
TA24 8YX

Contact: Adrian Forman
Tel/fax: 01643 862511

Distributor and agent for Bender
Books USA. Free illustrated book
list of German Third Reich new
books including many out of print
titles and Bowens classic Iron Cross
book, plus Forman's guide series
on German Third Reich awards
and documents (with values).

**Aldershot Military History
Trust**
Evelyn Woods Road, Queens
Avenue, Aldershot, Hampshire
GU11 2LG

American Western Bookshelf
The Glen, Vicarage Road, Bude,
Cornwall EX23 8LT

Andrew Burroughs Books
34 St Martins, Stamford,
Lincolnshire PE9 2LJ

Contact: Andrew Burroughs
Tel/fax: 01780 51363

Naval and military. Specialities:
20th century British naval and
military history, in particular the
Second World War. Catalogues
issued.

Antik 13
Pieperstrasse 13, Bochum
D-44789, Germany

Contact: Peter J Nachtigall
Tel: (0049) 234 300289
Fax: (0049) 234 330919

Antiquarian and second-hand
books on militaria and weapons;
Osprey Men at Arms series, V
S-Books; also 20th century
collectables including militaria.

**Antiquariat Buchvertrieb &
Verlag**
Oselbachstrasse 72, 6660
Zweibrucken, Germany

Contact: Heinz Nickel-Verstand

Argus Books
Eddington Hook Ltd, 406 Vale
Road, Tonbridge, Kent TN9 1XR

Armchair Auctions
98 Junction Road, Andover,
Hampshire SP10 3JA

Contact: George Murdoch
Tel/fax: 01264 362048

Holds monthly postal auctions
specialising in 1914-18 military
and aviation books, relics and
ephemera. All guaranteed to be in
stock. Write/phone/fax for free
catalogue.

Arnold Busck
49 Kobmagerade 1150,
Copenhagen K, Denmark

Articles of War Ltd
8806 Bronx Avenue, Skokie,
Illinois 60077-1896, USA

Contact: Robert Ruman
Tel: (001) 847 674 7445
Fax: (001) 847 674 7449
Email: warbooks@aol.com

New and second-hand military
history books, all eras ancient to
modern; over 20,000 titles.
Catalogues issued; open Tues –
Sat. Website: http://www.
sonic.net/~bstone/articles.

Athena Books
34 Imperial Crescent, Town
Moor, Doncaster, Yorkshire (S.)
DN2 5BU

Contact: Les Thomas
Tel: 01302 322913
Fax: 01302 730531
*Email: athenamilitarybooks
@btinternet.com*

UK's largest stockists of military
history, 3000 BC to WW3.
Monthly catalogue of over 1,000
books. SAE for sample. 50,000
books in stock. Normal shop
hours.

Aviation Bookshop
656 Holloway Road, London
N19 3PD

Tel/fax: 0171 272 3630

Large new and second-hand stock
of books, magazines, posters,
videos. Speciality: aviation in all
its aspects.

Aviation Shop
Kajanuksenkatu 12A, 00250
Helsinki, Finland

Barbarossa Books
West Beynon House, 242 High
Street, Bromley, Kent BR1 1PQ

Contact: Russell Hadler
Tel/fax: 0181 325 8646

Specialists in military, aviation,
naval, modelling books, videos
and magazines. Titles from
Russia, Poland, France, USA,
Japan. Free quarterly catalogue
with SAE/IRC. Mail order to the
world.

Blitzkrieg Books
15 Woodlands Road, Alum Rock,
Saltley, Birmingham, Midlands
(W.) B8 3AG

Contact: J Bentley
Tel/fax: 0121 422 6814

Specialists in rare, out-of-print
and second-hand books and
magazines. Military, aviation,
nautical, wargaming, modelling.
Send A4 SAE for catalogue.
Visitors by appointment.

The Book Room
Post Office, Twyning,
Gloucestershire GL20 6DF

Contact: Kenneth Fergusson
Tel/fax: 01684 274641

Opening times: Monday to
Saturday 10.00am-5.00pm. Good
general stock. Specialities:
Aviation and Military.

Books International
101 Lynchford Road,
Farnborough, Hampshire
GU14 6ET

Tel: 01252 376564
Fax: 01252 370181

Brentano's
37 Avenue de L'Opera, 75002
Paris, France

Bufo Books
32 Tadfield Road, Romsey,
Hampshire SO51 5AJ

Tel/fax: 01794 517149

Dealers in second-hand and out of
print military books, by post and
at book fairs. Stock mainly 20th
century related. Catalogues issued
irregularly.

Bushwood Books
6 Marksbury Avenue, Kew
Gardens, Surrey TW9 4JF

Contact: Richard Hansen
Tel: 0181 392 8585
Fax: 0181 3929876
Email: bushwd@aol.com

Exclusive UK distributor for
Schiffer Publishing military list,
and also distribute for S S
Fedorowicz.

Buttercross Books
2 The Paddock, Bingham,
Nottinghamshire NG13 8HQ

Tel/fax: 01949 837147

Small stock of military books.
Specialities: Napoleonic era and
World War I. Facsimile editions of
rare Napoleonic items published
in limited editions. Two
catalogues published per year.

Librairie Armes et Collections
19 Avenue de la Republique,
75011 Paris, France

Specialist bookshop for all
military publications; the retail
branch of the publishers of the
leading French magazines
'Militaria', 'Tradition', 'RAIDS',
etc. Also stocks a wide range of
outdoor clothing, survival
equipment, etc.

Librairie Uniformologique Internationale
111 Avenue Victor Hugo, Galerie
Argentine, 75116 Paris, France

Contact: Thierry Lecourt

Libros Reyes
Dato 1, 50005 Zaragoza, Spain

Contact: Angel Marti
Tel: (0034) 76 219443
Fax: (0034) 76 230179
Email: l.reyes@arrakis.es

Specialists in Spanish military
history, Spanish air force, navy,
uniforms and weapons, all
periods. Send for free catalogue.

La Maison du Livre Aviation
75 Boulevard Malesherbes,
75008 Paris, France

Major Book Publications
Hougoumont, Maxworthy,
Launceston, Cornwall PL15 8LZ

Contact: H.W. or C.P. Adams
Tel/fax: 01566 781422

Military research material. Books
and our unique, comprehensive
archival data sheet system. 'British
Army at Waterloo' identifies
regimental arrivals, departures,
dates, locations, formations,
commanders, strengths, losses
plus Allied Army of Occupation
details. SAE essential.

Maritime Books
Lodge Hill, Liskeard, Cornwall
PL14 4EL

Contact: Roger May
Tel: 01579 343663
Fax: 01579 346747

Specialist publishers of Royal
Naval books & bookseller of RN
titles. Over 100,000 books in stock.
Mail order only; free catalogues.

Mark Weber
35 Elvaston Place, London
SW7 5NW

Tel: 0171 225 2506
Fax: 0171 581 8233
Email: wscbooks@mail.bogo.co.uk

Private premises: postal business.
2,000 books exclusively in two
specialities: official war histories
(HMSO & Commonwealth); and
books by/about Winston
Churchill.

Martin Gladman
235 Nether Street, Finchley,
London N3 1NT

Tel/fax: 0181 343 3023

Large stock of Military /
Naval/Aviation within a very
large history stock. 'Wants' lists
welcomed. Individual volumes or
collections purchased.

McGowan Book Co
PO Box 16325, Chapel Hill,
North Carolina 27516-6325, USA

Contact: R Douglas Sanders
Tel: (001) 919 968 1121
Fax: (001) 919 968 1169

We offer the finest in rare and
out-of-print American Civil War
first editions. Send US $3.00 or
equivalent for subscription to our
next three catalogues.

Michael Haynes
46 Farnaby Road, Bromley, Kent
BR1 4BJ

Specialist in books on the
American Civil War and
American Frontier. Occasional
listings on Colonial, Revolution,
Texas etc. Mail order only. List
free on request.

Midland Counties Publications
Unit 3, Maizefield, Hinckley,
Leicestershire LE10 1YF

Contact: Mike Everton
Tel: 01455 233747
Fax: 01455 233737
Email: 106371.573
@compuserve.com

Distributors of aviation, military,
railway, spaceflight and
astronomy books, magazines and
videos. Large selections of titles
in stock for mail order worldwide.
Free illustrated catalogue.

Mil-Art
41 Larksfield Crescent,
Dovercourt, Harwich, Essex
CO12 4BL

Tel/fax: 01255 507440

Producers of metal military figure
kits in 54mm, 80mm and 100mm
scales. Also retailer of new and
out of print military books.

Militaria
c/Bailen 120, 08009 Barcelona,
Spain

Contact: Xavier Andrew
Tel/fax: (0034) 3 2075385

Buy, sell and exchange militaria,
medals, edged weapons, insignia,
headgear etc. Also large selection
of military books in stock.

Military and Aviation Book Society
Guild House, Farnsby Street,
Swindon, Wiltshire SN99 9XX

Tel: 01793 512100
Fax: 01793 616789

A book club for military and
aviation enthusiasts ancient and
modern (membership UK only).
Books sold by post at reduced
prices.

Military Books
16 Conway Road, Wimbledon,
London SW20 8PA

Contact: David J Harrison
Tel/fax: 0181 946 3219

Second-hand and new books on
the Great War 1914-18, postcards,
magazines, ephemera, battlefield
items, maps, Michelin guides, etc.
Visitors welcome seven days a
week, morning, noon and night.
Easy parking. Can collect from
Wimbledon Station, District Line.
Free book search service.

Military Books
Low Field, Newby, Penrith,
Cumbria CA10 3HB

Contact: Major J R McKenzie
Tel/fax: 01931 715253

Purchase and sale of military
books on land warfare. Please
send SAE for catalogue. Fair
prices paid for books in good
condition. Free specialist
booksearch.

Military Bookworm
PO Box 235, London SE23 1NS

Contact: David W Collett
Tel/fax: 0181 291 1435

British regimental and divisional
histories; campaigns; memoirs;
school registers and rolls of
honour; biographies. Postal
business only; regular catalogues
issued, subscription UK £10.00,
overseas £14.00.

Military History Bookshop
77-81 Bell Street, London
NW1 6TA

Contact: Jonathan Prickett
Tel: 0171 723 4475
Fax: 0171 723 4665

Specialist in military history.
Large stock of new and
second-hand books including
many imported titles, on
uniforms, weapons, unit
histories, biographies, battles,
campaigns, espionage and
terrorism. Annotated catalogues
issued, each of over 1000 titles,
which often include videos and
computer simulations: 1. General
military. 2. American military
achievement, from the
Revolution to the Gulf.
3. German and European Axis
World War II Armed Forces.
£1.00 per catalogue. Visa and
Mastercard accepted. Visitors

welcome between
10.00am-1.00pm Tues to Fri
inclusive. The hours are different
at Christmas and other public
holidays, so it is always advisable
to telephone prior to calling.

Military Parade Bookshop
The Parade, Marlborough,
Wiltshire SN8 1NE

Contact: Peter Kent
Tel: 01672 515470
Fax: 01980 630150

Military warfare from the
Crusades to the Gulf. Specialises
in Regimental histories,
Napoleonic, World War I & World
War II. Over 5,000 titles available;
send SAE for catalogue. Open
Mon-Sat, 10.00am-5.00pm.

Military Services
87 Ellacombe Road, Longwell
Green, Bristol, Avon BS15 6BP

Contact: Terry Walsh
Tel/fax: 0117 9324085

New and second-hand military
books and magazines; mail order,
list available; book-finding
service. Send SAE for current
listing.

Military Subjects
2 Locks Road, Locks Heath,
Southampton, Hampshire
SO3 6NT

Contact: C Pearce
Tel/fax: 01489 572582

Books, models and militaria.
Private premises; visitors by prior
appointment only, please.

Model & Hobby
Fredriksborgade 23, DK-1360
Copenhagen, Denmark

Motor Books
33 St Martins Court, London
WV2N 4AL

Tel: 0171 836 5376
Fax: 0171 497 2539

Real specialists in military,
maritime, motoring, aviation and
railways. Worldwide mail order
service, credit cards accepted.
Booklists for each specialisation
available. Very close to Leicester
Square underground station.
Large stocks of new books and
video cassettes. Open Mon-Fri
9.30am-5.30pm (Thurs to
7.00pm), Sat 10.30am-1.00pm
2.00pm-5.30pm. Branches: 10,
Theatre Square, Swindon, SN1
1QN (01793 523170). 8 The
Roundway, Headington, Oxford
OX3 8DH. (01865 66215).
Business hours vary.

Motor Books
10 Theatre Square, Swindon,
Wiltshire SN1 1QN

Contact: Janet Terry
Tel: 01793 523170
Fax: 01793 432070

Specialist bookshop for all
military publications. Large range
of military figures. Worldwide
mail order service.

Oppenheim Booksellers
7/9 Exhibition Road, South
Kensington, London SW7

Tel/fax: 0171 584 4143

**Opportunity Book
Distributors**
PO Box 62514, Sharonville, Ohio
45262, USA

Contact: Robert Wagner
Tel: (00) 513 554 1162
Fax: (001) 513 554 1162
*Email: rrwagner
@www.earthlink.net*

Mail order bookseller specializing
in toy soldier/model soldier
books, and the painting and
modelling of toy and model
soldiers. List free with large SAE.

Outdoorsman's Bookstore
Unit 27, Monument Industrial
Park, Chalgrove, Oxfordshire
OX44 7RW

Contact: Alan Badger
Tel: 01865 400256
Fax: 01865 400257

Selected books on military,
firearms and related subjects,
from across the world, ensure a
diverse range of new books
available by mail order.

Paul Meekins Books
34 Townsend Road, Tiddington,
Stratford upon Avon,
Warwickshire CV37 7DE

Contact: Paul Meekins
Tel/fax: 01789 295086

Large stock, Roman to WWII.
Specialising in English Civil War
and Napoleonic subjects.
Second-hand and new. 3 x 1st
class stamps for catalogue. Private
premises.

Pelta
16 Swietokrzyska Street, 00
Warsaw 050, Poland

Contact: Mark Machala
Tel: (0048) 2227661
Fax: (0048) 22269186

Leading distributor of Polish and
Russian books on aviation,
vehicles, AFVs, naval and

militaria subjects (e.g. uniforms),
covers all historical periods. ALSO
military models, plastic kits,
vacuforms, figures (30mm, 54mm)
painted and unpainted. The
widest selection in the world,
available for trade and mail order
world wide. Free catalogues and
super prices.

Peter A Heims Ltd
274 Kingston Road, Leatherhead,
Surrey KT22 7QA

Contact: Peter A Heims

'Parachute Wings' by Bragg &
Turner – £15.00 or equivalent,
postage paid, from above address.

Peter de Lotz Books
20 Downside Crescent, Belsize
Park, London NW3 2AP

Tel: 0171 794 5709
Fax: 0171 284 3058

Out of print books on all aspects
of military, naval, aviation
warfare. Catalogue and search
only. Publisher of Hackett 'South
African War Books Bibliography'
£65.

Phillip Austen
23 Westgate, Sleaford,
Lincolnshire NG34 7PN

British military history books.
Postal business only. Catalogues
issued. Your particular
requirements are welcome.

Pocketbond Ltd
PO Box 80, Welwyn,
Hertfordshire AL6 0ND

Contact: Phillip Brook
Tel: 01707 391509
Fax: 01707 327466

Exclusive importer for: Squadron
Signal Publications and Squadron
products: Testors, Glencoe, Tauro,
Hobbycraft, AFV Club and Emhar
plastic kits and Imex Civil War
figures.

Preussisches Bücherkabinett
Knesebeckstrasse 88, 10623
Berlin, Germany

Contact: Stefan Muller
Tel: (0049) 303130802
Fax: (0049) 303131180

Specialists in military books from
around the world. Mail order
catalogue available. Shipping
anywhere.

Prospect Books
18 Denbigh Street, Llanrwst,
Gwynedd LL26 0AA, Wales

Tel/fax: 01492 640111

Rare books on firearms and edged
weapons bought and sold; send
for catalogue.

Quartermasters Stores
17-19 West Wycombe, High
Wycombe, Buckinghamshire
HP11 2NF

R M England
4 Baker House, Darien Road,
Battersea, London SW11 2EQ

Tel/fax: 0171 738 1304

Old newspapers, books and
prints; wide selection of original
newspapers, books and
newspaper prints on Napoleonic
and Victorian campaigns.

**Railway Book and Model
Centre**
The Roundway, Headington,
Oxford, Oxfordshire OX3 8DH

Tel/fax: 01865 66215

Ray Westlake Military Books
53 Claremont, Malpas, Newport,
Gwent NP9 6PL, Wales

Tel: 01633 854135
Fax: 01633 821860

A mail order book firm offering a
service to historians, collectors
and modellers. Holding more
than 5,000 new and second-hand
military books in stock, we are
able to despatch orders within 24
hours of receipt. Orders can be
taken by post or telephone
(8.00am-8.00pm including
weekends). Quarterly lists are
published – send £1.00 in stamps.
The Ray Westlake Unit Archives
hold files dealing with the
histories, uniforms, badges and
organisation of some 6,000 units of
the British Army.

Read & Relax Books
28 High Street, Kinver,
Stourbridge, Midlands (W.) DY7
6HF

Contact: Ray Jones
Tel/fax: 01384 872596

Militaria reading for all.

Robin Turner
30 Great Norwood Street,
Cheltenham, Gloucestershire
GL50 2BH

Tel/fax: 01242 234303

Small stock of military books (pre
1914). Specialising in Napoleonic
period. Lists/catalogues issued 2
or 3 times a year.

Robinson Imports
25 Princetown Road, Bangor,
Co.Down BT20 3TA, N. Ireland

Contact: Cameron Robinson
Tel/fax: 01247 472860

Importer of RAFM 25mm figures
& equipment. SYW, French &
Indian Wars, ACW; also ACW
Society Magazine 'Zouave', ACW
books and related items.

Royal Air Force Museum
Grahame Park Way, Hendon,
London NW9 5LL

Tel: 0181 205 2266
Fax: 0181 200 1751

Britain's National Museum of
Aviation, displays 70 full size
aircraft. Open daily. Extensive
library research facilities
(weekdays only). Large free car
park. Licensed restaurant.
Souvenir shop with extensive
range of specialist books and
model kits.

Selous Books Ltd
40 Station Road, Aldershot,
Hampshire GU11 1HT

Contact: Trevor Hudson
Tel: 01252 333611
Fax: 01252 342337
Email: 100307,1735@compuserve

Specialist book dealers in military
history, historic travel and
ethnographic books. Subjects
covered include military histories,
regimental histories, uniforms,
insignia, medal rolls, firearms and
equipment.

Spink & Son Ltd
5 King Street, St James's, London
SW1 6QS

Contact: David Erskine-Hill
Tel: 0171 930 7888
Fax: 0171 839 4853

Spink stock reference books for
Orders, Decorations and Medals
of the World including their
standard price guide and 'British
Battles and Medals', the definitive
work on campaign awards. Please
contact the Book Department.
Open Monday to Friday, 9am to
5.30pm.

Stenvall's
Foreningsgatan 12, S-21144
Malmo, Sweden

Tel: (0046) 40 127703
Fax: (0046) 40 127700

Sweden's best stocked bookshop
for military and transport subjects.
Mail order catalogue issued
periodically. Shop open Mon-Fri
9am-6pm.

Stephen E Tilston
37 Bennett Park, Blackheath,
London SE3 9RA

Tel/fax: 0181 318 9181

Second-hand and out of print
military, maritime and naval
books bought and sold; WWI a
speciality. Lists issued;
booksearch service; postal
business only.

Steven J Hopkins
Court Farm, Kington, Flyford
Flavell, Hereford & Worcester
WR7 4DQ

Contact: Steven Hopkins
Tel/fax: 01386 793427

Napoleonic and Victorian
campaign histories, particularly
African wars. Send SAE for
catalogue. Good military books
and Zulu memorabilia purchased.

T R Robb
17 Thorney Bay Road, Canvey
Island, Essex SS8 OHG

Contact: Terry Robb
Tel: 01268 696054
Fax: 01268 681068
Email: terry.trrobb
@pop3.hiway.co.uk

Airgun tuning books, range
targets and airgun tuning parts,
tuning kits.

Terence Wise Military Books
Pantiles, Garth Lane, Knighton,
Powys LD7 1HH, Wales

Contact: Terence Wise

Mail order only. Military history,
3000 BC to AD 1990, specialising
in regimental histories. Catalogues
four per year. Also occasional
catalogues model soldiers,
wargame figures.

Thierry Lecourt
31 Rue Berthollet, 94110 Arcueil
94110, France

Tel/fax: (0033) 47353158

Tom Donovan Military Books
52 Willow Road, Hampstead,
London NW3 1TP

Tel/fax: 0171 431 2474

Private premises, appointment
necessary. Printed works,
documentation and manuscript
material depicting all aspects of
the Great War.

Tony Gilbert Antique Books
101 Whipps Cross Road,
Leytonstone, London E11 1NW

Contact: A S Gilbert
Tel/fax: 0181 530 7472

Storeroom – appointment
necessary. Naval, Military and
Aviation books, antiquarian and
second-hand; all periods,
especially Nelsonian, Napoleonic,
both World Wars. 'Wants' lists
welcome; annual catalogue.

Toytub
100a Raeburn Place, Edinburgh,
Lothian EH4, Scotland

Tradition H Zorn
Bettenfeld 21, 91541
Rothenburg, Germany

Tel/fax: (0049) 9861261 1

Traeme Selection
62 Boulevard Jean-Jaures, 92100
Boulogne, France

Traplet Publications Limited
Traplet House, Severn Drive,
Upton upon Severn, Hereford &
Worcester WR8 OJL

Tel/fax: 01684 594505

**Turner Donovan Military
Books**
1132 London Road, Leigh on Sea,
Essex SS9 2AJ

Contact: Brian Turner
Tel/fax: 01702 78771

Tuttostoria
Via S Sonnino 34, 43100 Parma,
Italy

Under Two Flags
4 Saint Christopher's Place,
London W1M 5HB

Contact: Jock Coutts
Tel/fax: 0171 935 6934

Stockist of toy soldiers, model
kits, military books, painted
figures & dioramas. Open
Mon-Sat, 10.00am-5.00pm.

Universal Soldier
10 Old Rectory Close, Instow,
Bideford, Devon EX39 4LY

V S-Books
PO Box 20 05 40, D-44635
Herne, Germany

Contact: Torsten Verhülsdonk
Tel: (0049) 2325 73818
Fax: (0049) 201 421396

Mail order supplier of German
and some English books on
militaria and weapons, also some
antiquarian titles.

**Van Nieuwenhuijzen
Modelshop**
Oude Binnenweg 91, 3012 JA
Rotterdam, Netherlands

Contact: Mike Lettinga
Tel: (0031) 10 4135923
Fax: (0031) 10 4141324

Victor Sutcliffe
36 Parklands Road, London
SW16 6TE

Tel: 0181 769 8345
Fax: 0181 769 6446

Opening times: at any time but by
appointment only. About 1500
books on Land Warfare before
1914. Specialities: Napoleonic and
British Colonial.

W E Hersant Ltd
17 The Drive, High Barnet,
Hertfordshire EN5 4JG

Contact: Nancy Kirkham
Tel/fax: 0181 440 6816
Email: herbooks@dircon.co.uk

We specialise in military history
books by mail order or from our
trade stand at various
wargaming/modelling shows.
Period lists on request.

Warrior Videos
38 Southdown Avenue, Brighton,
Sussex (E.) BN1 6EH

World War Bookfairs
Oaklands, Camden Park,
Tunbridge Wells, Kent TN2 5AE

Contact: Tim Harper
Tel/fax: 01892 538465

Organisers of six annual specialist
Military, Naval and Aviation
bookfairs in London, York,
Marlborough and Tunbridge
Wells.

World War I Books
Beeches House, 53 Park Hill,
Carshalton Beeches, Surrey SM5
3SE

Contact: Alan Jeffreys
Tel/fax: 0181 401 2108

Second-hand books on all aspects
of the First World War.
Catalogues issued; 'wants' list
welcomed; postal business only.

World War II Books
PO Box 55, Woking, Surrey
GU22 8HP

Contact: Graham Palmer
Tel: 01483 722880
Fax: 01483 721548

This is a mail order business. 12
catalogues issued each year.
'Wants' list available.

HISTORICAL SOCIETIES

1066 The Medieval Society
PO Box 1163, Parramatta, New South Wales 2124, Australia

Contact: Anne Davey
Tel: (02) 95191803
Fax: (02) 92507816
Email: adavey@soh.nsw.gov.au

Through re-enactment we aim to reproduce authentically the social and combat activities of the Kingdom of Briton, 960–1066 A.D.

1940 Association
43 The Drive, Ilford, Essex IG1 3HB

Contact: Michael Conway
Tel/fax: 0181 554 8169

The Association brings together all those interested in Britain's Home Front 1939-45. Membership includes individuals, museums, libraries, specialist groups, etc. Magazine.

95th Rifles Living History Group
10 Park Hill, Awsworth, Nottingham, Nottinghamshire NG16 2RD

Contact: William Whitlam

Educational displays of weapons, equipment & tactics of riflemen employed by His Majesty King George III. Private lectures/publicity venues by appointment. Beware of imitations.

A C M N
11 Rue Alexis Quirin, (Bat.Al), 94350 Villiers sur Marne, France

Contact: Daniel Poisson

Aldershot Militaria Society
1 Littlefield Gardens, Ash, Aldershot, Hampshire GU12 6LN

Contact: H W Glover
Tel/fax: 01252 21931

AMS meets last Thursday each month at Aldershot Library at 7.30pm; visitors welcome. Interests include medals, insignia, weapons, and all aspects of military history.

American Civil War Round Table (UK)
41 Templemere, Oatlands Drive, Weybridge, Surrey KT13 9PA

Contact: Paul Pilditch
Tel/fax: 01932 846150

Society for the study of the military, naval and civil history of the American Civil War 1861-65. Formed 1953; affiliated to CWRT Associates, USA. Meetings, talks, newsletter, library. Membership for amateur/professional historians/researchers £5 p.a. UK, £6 overseas.

American Civil War Society
PO Box 52, Brighouse, Yorkshire (W.) HD6 1JQ

Contact: Chris Wood
Tel/fax: 01625 265226

For 'living history' and re-enactment country-wide.

Arms & Armour Society
PO BOx 10232, London SW19 2ZD

Contact: A Dove

The society promotes the study, collection and conservation of arms and armour from earliest times to the present, by meetings, lectures, visits, newsletters and journals.

Army Records Society
Dept of History, University of Luton, Luton, Bedfordshire LU1 3AJ

Contact: Professor I F W Beckett

Publishes annual volume illustrating some aspect of the Army's past. Subscription only £15.00 per annum, although the volumes retail commercially at £40.00 or more.

Association of Friends of Waterloo Committee
2 Coburn Drive, Four Oaks, Sutton Coldfield, Midlands (W.) B75 5NT

Contact: John S White
Tel/fax: 0121 308 4103

The Association acts in support of the Belgian Waterloo Committee, conducting research, promoting appreciation and preserving the battlefield. Programme of meetings, lectures and visits; three journals annually.

Battlefields Trust
Meadow Cottage, 33 High Green, Brooke, Norfolk NR15 1HR

Contact: Michael Rayner
Tel/fax: 01508 558145

The Trust has been formed to help save battlefield sites from destruction and to preserve them for posterity as educational and heritage resources.

British Militaria Collectors Community
Keplerstrasse 12, 91056 Erlangen, Germany

Contact: Klaus Stuebiger
Tel: (0049) 9131 991446
Fax: (0049) 9131 991403

British Militaria Collectors' Community of Germany established 1991. Members throughout Germany. Main focus on 19th/20th century. Regular newsletter published. New members welcome.

British Plate Armour Society
82 Skinner Street, Poole, Dorset BH15 1RJ

Contact: David Barnes

Charles II Society
The Whitemoor, Codsall Wood, Wolverhampton, Staffordshire WV8 1RA

Contact: M Whittaker
Tel/fax: 01902 850363

A society formed to promote interest, at all levels, in the life and times of Charles II in particular, and the Stuart dynasty in general.

Chute and Dagger/ Europe
Bordes de Riviere, 31210 Montrejeau, France

Contact: A P Gaudet

We are 250 collectors of elite unit insignia from around the world. Subscription $19 US or equivalent for non-Europeans, 75 French francs or equivalent for Europeans. For a free sample copy of Chute and Dagger/ Europe illustrated newsletter, apply in writing.

Chute and Dagger/ UK
21 Old Brickfield Road, Aldershot, Hampshire GU11 3UE

Contact: Roy Turner
Tel/fax: 01252 25386

Parachutists and Special Forces insignia collectors group; by subscription.

Commonwealth Forces History Trust
37 Davis Road, Acton, London W3 7SE

Contact: Shamus Wade
Tel/fax: 0181 749 1045

Provides a history of the Defence Forces of the British Commonwealth and Empire from 1066 to 1945. Information on 10,380 different units from 82 countries.

Corps of Drums Society
62 Gally Hill Road, Church Crookham, Hampshire GU13 OR4

Contact: Reg Davies
Tel/fax: 01252 614852

The Corps of Drums Society for anyone who plays military drum or fife, or who wishes to study their history and preserve their future.

Crimean War Research Society
4 Castle Estate, Ripponden, Sowerby Bridge, Yorkshire (W.) HX6 4JY

Contact: David Cliff
Tel/fax: 01422 823529

The Society encourages research into every facet of the war; and has a quarterly journal 'The War Correspondent'. New members are always welcome.

Crown Imperial
37 Wolsey Close, Southall, Middlesex UB2 4NQ

Contact: Lt Cdr Maitland Thornton
Tel/fax: 0181 574 4425

This society was formed in 1973, to study traditions and regalia of the forces of the crown. Four journals circulated annually.

Fairfax Historical Associates

The Whitemoor, Codsall Wood, Wolverhampton, Staffordshire WV8 1RA

Contact: M Whittaker
Tel/fax: 01902 850363

Freelance historians specialising in period 1485–1901. History workshops for schools. Lectures and courses for adults. Costumed recreations for museums and historic sites. Details available on request.

Fortress Study Group

The Severals, Bentleys Road, Market Drayton, Shropshire TF9 1LL

Contact: B C Lowry
Tel/fax: 01630 653433

An international group, founded in 1975, to promote public knowledge and awareness of fortifications since the 16th century. Publish journal and thrice yearly newsletter.

Heraldry Society

44-45 Museum Street, London WC1A 1LY

Contact: Mrs Marian Miles
Tel/fax: 0171 430 2172

The Society exists to increase and extend interest in, and knowledge of Heraldry, Armory, Chivalry and Genealogy and is a registered Educational Charity.

Holts' Tours (Battlefields & History)

15 Market Street, Sandwich, Kent CT13 9DA

Contact: John Hughes-Wilson
Tel: 01304 612248
Fax: 01304 614930
Email: www.battletours.co.uk

Europe's leading military historical tour operator, offering annual world-wide programme spanning history from the Romans to the Falklands War. Holts' provides tours for both the Royal Armouries Leeds and the IWM. Every tour accompanied by specialist guide-lecturer. Send for free brochure.

Indian Military Historical Society

37 Wolsey Close, Southall, Middlesex UB2 4NQ

Contact: Lt Cdr Maitland Thornton
Tel/fax: 0181 574 4425

Formed in 1983 to study the military history of the Indian sub-continent. Subjects include medals, uniforms, insignia of all forces including police. Four journals are published annually.

The Keep Military Museum

The Keep, Bridport Road, Dorchester, Dorset DT1 1RN

Contact: Len Brown
Tel: 01305 264066
Fax: 01305 250373

Covers Devonshire Regiment, Dorset Regiment, The Devonshire and Dorset Regiment, Queens Own Dorset Yeomanry, The Royal Devonshire Yeomanry, Royal North Devonshire Hussars, Dorset Yeomanry and 94th Field Regiment, RA.

Kuratorium zur hist. Waffensammlungen

Staudenweg 28, 44265 Dortmund, Germany

Linnet the Seamstress

19 Cowper Close, Mundesley, Norfolk NR11 8JS

Tel/fax: 01263 721574

15th century costume, eg. shirts, braies, hose, jerkins, kirtles, gowns. Personal attention, reasonable prices. List on request. Other periods considered. Accuracy & hand finishing paramount. 'Living history' & talks on request. Member of The Company of the White Lion.

Medieval Archery Society of Australia

Flat 4/22, Highbury Grove, Kew, Victoria 3101, Australia

Contact: Stephen Wyley

Medieval Combat Society

Flat 2, 93 Surbiton Road, Kingston, Surrey KT1 2HW

Contact: David Debono
Tel/fax: 0181 974 8101
Email: david @nouarltd.demon.co.uk

We are a society made up from all ages and walks of life and recreate the period of Edward III, one of England's greatest kings.

Medieval Society of England

7 Markstakes Corner, South Chailey, Sussex (E.) BN8 4BP

Contact: Dr K Fry
Tel/fax: 01273 400037

Research into all aspects of the middle ages. Lectures to societies and schools. Meetings once a year. Postal library. SAE membership details.

Mercia Military Society

17 Barne Close, Nuneaton, Warwickshire CV11 4TP

Contact: Joe Lawley

Meets at 7.30pm on the second Monday in every month, at Nuneaton Chilvers Coton Liberal Club, for lectures/discussions on military subjects. Membership £5.00 p.a.

Military Heraldry Society

37 Wolsey Close, Southall, Middlesex UB2 4NQ

Contact: Lt Cdr W Maitland Thornton
Tel/fax: 0181 574 4425

The society specialises in researching and exchanging cloth military insignia worldwide. These include formation signs, shoulder titles and other cloth items. Four journals published annually.

Military Historical Society

National Army Museum, Royal Hospital Road, London SW3 4HT

Contact: Robin Hodges
Tel: 01980 630613
Fax: 01980 615305

Military history of the Forces of the Crown. The society for collectors and researchers. 1,200 members receive quarterly journal and annual reference book. Subscription £10 pa.

Military History Society of Ireland

c/o University College Dublin, Newman House, 88 St Stephens Green, Dublin 2, Ireland

Contact: F G Thompson

Naval Historical Colls & Research Assoc

Flat 3, 19 Royal Crescent, London W11 4SL

Contact: David Rolfe
Tel/fax: 0171 602 6866

For those who research or collect Naval subjects or with a general interest in Naval history, from the Nelson era to Desert Storm.

New England Medieval Arts Society

201 Mann Street, Armidale, New South Wales 2350, Australia

Contact: Adam Cafarella

The Normandy Arnhem Society

22 Cousin Lane, Illingworth, Halifax, Yorkshire (W.) HX2 8AF

Contact: David P Mitchell
Tel/fax: 01422 256891

The Normandy Arnhem Society is a living, breathing museum of remembrance dedicated to keeping history alive by recreating, as accurately as possible, the life and times of German and British soldiers during the latter half of WWII. Comprises 9./SS-Pz.Gr.Rgt.20 and 9 Field Company RE. Specialises in 'living history' displays for museums.

North Eastern Militaria Society

2 Meadowgate, Eston under Nab, Middlesbrough, Cleveland TS6 9JB

Contact: Amanda Luker
Tel/fax: 01642 454609

Meetings first Thurs each month at Eaglescliffe Hotel, Stockton. Varied interests – raffles – quarterly newsletter. New members welcome.

Ordnance Society

3 Maskell Way, Southwood, Farnborough, Hampshire GU14 0PU

Contact: Ian McKenzie
Tel/fax: 01252 521201

Society for the furthering of research into ordnance, artillery and all related subjects.

Pa Engliscan Gesipas

Gerefa, 38 Cranworth Road, Worthing, Sussex (W.) BN11 2JF

Contact: Janet Goldsbrough-Jones
Tel/fax: 01903 207485

Society for the study of Dark Ages subjects.

Police Insignia Collectors Association
43 Hunters Way, Saffron Walden, Essex CB11 4DE

Contact: Steve Daly

The Association is the only organisation in mainland Britain devoted entirely to the study and collection of police insignia world-wide. Annual fees £10 (UK & BFPO), £12 (Europe), £15 (rest of world); initial joining fee £5.

Regia Anglorum (Admin body)
9 Durleigh Close, Headley Park, Bristol, Avon BS13 7NQ

Contact: J K Siddorn
Tel/fax: 0117 9646818

An international society of over five hundred people, dedicated to the authentic re-creation of Vikings, Saxons and Normans (approximately 950 to 1066). We have a very extensive working 'living history' & craft exhibit. Early music and many crafts are available. Our wooden 'Viking' ship replica is over forty feet long, with its own experienced, costumed & equipped crew. Battle re-enactments involving over 200 people are available, as are KS2 school visits. Branches everywhere!

Richard III Society
4 Oakley Street, Chelsea, London SW3 5NN

Contact: E M Nokes
Email: enokes@rcpsych.ac.uk

The Society promotes research and interest in the field of 15th Century history in general, and the life of King Richard III in particular.

Roman Research Society
8 Leechmere Way, Ryhope, Sunderland, Tyne & Wear SR2 0DH

Contact: Eddie Barrass
Tel/fax: 0191 523 6377

Royal Marines Historical Society
2A Seaview Road, Drayton, Portsmouth, Hampshire PO6 1EN

Contact: Major A P Whitehead

Royal Navy Enthusiasts Society
5 Midways, Stubbington, Fareham, Hampshire PO14 2DA

Contact: D J Maxted

Interested in the Royal Navy, its customs and traditions? Do you collect tallies, crests, badges, histories, literature? Then join the RNES.

La Sabretache
7 Rue Guersant, 75017 Paris, France

Tel/fax: (0033) 1 45726410

Long-established society for specialists of military history, uniforms and tin soldiers. Quarterly bulletin issued. Membership carries free admission to Paris army and navy museums.

Salisbury Militaria Society
Red Lion Hotel, Milford Street, Salisbury, Wiltshire SP1 2AN

Contact: M Maidment
Tel: 01722 323334
Fax: 01722 325756

Meets third Wednesday of each month at above address for talks and discussion on all aspects of military history and memorabilia.

Scottish Military Club
103 Taylor Street, Methil, Fife KY8 3AY, Scotland

Contact: Alexander Ferrier
Tel: 01333 429692
Fax: 01333 429244

The aims of the club are to encourage wargaming, modelling, art, re-enactment and general interest in military history.

Scottish Military Historical Society
4 Hillside Cottages, Glenboig, Lanarkshire ML5 2QY, Scotland

Contact: Tom Moles

The Society exists to encourage the study of Scottish military history and publishes its own illustrated journal, covering the collecting of badges, headdress, uniforms, medals, photographs, postcards, prints, watercolours, equipment, pistols, powderhorns etc. The SMHS has a world-wide membership.

Société des Amis du Musée de l'Armée
Musee de l'Armee, Hotel National des Invalides, 75007 Paris, France

Contact: Mlle Pierron

Society for Army Historical Research
c/o National Army Museum, Royal Hospital Road, London SW3 4HT

Contact: The Hon Secretary

The senior British military historical society, founded in 1921. Publishes quarterly Journal covering history of British and Commonwealth land forces; annual subscription £22.00 (UK membership).

Society of Archer-Antiquaries
61 Lambert Road, Bridlington, Yorkshire (E.) YO16 5RD

Contact: Douglas Elmy

Society of Friends of the National Army Museum
c/o National Army Museum, Royal Hospital Road, London SW3 4HT

Contact: Derek A Mumford
Tel/fax: 0171 730 0717

The Society of Friends of the National Army Museum assist the Museum in the acquisition of significant militaria; and members enjoy lectures, private views, battlefield and Army establishment excursions, and newsletters. Annual subscription £8.00; contact Secretary/Treasurer above.

Soke Military Society
45 Warwick Road, Walton, Peterborough, Cambridgeshire PE4 6DE

Contact: Roger Negus
Tel/fax: 01733 578842

Varied membership of historians, collectors, modellers and wargamers. Monthly meeting every second Wednesday at the Peterborough Museum, with guest speaker. A monthly newsletter is also distributed. Subscription £3.00.

South & Central American Military Society
27 Hallgate, Cottingham, Yorkshire (E.) HU16 4DN

Contact: Terry D Hooker
Tel/fax: 01482 847068
Email: scamhs@aol.com

Publishers of quarterly journal 'El Dorado' covering from pre-Columbian times to present day; English and Spanish text, with illustrations and book reviews, all areas of military history.

Victorian Military Society
20 Priory Road, Newbury, Berkshire RG14 7QN

Contact: Dan Allen
Tel/fax: 01635 48628

The leading society in this field promotes research into all aspects of military history 1837-1914, wargames, re-enactment, etc; and publishes a quartery journal.

The Vikings (N F P S) (Admin body)
2 Stanford Road, Shefford, Bedfordshire SG17 5DS

Contact: Sandra Orchard
Tel/fax: 01462 812208

As the original re-enactment society, The Vikings – Norse Film and Pageant Society – captures the atmosphere of the Dark Ages with a unique blend of authenticity and humour. Our Vikings, Celts, Saxons and Normans encapsulate the whole flavour of Medieval life, from the simple craftsman to the professional warrior. We have performed at home and abroad, and with equal ease for film crews and schoolchildren. Contact your local Hird now.

Welsh Maritime Association
257 Clydach Road, Morriston, Swansea, Glamorgan (W.) SA6 6QJ, Wales

Contact: Robert Morgan
Tel/fax: 01792 797185

Formed in 1982 to promote and encourage all aspects of maritime research and hobbies in Wales. Corresponding members throughout Europe and the world.

Western Front Association (1914-1918)
PO Box 1914, Reading, Berkshire RG4 7YP

Formed to perpetuate the memory of the courage and comradeship of all who served in France and Flanders. Open to members of all ages and both sexes.

WWII Railway Study Group
17 Balmoral Crescent, West Molesey, Surrey KT8 1QA

Contact: Greg Martin

Group for promotion and research into all aspects of railways during the War. A bi-monthly bulletin is published. Send SAE for details.

ILLUSTRATORS, ARTISTS & DISPLAY FIGURES

A Buttery
6 Gypsy Lane, Oulton, Leeds,
Yorkshire (W.) LS26 8SA

Admiralty House Publications
PO Box 6253, Los Osos,
California 93412, USA
Contact: Beth Queman
Tel: (001) 805 534 9723
Fax: (001) 805 534 9127
Email: lbj.greene@thegrid.net
Computer generated maps and
artwork for both books and
games. All military topics.

Alfio Serafini
15 Oak Close, Kilamarsh,
Sheffield, Yorkshire (S.) S31 8FB
Tel/fax: 0114 2482022
Specialist in equestrian and
Napoleonic oil paintings and
unique hand-made figures, any
size or configuration. Regimental
and private commissions
undertaken.

Alix Baker
Exmoor House, Castle Hill,
Brenchley, Kent TN12 7BL
Contact: Alix Baker
Tel/fax: 01892 722866
Postcards, prints, paintings. Many
regiments in stock. Internationally
collected artist working full-time
for British Army units. See
advertisement for details in
Paintings, Prints & Postcards
section.

Angus McBride
c/o Scorpio Gallery, PO Box 475,
Hailsham, Sussex (E.) BN27 2SH
Contact: Hilary Hook

Bryan Fosten
5 Ross Close, Nyetimber,
Bognor Regis, Sussex (W.)
PO21 3JW
Contact: Bryan Fosten

C & C Military Fine Art
98 Wycombe Road, Marlow,
Buckinghamshire 5L7 3JE
Contact: Mark Churms
Tel: 01628 483985
Fax. 01628483985
Military and equestrian art,
original oil paintings by Mark
Churms (commissions and image
licences available). Full range of
Cranston Fine Arts prints.
Visa/Mastercard accepted.

Christopher Collingwood
1 Barton Cottges, Monkleigh,
Bideford, Devon EX39 5JX
Tel/fax: 01805 623023
England's foremost military artist
– work includes ECW, American
Civil War, Roman and Norman,
Medieval, Jacobite, Napoleonic,
First and Second World Wars.
Private commissions undertaken.

Clive Farmer
6 Churchway, Faulkland, Bath,
Somerset BA3 5US
Tel/fax: 01373 834752
Military artist. Accurately
researched illustrations. Staff artist
to the Crimean War Research
Society. Prints and paintings
always in stock. Commissions
undertaken

David Cartwright
Studio Cae Coch Bach,
Rhosgoch, Anglesey, Gwynedd
LL66 0AE, Wales
Contact: Sara Cartwright
Tel/fax: 01407 710801
Military artist specializing in
Napoleonic and Crimean scenes
on canvas. Commissions
undertaken. Contact for further
details of originals and limited
edition prints.

David Rowlands
6 Saville Place, Clifton, Bristol,
Avon BS8 4EJ
Contact: David Rowlands
Tel/fax: 0117 9731722
Military artist. Military prints.

Douglas N Anderson
37 Hyndland Road, Glasgow,
Strathclyde G12 9UY, Scotland
Contact: Douglas N Anderson
Tel/fax: 0141 339 8381
Professional artist specialising
(most media) in military and
historic costume, male and female,
particularly of Scotland. Wide
knowledge of Scottish/Highland
regiments and Highland dress
ancient and modern.

Gallery Militaire
1 Holstock Road, Ilford, Essex
IG1 1LG
Contact: Rodney Gander
Tel: 0181 478 8383
Fax: 0181 533 4331
Fine and investment art dealers
and publishers, supplying original
paintings, limited edition prints,
reproductions, plates and
postcards. All types of framing
and art commissions undertaken.
European dealers and distributors
for major military artists. Gallery
viewing by appointment. Mail
order. Large A4 illustrated
catalogue £3.00 UK, £4.00 Europe
$10.00 USA Airmail.

Heinz Rode
Karl Marx-Allee 141, 10243
Berlin, Germany

Ian White Models
238 Taunton Avenue, Whitleigh,
Plymouth, Devon PL5 4EW
Contact: Ian White
Tel/fax: 01752 768507
Completely new figures. Large
scale (1:5.5) historical and modern
military uniform figurines.
Available complete or as kits.
Resin and white metal. Easy
assembly. Larger commissions
accepted.

Inkpen Art Productions
12 Westdene Crescent,
Caversham, Reading, Berkshire
RG4 7HD

John Wayne Hopkins
Dept.MD, Gatooma, 58 Queen
Victoria Road, Llanelli, Carms
SA15 2TH, Wales
Tel/fax: 01554 750761
Artist specialising in 20th century
military and aviation paintings in
oils. Regularly commissioned by
British Army. Rhodesian Fireforce
and Army Air Corps limited
edition prints available.
Commissions accepted. Presently
researching Dark Ages and
Medieval periods and Battle of
Bosworth for a series of Welsh
history paintings. SAE for free
details.

Kevin Lyles
24 Victoria Road, Berkhamsted,
Hertfordshire HP4 2JT

Mike Chappell
14 Downlands, Deal, Kent CT14
7XA

Military History Workshop Intl
cp 231, 36078, Valdagno, VI, Italy
Contact: Alessandro Massignani
Tel: (0039) 445 407968
Fax: (0039) 445 406894

**Morrison Frederick
(Tableaux)**
Studio 5D, 1 Fawe Street,
London E14 6PD
Contact: Jasper Lyon
Tel/fax: 0171 515 4110

Pan European Art
Lansbury Business Estate, Lower
Guildford Road, Knaphill,
Woking, Surrey GU21 2EP
Contact: Clive Jackson
Tel: 01483 799550
Fax: 01483 799660
Producers of fine art oil paintings,
primarily from customers'
photographs / artwork.
Specialists in the area of
uniformed personnel, often with
spouse and modes of
transportation that represent their
profession. Company also paint
special limited editions in oils or
watercolour of specific events in
military, regimental or public
service history. Commission a fine
work of art at affordable prices.

Patrice Courcelle
38 Avenue Des Vallons, B-1410
Waterloo B-1410, Belgium

Contact: Patrice Courcelle
Tel/fax: (0033) 322 354 3607
Email: courcelle@linkline.be

Illustrator and painter. Main periods: American Revolution, French Revolution & Napoleonic wars. Publisher of the plate series 'Ceux Qui Bravaient l'Aigle' and of the 'Waterloo' prints.

Paul Hannon
90 Station Road, Kings Langley, Hertfordshire WD4 8LB

Peter Dennis
28 The Park, Mansfield, Nottinghamshire NG18 2AT

Tel/fax: 01623 654402

All military subjects, with a particular interest in the American Civil War. Portrait reconstructions regularly published in 'Gallery' feature in 'Military Illustrated Past & Present'.

Pierre Conrad
Residence Mozart, 691 Avenue de la Liberation, 77350 La Mee sur Seine, France

R I G O / Le Plumet
Louannec, 22700
Perros-Guirrec, France

Richard & Christa Hook
158 Mill Road, Hailsham, Sussex (E.) BN27 2SH

Richard Kernick
Wrecclesham Road, Farnham, Surrey GU10 4PS

Tel/fax: 01252 733255

Scorpio Gallery
PO Box 475, Hailsham, Sussex (E.) BN27 2SH

Contact: Hilary Hook

Agents for resale of original paintings by leading illustrators, eg Richard Hook, Angus McBride, Gerry Embleton etc.

Simon McCouaig
4 Yeomans Close, Stoke Bishop, Bristol, Avon BS9 1DH

Time Machine GA
La Chaine 15, 2515 Preles, Switzerland

Contact: Gerry Embleton
Tel: (0041) 32 315 2393
Fax: (0041) 32 315 1793

Realistic, accurately costumed life-size mannequins for museum and other exhibition settings. Specialist in military and other historical costumes, arms and armour; research for TV and film productions, etc., to the most exacting standards. See our work in the National Army Museum, London (Napoleonic and 19th century galleries).

William Sumpter Models
Barton End House, Bath Road, Nailsworth, Gloucestershire GL6 0QQ

Contact: Geoff Sumpter
Tel/fax: 01453 833883

Models, props, themed special effect dioramas amd murals for museums, film etc. Tiny scale to life-size and beyond. Quality work by a team of specialist staff, we work within your budget and to your deadline.

LIBRARIES & RESEARCH SERVICES

Admiralty Research Establishment
Southwell, Portland, Dorset DT5 2JS

Contact: A E G Wilson
Tel/fax: 01305 820381

Airborne Forces Museum
Browning Barracks, Aldershot, Hampshire GU11 2BU

Contact: Diana Andrews
Tel/fax: 01252 349619

The largest collection of airborne weapons and equipment ever assembled. A concise history of British Airborne Forces from their WWII formation to the present day.

Andrew Mollo
10 Rue Jean Juares, 03320 Lurcy-Levis, France

Tel/fax: (0033) 70679183

Military consultant to film and TV production companies; author and collector, specialising in German and Russian subjects; publisher of Historical Research Unit books.

Army Museums Ogilby Trust
58 The Close, Salisbury, Wiltshire SP1 2EX

Contact: Antony Makepeace-Warne
Tel: 01722 332188
Fax: 01722 334211

Registered charity (No.250907) supporting all Regimental and Corps Museums of the British Army; focus on research and education; publishers of military reference works.

B D I C – Musée d'Histoire
Hotel National des Invalides, 75007 Paris, France

Tel: (0033) 144425491
Fax: (0033) 144189384
Email: mhc_bdic@club_internet.fr.

Important documentary, art and photo archive, both World Wars; limited space so partial displays only, though occasional special exhibitions. Closed Sun, Mon, & August.

Britannia Royal Naval College
Dartmouth, Devon TQ6 0HJ

Contact: R J Kennell
Tel: 01803 837279
Fax: 01803 837015

Confederate States Navy
HQ Redoubt Fortress, Royal Parade, Eastbourne, Sussex (E.) BN20 8BB

Tel/fax: 01323 410300

Archives and library. 1858–1870. Great Britain and America. Navies and blockade running. Artifacts. Lectures. 'Living history' section with howitzer. Secretary call 01273 4000037.

Duke of Cornwall's Light Infantry Museum
The Keep, Bodmin, Cornwall PL31 1EG

Contact: Hugo White
Tel/fax: 01208 72810

History of the Duke of Cornwall's Light Infantry, and the local Cornish forces. Excellent display of uniforms, medals and weapons. Very comprehensive reference library.

Helion & Company
26 Willow Road, Solihull, Midlands (W.) B91 1UE

Contact: Duncan Rogers
Tel: 0121 705 3393
Fax: 0121 711 1315

Expert bibliographic research for all periods of military history.

International Institute for Strategic Studies
23 Tavistock Street, London WC2E 7NQ

Contact: Hilary Oakley
Tel: 0171 379 7676
Fax: 0171 836 3108
Email: oakley@iiss.org.uk

Open 10am-5pm Mon to Thurs; 10am-12pm Fri; no appointment necessary; entry fee £2.00 per day (students), £5.00 per day all other categories. Subject coverage: security, arms control, international relations. Services: research facilities incl. 6,000 books, 12,000 pamphlets, press library, CD-Roms and on-line services. Enquiries by phone or letter, please.

The Keep Military Museum
The Keep, Bridport Road, Dorchester, Dorset DT1 1RN

Contact: Len Brown
Tel: 01305 264066
Fax: 01305 250373

Covers Devonshire Regiment, Dorset Regiment, The Devonshire and Dorset Regiment, Queens Own Dorset Yeomanry, The Royal Devonshire Yeomanry, Royal North Devonshire Hussars, Dorset Yeomanry and 94th Field Regiment, RA.

Major Book Publications
Hougoumont, Maxworthy, Launceston, Cornwall PL15 8LZ

Contact: H.W. or C.P. Adams
Tel/fax: 01566 781422

Military research material. Books and our unique, comprehensive archival data sheet system. 'British Army at Waterloo' identifies regimental arrivals, departures, dates, locations, formations, commanders, strengths, losses plus Allied Army of Occupation details. SAE essential.

Musée de l'Armée
Hotel National des Invalides, 75007 Paris, France

Tel/fax: (0033) 145559230

The French national museum of military history; not to be missed by any Paris visitor.

National Army Museum Reading Room
Royal Hospital Road, Chelsea, London SW3 4HT

Contact: Linda Washington
Tel: 0171 730 0717
Fax: 0171 823 6573

About 30,000 books; over 200 current historical and regimental journals; archives; prints and drawings; photographs. Normally open Tues-Sat, 10am-4.30pm, to holders of readers' tickets, for which application must be made in advance on forms available from the Museum's Department of Printed Books.

National Maritime Museum
Park Row, Greenwich, London SE10 9NF

Tel/fax: 0181 858 4422

Britain's national museum of naval and maritime history, set in magnificent surroundings by the Thames at Greenwich. Historic exhibits, models, art collection, documentary and pictorial archives. Open September to April, Mon-Sat 10am-5pm, Sun 12am-5pm; May-October, closes 6pm.

Positive Proof
17 Argyll St, Shrewsbury, Shropshire SY1 2SF

Tel/fax: 01743 240960

Specialist photographic service with comprehensive photo-index who will endeavour to locate and photograph any 1914–18 war grave situated within the Ypres Sallent, Belgium.

Prince Consort's Army Library
Knolly's Road, Aldershot, Hampshire GU11 1PS

Contact: P H Vickers
Tel/fax: 01252 24431 X328

Public Record Office
Ruskin Avenue, Kew, Richmond, Surrey TW9 4DU

Contact: The Keeper
Tel: 0181 876 3444
Fax: 0181 878 8905

The national archive; contains records of the British Army and its personnel from 1660, naval archives, etc. Open Mon-Fri, 9.30am-5pm, admission free.

Ray Westlake Military Books
53 Claremont, Malpas, Newport, Gwent NP9 6PL, Wales

Tel: 01633 854135
Fax: 01633 821860

A mail order book firm offering a service to historians, collectors and modellers. Holding more than 5,000 new and second-hand military books in stock, we are able to despatch orders within 24 hours of receipt. Orders can be taken by post or telephone (8.00am-8.00pm including weekends). Quarterly lists are published – send £1.00 in stamps. The Ray Westlake Unit Archives hold files dealing with the histories, uniforms, badges and organisation of some 6,000 units of the British Army.

Royal Air Force Museum
Grahame Park Way, Hendon, London NW9 5LL

Tel: 0181 205 2266
Fax: 0181 200 1751

Britain's National Museum of Aviation, displays 70 full size aircraft. Open daily. Extensive library research facilities (weekdays only). Large free car park. Licensed restaurant. Souvenir shop with extensive range of specialist books and model kits.

Royal Army Medical College
Millbank, London SW1P 4RJ

Contact: P R R Gibbs
Tel/fax: 0171 834 9060 X220

Royal College of Defence Studies
Seaford House, 37 Belgrave Square, London SW1X 8NS

Contact: Cdr T Binney
Tel/fax: 0171 235 1091 X210

Royal Engineers Library
Brompton Barracks, Chatham, Kent ME4 4UG

Tel/fax: 01634 822416

Founded in 1813, this remarkable archive contains printed books, manuscripts, maps, plans and photographs. A unique source for scholars and family history. Open by appointment.

Royal Marines Museum
Southsea, Portsmouth, Hampshire PO4 9PX

Contact: Jorj Jarvie
Tel: 01705 819385
Fax: 01705 838420

The history of the Royal Marines from 1664 to the present. Open 7 days a week. Library and archives by arrangement.

Royal Military Academy Sandhurst
Central Library, Camberley, Surrey GU15 4PQ

Contact: J W Hunt
Tel/fax: 01276 63344 X367

Royal Naval College
Greenwich, London SE10 9NN

Contact: D Male
Tel/fax: 0181 858 2154

Sunset Militaria
Dinedor Cross, Herefordshire HR2 6PF

Contact: David Seeney
Tel/fax: 01432 870420

Military research. Medal card and computer check 1914 1922 £4.00 + SSAE when number, regiment known; £6.00 + SSAE when details vague. For details of this and other research, all periods, send SSAE.

T R Military Search
65A Wixs Lane, London SW4 0AH

Contact: Elizabeth Talbot Rice
Tel/fax: 0171 228 5129

Expert research into the military records of the British, the EIC's and the Indian army held by the Public Record Office, the Oriental and India Office Library and the National Army Museum.

Technical Info Centre Royal Engineers
Drayton Camp, Barton Stacey, Winchester, Hampshire SO21 3NH

Contact: J M Seaman
Tel/fax: 0126 472 372 X366

Time Machine GA
La Chaine 15, 2515 Preles, Switzerland

Contact: Gerry Embleton
Tel: (0041) 32 315 2393
Fax: (0041) 32 315 1793

Realistic, accurately costumed life-size mannequins for museum and other exhibition settings. Specialist in military and other historical costumes, arms and armour; research for TV and film productions, etc., to the most exacting standards. See our work in the National Army Museum, London (Napoleonic and 19th century galleries).

V S-Books
PO Box 20 05 40, D-44635 Herne, Germany

Contact: Torsten Verhülsdonk
Tel: (0049) 2325 73818
Fax: (0049) 201 421396

Research and consultancy to the media/theatre/museums on weapons, costume and uniforms.

Uniforms & Medals Research Service
Phoenix Enterprises (Crewe), 12 Bude Close, Crewe, Cheshire CW1 3XG

Contact: Megan Robertson
Tel/fax: 01270 500 565

Research into military and other uniforms, medals, insignia, ceremonial &c. of all nations, all periods, for modellers, re-enactors, historians, film-makers, museums. SAE for individual quote.

Warship Preservation Trust
HMS Plymouth, Dock Road, Birkenhead, Merseyside PL41 1DJ

Tel: 0151 650 1573
Fax: 0151 650 1473

Historic warships at Birkenhead – Falklands War veterans, the former Royal Navy frigate 'Plymouth' and submarine 'Onyx' are open to the public daily from 10am. New – German U Boat 534 now open.

PHOTOGRAPHIC & VIDEO

Belle & Blade Distributors (Video)
124 Penn Avenue, Dover, New Jersey 07801, USA

Contact: Steve Mormando
Tel/fax: (001) 201 328 8488

All videos ever printed are available! War, history, swashbuckling, classics, Westerns, horror – over 20,000 NTSC films in stock. Also swords, knives, books and toys from Belle & Blade Distributors; catalogues $10.00 US.

Black & White Photo Services
19 Butterworth Street, Chadderton, Lancashire OL9 OJL

Contact: Derek Denton
Tel/fax: 0161 620 4064

Old black and white photographs copied, military and other quality hand prints, reasonable rates. Send SAE for price list or telephone for details. All work returned recorded delivery.

Brian V Thomas
8 Russell Drive, Wollaton, Nottingham, Nottinghamshire NG8 2BH

Tel/fax: 0115 9281451

Brian V.Thomas offers a service of photographing individual war graves in Northern Europe, and research into locations of war graves anywhere in the world.

Chris Honeywell
c/o Collections Pic. Library, 13 Woodberry Crescent, London N10 1PJ

Contact: Brian Shuel
Tel: 0181 883 0083
Fax: 0181 883 9215
Email: collections@btinternet.com

This library holds a large selection of photographer Chris Honeywell's colour photographs of English Civil War re-enactment and reconstruction subjects, as published in his 'The English Civil War Recreated in Colour Photographs' (Windrow & Greene, 1993).

Classic Images
PO Box 1863, Charlbury, Chipping Norton, Oxfordshire OX7 3PD

Contact: Jef Savage
Tel/fax: 01608 676635

American Civil War specialists for 'Gettysburg' – the epic motion picture; re-enactment videos, books, prints, music, newspapers, maps, clothing patterns, toy soldiers and product search service.

Duke Marketing Ltd
PO Box 46, Milbourn House, St Georges Street, Isle of Man IM99 1DD

Tel: 01624 623634
Fax: 01624 629745
Email: duke.video@enterprise.net

Illustrated Entertainment
14 Overfield Road, Stopsley, Luton, Bedfordshire LU2 9JU

Contact: Paul Chamberlain
Tel/fax: 01582 616674

Photography for re-enactors. Copywork onto transparency for lectures a speciality.

International Historic Films, Inc.
Dept MM, Box 29035, Chicago, Illinois 60629, USA

Contact: Peter Bernotas
Tel: (001) 773 927 9091
Fax: (001) 773 927 9211
Email: intrvdeo@2x.netcom.com

Military, political and social history on videocassette. Over 600 original newsreels, documentaries and feature films including Nazi and Soviet propaganda films. All titles on VHS and all world TV standards; write ot FAX for free catalogue.

Mayhem Photographics (History in Camera)
Basement Flat, 10 Alexandra Road, St Leonards on Sea, Sussex (E.) TN37 6LE

Contact: Dick Clark

Military Features & Photos
20 Blamire Drive, Binfield, Bracknell, Berkshire RG42 4UN

Contact: John Norris
Tel/fax: 01344 424254

The agency supplies military pictures & historic prints for a variety of purposes including audio-visual presentations. Commissions also undertaken. Special re-enactor photographic files.

Military Scene (Photo Library)
28B Mill Street, Ottery St.Mary, Devon EX11 1AD

Contact: Bob Morrison
Tel/fax: 01404 814164

Library holding over 40,000 transparencies of military subjects for use by publishers and defence manufacturers. Includes images from Operations Granby, Desert Sabre, Hanwood and Grapple.

Photo Press Defence Pictures
Glider House, 14 Addison Road, Plymouth, Devon PL4 8LL

Contact: David Reynolds
Tel/fax: 01752 251271

A unique source of military photography, covering all areas of the Armed Forces, which is backed up by an innovative computer system which can pinpoint the best image to suit the client's requirements. Our images span from the Falklands to the streets of Ulster, and the Gulf, where Photo Press photographers were among the first cameramen into liberated Kuwait City.

Positive Proof
17 Argyll St, Shrewsbury, Shropshire SY1 2SF

Tel/fax: 01743 240960

Specialist photographic service with comprehensive photo-index who will endeavour to locate and photograph any 1914–18 war grave situated within the Ypres Salient, Belgium.

Running Wolf Productions (Video etc)
Crogo Maws, Corsock, Castle Douglas, Kirkcudbrightshire DG7 3DR, Scotland

Contact: M Loades
Email: mike @runningwolf.softnet.co.uk

Video production company – suppliers of 'Archery – Its History and Forms' and 'The Blow by Blow Guide to Swordfighting'. Mike Loades also available for lectures, workshops and fight arranging assignments.

V S-Books
PO Box 20 05 40, D-44635 Herne, Germany

Contact: Torsten Verhülsdonk
Tel: (0049) 2325 73818
Fax: (0049) 201 421396

Transparencies and prints of modern military and police subjects for illustration and advertising; also re-enactments of various periods; some historical material; commissions undertaken.

Warrior Videos
38 Southdown Avenue, Brighton, Sussex (E.) BN1 6EH

MAGAZINES

39/45 Magazine
Editions Heimdal, BP 320,
Chateau de Damigny, F-14400
Bayeux, France

Contact: Bernard Paich
Tel: (0033) 2 31516868
Fax: (0033) 2 31516860

A B M Magazine
96 Rue de Paris, 92100
Boulogne, France

Contact: Nicolas Draeger

Air International
PO Box 100, Stamford,
Lincolnshire PE9 1XQ

Contact: Malcolm English
Tel: 01780 755131
Fax: 01780 757261
Email: english
@keymags.demon.co.uk

Aircraft Illustrated
Ian Allan, Terminal House,
Shepperton, Middlesex
TW17 8AS

Contact: Allan Burney

Antenna
Forest Publishing, 15 Welland
Close, New Road, St. Ives,
Huntingdon, Cambridgeshire

Tel: 01480 300661
Fax: 01480 62286

Armees d'Aujourd'hui
6 Rue St Charles, 75015 Paris,
France

Contact: Bruno Nielly

The Armourer Magazine
Beaumont Publishing Ltd, 25
Westbrook Drive, Macclesfield,
Cheshire SK10 3AQ

Contact: Irene Moore
Tel/fax: 01625 431583
Email: editor
@armourer.u_net.com

The Armourer magazine for
militaria collectors is published
bi-monthly. Written by collectors
for collectors and sold at militaria
fairs and outlest throughout the
UK, the magazine has subscribers
in 24 countries. Each issue
contains hundreds of contacts for
buying and selling plus
informative articles on topics such
as: medals, ordnance, insignia,
uniforms, Third Reich awards, the
Western Front, British battles, the
Napoleonic Wars etc. A unique
feature is photographs of items on
dealer's stands taken at militaria
fairs with prices and contact

numbers. A comprehensive list of
UK arms fairs is also included. For
a current copy please send £2.50
(£3.50 outside the UK). Cheques
payable to Beaumont Publishing
Ltd. Credit cards accepted. The
Armourer also has an extensive
Internet site at http://www.
armourer.u-net.com.

**Army Medical Services
Magazine**
Combined Services Publications,
PO Box 4, Farnborough,
Hampshire GU14 7LR

Contact: Joan Hare
Tel: 01252 515891
Fax: 01252 517918

Military medical journal.

**Army Quarterly & Defence
Journal**
1 West Street, Tavistock, Devon
PL19 8DS

Tel: 01822 613577
Fax: 01822 612785

Arrowhead
39 St Stephen Road, Bridlington,
Yorkshire (E.) YO16 4DW

Contact: R C Brown

Call to Arms
7 Chapmans Crescent, Chesham,
Buckinghamshire HP5 2QU

Contact: Duke Henry Plantagenet
Tel/fax: 01494 784271

The Worldwide Directory of
Historical Re-enactment Societies
and Traders. Published twice a
year with continuous updating
service – fax and email supported
– entry in our listings is free. For
sample copy send £2.50 (UK),
£3.00 (airmail Europe), or £3.50
(airmail rest of world) – sterling
only. Listings are uniquely
annotated with society size /
activity data. Also contains high
quality articles of news, research
and development. Free and
friendly advice on-line to
subscribers, a 'to-your-door'
update service planned. W W
Web pages planned. Email for
details. To find out anything
about Historical Re-enactment
and Living History, first you buy
'Call to Arms'. Get your copy
now. Email: duke@calltoarms.com

Casus Belli
Excelsior Publications, 1 Rue du
Colonel Pierre Avia, 75015 Paris,
France

Cibles
12 Rue Duguay-Trouin, 75006
Paris, France

Contact: Eric Bondoux

Classic Arms & Militaria
Peterson House, Northbank,
Berryhill Industrial Estate,
Droitwich, Hereford &
Worcester WR9 9BL

Tel/fax: 01905 795564

Deutsches Waffen-Journal
Schmollerstrasse 31, 74523
Schwabisch Hall, Germany

Tel: (0049) 791494515
Fax: (0049) 791404505
Email: dwj@schwend.de

Europe's premier magazine for
collectors, hunters and marksmen.

Editions Heimdal
Chateau de Damigny, 14400
Bayeux, France

Contact: Georges Bernage

Figurines
Histoire & Collections, 5 avenue
de la Republique, 75011 Paris,
France

Contact: Jean-Marie Mongin

Firm & Forester
Combined Services Publications,
PO Box 4, Farnborough,
Hampshire GU14 7LR

Contact: Sue Trotter
Tel: 01252 515891
Fax: 01252 517918

First Empire Ltd
1st Floor, 21 Whitburn Street,
Bridgnorth, Shropshire WV16 4QN

Contact: David Watkins
Tel: 01746 765691
Fax: 01746 768820
Email: enquiries
@firstempire.ltd.uk

Publishers of the Napoleonic
magazine 'First Empire'. Also
available: books, prints, wargames
figures & software. Send SAE for
details and subscription
information. WWW. First Empire.
Ltd.UK.

Forces News
Mandrake Marketing, 25-26
Market Place, Wisbech,
Cambridgeshire PE13 1HE

Tel/fax: 01945 5975

The Formation Sign
Military Heraldry Society, 124
Stoneleigh Park Road, Ewell
Epsom, Surrey KT19 0RG

Contact: F L Brown

Society for study and collecting of
world-wide cloth insignia,
established 1951. Quarterly
journal 'The Formation Sign'
mailed to members. See MHS
entry in Section 4.

Forum
Forest Publishing, 49-55 Fore
Street, Ipswich, Suffolk IP4 1JL

Tel/fax: 01473 55069

The Garter
Redcote, 228 Sydenham Road,
Croydon, Surrey CR0 2EB

Contact: Frances E Tucker
Tel/fax: 0181 684 1095

The directory that supports
costuming. The first source book
containing a host of fabric shops,
booksellers, costumers, accessory
makers, materials suppliers,
associated societies, pattern
makers etc.

The Globe and Laurel
HMS Excellent, Whale Island,
Portsmouth, Hampshire PO2 8ER

Tel: 01705 651305
Fax: 01705 547212
Email: GL97@aol

The journal of the Royal Marines.
Published bi-monthly, 12000
copies are sent all over the world,
keeping past and present RMs up
to date.

Green Howards Gazette
Combined Services Publications,
PO Box 4, Farnborough,
Hampshire GU14 7LR

Contact: Sue Trotter
Tel: 01252 515891
Fax: 01252 517918

Guards Magazine
HQ London District, Horse
Guards, Whitehall, London
SW1A 2AX

Tel: 0171 414 2271
Fax: 0171 414 2207

Gunner
Combined Services Publications,
PO Box 4, Farnborough,
Hampshire GU14 7LR

Contact: Sue Trotter
Tel: 01252 515891
Fax: 01252 517918

ESSENTIAL AMMUNITION FOR ALL MILITARY ENTHUSIASTS

MILITARY MODELLING

The World's best selling military modelling magazine. *The definitive magazine for all modellers, with editorial that is respected World-wide. From kit reviews to detailed constructional advice, vehicles to toy soldiers, there is no other magazine casting such a expert eye over the hobby. The stunning photographs alone make this magazine a pleasure to own.*

PRACTICAL WARGAMER

The UK's most popular **Wargames** magazine. *Embracing all aspects of the hobby, Practical Wargamer makes a great read for wargamers of all ages and experience. The widely respected coverage includes factual, historical and fantasy board games; role playing, computing and modelling, together with a full round up of all club activity, competitions and events.*

REGIMENT

The military heritage collection. *Each issue gives superb, in-depth coverage of the history of an individual military unit or armed force from around the World, charting its progress from inception to the present day. Using many previously unreleased facts and photographs, with the full support of the regiments involved, this builds into an invaluable source for historians, hobbyists and model makers.*

Available from all good newsagents.

FOR OUR LATEST SPECIAL SUBSCRIPTION OFFERS,

PLEASE CALL OUR ENQUIRY LINE ON

01858-435322

REMEMBER, IT'S ALWAYS CHEAPER TO SUBSCRIBE!

Historex Agents
Wellington House, 157 Snargate
Street, Dover, Kent CT17 9BZ
Contact: Lynn Sangster
Tel: 01304 206720
Fax: 01304 204528

Verlinden Productions Modelling
Magazine is a bi-monthly magazine
distributed in the UK by Historex
Agents; devoted to Verlinden
products, it is a high-quality
colour-illustrated publication with
interesting articles on modelling
techniques, painting, and in-depth
studies of products.

Historia
18 Rue Neuve des Bois, 75011
Paris, France

**Honourable Artillery
Company Journal**
Combined Services Publications,
PO Box 4, Farnborough,
Hampshire GU14 7LR
Contact: Sue Trotter
Tel: 01252 515891
Fax: 01252 517918

I P M S (UK) Magazine
26 Sandygate Road, Marlow,
Buckinghamshire SL7 3A2
Contact: Edgar Brooks

I P M S (UK) Magazine Editor
16 Green Street, Greasbrough,
Rotherham, Yorkshire (S.) S61
4EF
Contact: Neil Robinson

**International Arms &
Militaria Collector**
PO Box 80, Labrador 4215,
Australia

Jet 48
Forest Publishing Ltd, Breckland
House, Church Walk, Mildenhall,
Suffolk
Tel/fax: 01638 715445

**Journal of the Royal Naval
Medical Service**
Combined Services Publications,
PO Box 4, Farnborough,
Hampshire GU14 7LR
Contact: Joan Hare
Tel: 01252 515891
Fax: 01252 517918
Military medical journal.

**Journal of the Royal Army
Medical Corps**
Combined Services Publications,
PO Box 4, Farnborough,
Hampshire GU14 7LR
Contact: Joan Hare
Tel: 01252 515891
Fax: 01252 517918
Military medical journal.

The Lancashire Lad
Combined Services Publications,
PO Box 4, Farnborough,
Hampshire GU14 7LR
Contact: Sue Trotter
Tel: 01252 515891
Fax: 01252 517918

Living History Register
21 Oak Road, Woolston,
Southampton, Hampshire
SO19 9BQ
Contact: Roger Emmerson
Tel/fax: 01703 442011 evening

This international magazine
covers all periods of re-enactment
and 'living history'; promoting
research and contact between
groups and interested parties; a
networking aid. Published
bi-annually - £1.95 each, £3 for
annual subscription, posted direct.
'The Register' developed from a
list of participants and their skills
into a magazine, written by
participants, to fill a need for all.

**London Scottish Regimental
Gazette**
Combined Services Publications,
PO Box 4, Farnborough,
Hampshire GU14 7LR
Contact: Sue Trotter
Tel: 01252 515891
Fax: 01252 517918

**M H Q (Military History
Quarterly)**
29 West 38th Street/16th Floor,
New York, NY 10018, USA
Contact: Edward M Strauss
Tel: (001) 212 398 1550
Fax: (001) 212 840 6790

Fully illustrated hardcover
magazine devoted to military
history from ancient times to
present day. One year
subscription $70.00 US. Back
issues from 1988 available.

Maquettes Modèles Actualités
Sevart, BP 3067, 78130 Les
Mureaux, France
Contact: Didier Lefèvre
Tel: (0033) 134748080
Fax: (0033) 134740405
Email: 106636.2506
@compuserve.com

Kit modellers' magazine covering
aircraft, ships, vehicles, and figures.

Mars and Minerva
Combined Services Publications,
PO Box 4, Farnborough,
Hampshire GU14 7LR
Contact: Sue Trotter
Tel: 01252 515891
Fax: 01252 517918

Medal News
Token Publishing Ltd, PO Box
14, Honiton, Devon EX14 9YP
Contact: Carol Hartman
Tel: 01404 831878
Fax: 01404 831895

Monthly magazine for the
collector of medals and the
military historian; regular
features, articles, book reviews,
competitions and healthy
advertising; classified advertising
and medal tracker service now
free to all subscribers.

The Men of Harlech
Combined Services Publications,
PO Box 4, Farnborough,
Hampshire GU14 7LR
Contact: Sue Trotter
Tel: 01252 515891
Fax: 01252 517918

Militaria
Histoire et Collections, 5 Avenue
de la Republique, 75011 Paris,
France
Contact: Philippe Charbonnier

The world's leading magazine for
20th Century uniform, insignia,
and equipment collectors and
enthusiasts. Colour illustrated
throughout, including many
expert photographic
reconstructions of the soldiers,
sailors, and airmen of the two
World Wars.

Military Illustrated
43 Museum Street, London
WC1A 1LY
Contact: Tim Newark
Tel: 0171 404 0304
Fax: 0171 242 0762

Monthly magazine covering all
periods and nationalities, strong
emphasis on uniform and
equipment history but also articles
on weapons, vehicles, medals, etc.

Military In Scale
Traplet Productions Ltd, Traplet
House, Severn Drive,
Upton-upon-Severn, Hereford &
Worcester WR8 0JL
Contact: John Cheyne
Tel: 01684 594505
Fax: 01684 594586
Email: traplet@dial.pipex.com

The most colourful and
informative monthly magazine
available to plastic modellers,
featuring tanks and AFVs, aircraft
and figures, plus the best reviews,
information and techniques. Web
site: http://www.
traplet.co.uk/traplet/.

Military Miniatures in Review
Ampersand Publishing Co Inc,
21045 Commercial Trail, Boca
Raton, Florida 33486, USA

Military Model Preview
PO Box 98865, Tacoma,
Washington 98498-0865, USA

Military Modelling
Nexus House, Boundary Way,
Hemel Hempstead,
Hertfordshire HP2 7ST
Contact: Ken Jones
Tel/fax: 01442 66551

The monthly magazine for
modellers, military enthusiasts
and wargamers of all persuasions.
This international magazine is a
trend-setter and forum for
everything military.

**Military Provost Staff Corps
Journal**
Combined Services Publications,
PO Box 4, Farnborough,
Hampshire GU14 7LR
Contact: Sue Trotter
Tel: 01252 515891
Fax: 01252 517918

Model & Collectors Mart
Aceville Publications, 97 High
Street, Colchester, Essex
CO1 1TH

Model Time
Via S Sonnino 34, 43100 Parma
43100, Italy

Museum Ordnance
PO Box 5884, Darlington,
Maryland 21034, USA

Nautical Magazine
Brown, Son and Ferguson Ltd.,
4-10 Darnley Street, Glasgow,
Strathclyde G41 2SD, Scotland
Tel/fax: 0141 429 1234

Naval Engineers Journal
Hunter's Moon, Exford,
Minehead, Somerset TA24 7PD
Contact: Bryan H Jackson
Tel: 01643 831695
Fax: 01643 831576

The Journal dealing with Naval
design, weapons, electronic
warfare, propulsion, construction,
in fact all that goes to make up the
fighting ship.

The Naval Review
32 West Street, Chichester,
Sussex (W.) PO19 1QS

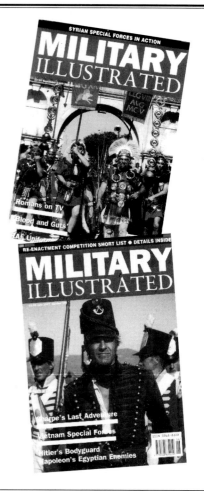
Nexus Special Interests
Nexus House, Boundary Way, Hemel Hempstead, Hertfordshire HP2 7ST

Contact: Ken Jones
Tel: 01442 66551
Fax: 01442 66998

Publishers of 'Military Modelling' the world's best selling military modelling magazine, and 'Practical Wargamer'. Organisers of Europe's premier military modelling show – 'Euromilitaire'. Publish numerous military modelling books.

Officers' Pensions Society
68 South Lambeth Road, Vauxhall, London SW8 1RL

Tel: 0171 820 9988
Fax: 0171 820 9948

Membership organisation protects pension interests of ex-Service people and gives members pension advice. Magazine is 'Pennant' – print run 37,000, biannual.

Partizan Press
816-818 London Road, Leigh on Sea, Essex SS9 3NH

Contact: David Ryan
Tel/fax: 01702 73986

Publishers of books, magazines, ECW Times, Age of Napoleon, Age of Empires; 18th Century and Renaissance Notes & Queries, Battlefields. Sole European distributor of major international wargaming magazines.

Pegasus
Combined Services Publications, PO Box 4, Farnborough, Hampshire GU14 7LR

Contact: Sue Trotter
Tel: 01252 515891
Fax: 01252 517918

Pireme Publishing Ltd
34 Chatsworth Road, Bournemouth, Dorset BH8 8SW

Contact: Iain Dickie
Tel: 01202 773490
Fax: 01202 512355

Publishers of 'Miniature Wargames' magazine – covers all periods of history and all theatres of conflict. Inexpensive military prints, booklets and 15mm Waterloo card model buildings, also available.

The Preston Kingsman
3 Valley View, Fulwood, Preston, Lancashire PR2 4HP

Contact: Peter Golding
Tel/fax: 01772 718357

Monthly newsletter of the King's and Manchester Regiments' Association. Events, news, views, pictures, poems, searchline, postbag, minutes, adverts, memorabilia, meetings. Annual subscription.

Q A R A N C Gazette
Combined Services Publications, PO Box 4, Farnborough, Hampshire GU14 7LR

Contact: Joan Hare
Tel: 01252 515891
Fax: 01252 517918

Military medical journal – Queen Alexandra's Royal Army Nursing Corps.

RAIDS
Histoire et Collections, 5 Avenue de la Republique, 75011 Paris, France

Leading monthly magazine of current military news, with colour photo features from today's best correspondents with the world's elite forces and front line reports. An English edition with partly different contents is published by Ian Allan Ltd – see Section 2, Book Publishers.

Royal Air Force News
MOD, Turnstile House, 98 High Holborn, London WC1V 6LL

Tel/fax: 0171 430 5020

The Royal Highland Fusilier
Combined Services Publications, PO Box 4, Farnborough, Hampshire GU14 7LR

Contact: Sue Trotter
Tel: 01252 515891
Fax: 01252 517918

Royal Logistic Corps Journal
Geerings of Ashford Ltd, Cobbs Wood House, Chart Road, Ashford, Kent TN23 1EP

Contact: Mark Ansett
Tel: 01233 633366
Fax: 01233 665713

The RLC Journal is circulated to the soldiers and officers of the Royal Logistic Corps – the largest corps in the British Army.

Royal Signals Journal
Combined Services Publications, PO Box 4, Farnborough, Hampshire GU14 7LR

Contact: Sue Trotter
Tel: 01252 515891
Fax: 01252 517918

S S A F A News
16-18 Old Queen Street, Westminster, London SW1

Tel/fax: 0171 518 1597

Scottish Military Historical Society
4 Hillside Cottages, Glenboig, Lanarkshire ML5 2QY, Scotland

Contact: Tom Moles

The Society exists to encourage the study of Scottish military history and publishes its own illustrated journal, covering the collecting of badges, headdress, uniforms, medals, photographs, postcards, prints, watercolours, equipment, pistols, powderhorns etc. The SMHS has a world-wide membership.

The Silver Bugle
Combined Services Publications, PO Box 4, Farnborough, Hampshire GU14 7LR

Contact: Sue Trotter
Tel: 01252 515891
Fax: 01252 517918

Soldier Magazine
Parsons House, Ordnance Road, Aldershot, Hampshire GU11 2DU

Contact: Chris Horrocks
Tel: 01252 347355
Fax: 01252 347358

Published by the Ministry of Defence, is the official magazine of the British Army and circulates worldwide. It is aimed at serving and retired soldiers and their families as well as civilian enthusiasts. It was the only British Service publication with combat reporting teams covering the Gulf War.

South & Central American Military Society
27 Hallgate, Cottingham, Yorkshire (E.) HU16 4DN

Contact: Terry D Hooker
Tel/fax: 01482 847068
Email: scamhs@aol.com

Publishers of quarterly journal 'El Dorado' covering from pre-Columbian times to present day; English and Spanish text, with illustrations and book reviews, all areas of military history.

Spartan Spirit
Forest Publishing, 15 Welland Close, St Ives, Cambridgeshire

Tel: 01480 300661
Fax: 01480 62286

The Stafford Knot
Combined Services Publications, PO Box 4, Farnborough, Hampshire GU14 7LR

Contact: Sue Trotter
Tel: 01252 515891
Fax: 01252 517918

Steel Masters
Histoire & Collections, 19 avenue de la Republique, 75011 Paris, France

Contact: Didier Chomette

Tank
Combined Services Publications, PO Box 4, Farnborough, Hampshire GU14 7LR

Contact: Joan Hare
Tel: 01252 515891
Fax: 01252 517918

Journal of the Royal Tank Regiment.

Tank TV
PO Box 9724, Wellington 6001, New Zealand

Contact: Peter Cooke

A4 format newsletter covering historical and current armoured vehicles.

Tankette
15 Berwick Avenue, Heaton Mersey, Stockport, Cheshire SK4 3AA

Contact: G E G Williams
Tel: 0161 432 7574
Fax: 0161 283 0283
Email: mafvahq@aol.com

The bi-monthly magazine of the Miniature AFV Association; contains plans, articles, reviews and photographs of military vehicles and equipment. Includes members' want ads and information.

The Thistle
Combined Services Publications, PO Box 4, Farnborough, Hampshire GU14 7LR

Contact: Sue Trotter
Tel: 01252 515891
Fax: 01252 517918

Token Publishing
Orchard House, Duchy Road, Heathpark, Honiton, Devon EX14 8YD

Contact: Philip Mussell
Tel: 01404 831878
Fax: 01404 831895

Publishers of Medal News, the world's only magazine devoted to medals and battles. Also The Medal Yearbook, the definitive guide to British orders and medals.

Tradition Magazine
25 Rue Bargue, 75015 Paris, France

Contact
Contact: Corrine Jarque
Tel: (0033) 140619767
Fax: (0033) 140619633

Monthly colour magazine of pre-1870 uniforms, weapons, military history 1st and 2nd Empire.

Trident
Medway Press Service Ltd, 58
High Street, Maidstone, Kent
ME14 1SY

Tel/fax: 01622 682026

Vae Victis
5 Avenue de la Republique,
75011 Paris, France

Contact: Theophile Monnier
Tel/fax: (0033) 110211820

The French wargame magazine
(boardgames, miniatures and
computer games). Also includes
military history articles on
strategy and tactics.

Warbirds Worldwide Ltd
PO Box 99, Mansfield,
Nottinghamshire NG19 9GN

Contact: Paul Cogan

Wargames Illustrated
Stratagem Publications Ltd, 18
Lovers Lane, Newark,
Nottinghamshire NH24 1HZ

Contact: Duncan McFarlane
Tel/fax: 01636 71973

Warship World
Lodge Hill, Liskeard, Cornwall
PL14 4EL

Contact: Roger May
Tel: 01579 343663
Fax: 01579 346747

Quarterly magazine covering both
the current Royal Navy and
Auxiliary scene and naval history.
Subscription £12 p.a. (£14
overseas). SAE for details.

WWII Military Journal
PO Box 28906, San Diego CA
92198, USA

Tel: 001 619 884 3483
Fax: 001 619 471 7406

A quarterly magazine concerned
with the history of WWII
(1931–1945).

A D Hamilton & Co.
7 St. Vincent Place, Glasgow,
Strathclyde G1 2DW, Scotland

Tel: 0141 221 5423
Fax: 0141 248 6019

Adrian Forman
PO Box 25, Minehead, Somerset
TA24 8YX

Tel/fax: 01643 862511 1

German WWII decorations, awards
and militaria (all GUARANTEED
original pre-1945), plus books. Free
sample illustrated catalogue and
book list on request.

Border Coins & Medals
29A Princess Street, The Square,
Shrewsbury, Shropshire SY1 1LW

Contact: P Druckers
Tel/fax: 01743 246150

Campaign medals, orders and
decorations. Largely British 1815
to date; strong emphasis on
1914–18. Large SAE for medal list.

Brandenburg Historica
342 A Winchester Street, Ste.
21, Keene, New Hampshire NH
03431, USA

Contact: Diane M Schreiber
Tel: (001) 603 352 1961
Fax: (001) 603 357 5364
Email: preussen@top.monad.net

Brandenburg Historica mail order
catalogue of books, military music
and militaria; Imperial, Third
Reich, DDR. Uniforms, insignia
and medals. Bi-monthly listings.
We ship worldwide. US$4.00.

Brian Clark Medals
16 Lothian Road, Middlesbrough,
Cleveland TS4 2HR

Tel/fax: 01642 240827

Medal lists issued eight or nine
times per year. Ribbons and
medal-related books also stocked.
Ring or send SAE for sample list.

Bryant & Gwynn Antiques
8 Drayton Lane, Drayton
Bassett, Staffordshire B78 3TZ

Contact: David Bryant
Tel/fax: 0121 378 4745

All types of military uniforms,
medals, swords and militaria
supplied. Also deactivated
military arms and muskets. Items
not stocked can be ordered and
traced.

C F Seidler
Stand G12, Grays Antique Mkt,
1-7 Davies Mews, Davies Street,
London W1V 1AR

Contact: Christopher Seidler
Tel/fax: 0171 629 2851

American, British, European,
Oriental edged weapons, antique
firearms, orders and decorations,
uniform items; watercolours,
prints; regimental histories, army
lists; horse furniture; etc. We
purchase at competitive prices
and will sell on a consignment or
commission basis. Valuations for
probate and insurance. Does not
issue a catalogue but will gladly
receive clients' wants lists. Open
Mon-Fri, 11.00am-6.00pm.
Nearest tube station Bond Street
(Central line).

C J & A Dixon Ltd
23 Prospect Street, Bridlington,
Yorkshire (E.)

Tel: 01262 676877
Fax: 01262 606600

Specialist dealers in war medals
and decorations. Dixons Gazette
contains over 1,000 priced medals
– £3 a copy, subscription 4 issues
UK £6, overseas £9.

**Central Antique Arms &
Militaria**
7 Smith Street, Warwick,
Warwickshire CV34 4JA

Contact: Chris James
Tel/fax: 01926 400554

Buys and sells antique guns,
swords, bayonets, helmets,
badges, British and German
medals and decorations, original
Third Reich militaria and
documents. Send 10 x 1st class
stamps for illustrated catalogue.
Shop open Mon-Sat,
10.00am-5.00pm. Exhibits at most
major fairs.

The Collector (Medals)
242 High Street, Orpington, Kent
BR6 0LZ

Contact: Michael Rochford
Tel/fax: 01689 890045

British campaign medals,
regimental cap badges and
postcards.

FOR SALE

Guaranteed original Nazi German Medals, Badges, Decorations etc,. British Fascist and other Axis related items. Please send £1 postal order/cheque/stamps for a list to:

**JAMIE CROSS
PO BOX 73 NEWMARKET
SUFFOLK CB8 8RY
TEL/FAX: 01638 750132
9AM-9PM**

*Over 15 years experience in the Third Reich Medal field.
Please telephone for a quote.
Wants list a speciality, please enquire for further details.*

Collectors Corner
51 The Old High Street,
Folkestone, Kent CT20 1RN

Contact: Clive Jennings

Collectors shop specialising in pre-1945 militaria, medals, insignia, helmets; also surplus, postcards, military books.

Collectors Market
London Bridge BR Station,
London Bridge, London EC4

Tel/fax: 0181 398 8065

Over 60 stands including medal dealers, badges and militaria on the concourse of London Bridge Main Line Station every Saturday.

Cotrel Medals
7 Stanton Road, Bournemouth,
Dorset BH10 5DS

*Contact: Peter Cotrel
Tel/fax: 01202 388367*

Mail-order company dealing in medals and related accessories. Send SAE for list. Callers by appointment only. Offers made on surplus medals. Mounting service available.

D M D Services
6 Beehive Way, Reigate, Surrey
RH2 8DY

Tel/fax: 01737 240080

David Hilton Medals
38 Chetwood Drive, Widnes,
Cheshire WA8 9BL

Tel/fax: 0151 424 0630

World medals bought and sold, four lists per year; subscription £2.50 inland, £6.00 overseas. Medals sold on commission basis. Phone for details.

Dunelme Coins & Medals
County Collectors Centre, 7 Durham Road, Esh Winning, Co.Durham DH7 9NW

*Contact: Peter G Smith
Tel: 0191 373 4446
Fax: 0191 373 6368*

Dealers in war and campaign medals (medal ribbons stocked), regimental badges, historical documents and autographed letters. Also coins and banknotes of the world. Open Mon-Sat 9.00am-5.00pm (closed Wed).

George Rankin Coin Company Ltd
325 Bethnal Green Road,
London E2

*Tel: 0171 729 1280
Fax: 0171 729 5023*

Buyers and sellers of all coins, banknotes, military and civilian medals, jewellery, scrap gold and silver, antiques and collectables.

Giuseppe Miceli Coin & Medal Centre
173 Wellingborough Road,
Northampton,
Northamptonshire NN1 4DX

*Contact: Guiseppe Miceli
Tel/fax: 01604 39776*

Gordons Medals
G12-13 Grays Antique Market,
1-7 Davies Mews, Davies Street,
London W1V 1AR

*Contact: R Gordon
Tel: 0171 629 2851
Fax: 017084 38987*

Specialising in campaign medals, German military collectables, Third Reich documents and awards. Comprehensive catalogue issued every two months – £8.00 UK, £16.00 overseas. Shop hours 11am-6pm.

Granta Coin & Antiquities Shop
23 Magdalene Street,
Cambridge, Cambridgeshire CB3 0AF

Tel/fax: 01223 61662

Militaria of the world bought and sold. Large stocks of cap badges, campaign and other medals, cloth badges, etc; also German militaria. All guaranteed genuine.

Great War Medals
22 Selborne Road, London
N14 7DH

*Contact: M A Law
Tel/fax: 0181 886 4120*

Regular mail-order listings of WWI medals and books. Items purchased or sold on commission. Research services for military and family historians. OMRS member. Visa/Mastercard facility. SAE for sample catalogue.

H & B Medals Ltd
38 Chetwood Drive, Widnes,
Cheshire WA8 9BL

*Contact: David Hilton
Tel: 0151 424 0630
Fax: 0151 495 1994*

Orders, medals and decorations bought and sold. 4 lists per year. Subscriptions £3.00 UK; £6 (sterling) overseas. Top prices paid for single items or collections.

III Arm Militaria
Hillbury, Kirkby-cum-Osgodby,
Lincolnshire LN8 3PE

*Contact: Duncan Lamb
Tel/fax: 01673 828081*

British Victorian medals and Soviet documented awards a speciality. A list is published periodically throughout the year; send SAE for free sample copy.

Jamie Cross
PO Box 73, Newmarket, Suffolk
CB8 8RY

*Contact: Jamie Cross
Tel/fax: 01638 750132*

Specialist dealer in Third Reich medals and badges. Axis-related awards also. Good selection always available. All guaranteed 100% original. Wants lists a speciality.

Jeremy Tenniswood
PO Box 73, Aldershot,
Hampshire GU11 1UJ

*Contact: Jeremy Tenniswood
Tel: 01252 319791
Fax: 01252 342339
Email: 100307,1735@compuserve*

Established 1966, dealing in collectable firearms civil and military, de-activated and for shooters; also swords, bayonets, medals, badges, insignia, buttons, headdress, ethnographica; and books. Regular lists of Firearms and Accessories; Medals; Edged Weapons; Headdress, Headdress Badges and Insignia; comprehensive lists Specialist and Technical Books. Office open 9am-5pm, closed all day Sunday. Medal mounting service.

John D Lowrey Militaria
PO Box 7928, Fremont,
California 94537-7928, USA

Tel/fax: (001) 510 657 7957

Medals and related books. 6 price lists per year (£2.50) includes US and foreign. Will consider exchanges. Accepts Visa/Mastercards/cheques.

Just Military
701 Abbeydale Road, Sheffield S7 2BE

Tel/fax: 0114 255 0536

Dealers in all types of military memorabilia and collectables. Full medal mounting and framing service including the supply of miniature and replacement medals.

Leroy's Miniature Medals
25 Stanfield Road, Talbot Park,
Bournemouth, Dorset BH9 2NL

*Contact: Michael Brachi
Tel/fax: 01202 519090*

Miniature medals, mainly early British Campaigns although also offers better quality modern medals. Also specialist in full size medals with fine quality, single British Campaign medals only. Free monthly lists available.

London Militaria Market
Angel Arcade, Camden Passage,
London N1

Tel/fax: 01628 822503

Saturdays only, 8am-2pm. About 35 stands dealing in badges, medals, arms, helmets, uniforms, regimental brooches, and many other military collectables.

M & W Coins
75 Church Road West, Walton,
Liverpool, Merseyside L45 UEE

M J & S J Dyas Coins/Medals
Sandy Hill Ind Estate, Unit 3, 42 Stratford Road, Shirley, Solihull,
Midlands (W.) B90 3LS

*Contact: Malcolm Dyas
Tel: 0121 733 2225
Fax: 0121 733 1533*

Medal dealers – but we also stock military books and associated supplies, and offer a research service.

Medal News
Token Publishing Ltd, PO Box 14, Honiton, Devon EX14 9YP

*Contact: Carol Hartman
Tel: 01404 831878
Fax: 01404 831895*

Monthly magazine for the collector of medals and the military historian; regular features, articles, book reviews, competitions and healthy advertising; classified advertising and medal tracker service now free to all subscribers.

Medal Society of Ireland
32 Clonmacate Road, Birches, Portadown, Co.Armagh, N. Ireland

Contact: Jonathan Maguire

Collector of medals and books relating to The Royal Dublin Fusiliers and their predecessors, research undertaken, very modest fees.

The Medals Yearbook
3rd Floor, 1 Old Bond Street, London W1X 3TD

Tel: 0171 499 5026
Fax: 0171 499 5023

The only up-to-date descriptive price guide to all British medals. Fully illustrated.

Michael's Medals & Miniatures
58A Stanfield Road, Talbot Park, Bournemouth, Dorset BH9 2NP

Tel/fax: 01202 512090

Militaria
c/Bailen 120, 08009 Barcelona, Spain

Contact: Xavier Andrew
Tel/fax: (0034) 3 2075385

Buy, sell and exchange militaria, medals, edged weapons, insignia, headgear etc. Also large selection of military books in stock.

Military Antiques
Shop 3, Phelps Cottage, 357 Upper Street, London N1 0PD

Contact: R Tredwin
Tel/fax: 0171 359 2224

In the heart of Camden Passage, long-established dealers in WWI & WWII uniforms, equipment, headgear, awards, edged weapons. All items original and carry a full money-back cover. Illustrated catalogue; wants lists welcome. Open Tues.-Fri., 11am-5pm; Sat, 10am-5pm; if travelling far, advisable to phone first.

Neate Militaria & Antiques
PO Box 3794, Preston St Mary, Sudbury, Suffolk CO10 9PX

Contact: Gary C Neate
Tel: 01787 248168
Fax: 01787 248363

Specialists in world-wide orders, decorations and medals. Catalogues available, five per annum: £6 UK, £10 rest of world. Stamped SAE for sample copy.

Norman W Collett
PO Box 235, London SE23 1NS

Tel/fax: 0181 291 1435

British and Commonwealth orders, decorations and medals. Postal business only. Regular catalogues issued – subscription UK £12, overseas £18. Access and Visa cards preferred.

Orders & Medals Research Society
123 Turnpike Link, Croydon CR0 5NU

Contact: N G Gooding
Tel/fax: 0181 680 2701

Researchers' and collectors' organisation. Quarterly journal, annual convention.

Peter Morris Militaria
1 Station Concourse, Bromley North BR Station, Bromley, Kent BR1 4EQ

Tel: 0181 313 3410
Fax: 0181 466 8502

Large stock of medals, militaria, cap badges; medal groups mounted. Retail sales lists issued quarterly; valuations given. Open Mon, Tues, Thurs, Fri, 10am-6pm; Sat, 9am-2pm.

Positive Proof
17 Argyll St, Shrewsbury, Shropshire SY1 2SF

Tel/fax: 01743 240960

Specialist photographic service with comprehensive photo-index who will endeavour to locate and photograph any 1914–18 war grave situated within the Ypres Salient, Belgium.

Poussières d'Empires SARL
33 Rue Brezin, 75014 Paris, France

Poussières d'Empires sells badges, orders, decorations and military items, particularly French Indochina and other Colonial. Open 11am-7pm; catalogue available for mail order service; Visa and Access accepted.

R H Smart
39 Mount Lane, Kirkby-la-Thorpe, Sleaford, Lincolnshire NG34 9NR

Contact: Roy Smart
Tel/fax: 01529 304236

Medal ribbons, specialising in former Soviet Union, United Nations, Imperial Germany and GB. Full size and miniature medals. Medal mounting service.

Raymond D Holdich
7 Whitcomb Street, Trafalgar Square, London WC2H 7HA

Tel: 0171 930 1979
Fax: 0171 930 1152

Selection of British campaign and gallantry medals, cap badges, Third Reich items, orders, medals and decorations from around the world. Open Mon-Fri, 10.00am-5.30pm.

Romsey Medals
5 Bell Street, Romsey, Hampshire SO51 8GY

Tel: 01794 512069
Fax: 01794 830332

Service Commemoratives Ltd
PO Box 173, Dromana, Victoria 3936, Australia

Tel/fax: (0061) 359 810201

Military service commemorative medals and a range of over 100 different area and specialist clasps/bars available for: combatant service; sea service; foreign service; army service; aviation service; volunteers; national defence; Antarctica service; and POW liberation 1945.

Southern Medals
16 Broom Grove, Knebworth, Hertfordshire SG3 6BQ

Contact: John Williams
Tel: 01438 811657
Fax: 01438 813320

Dealers in orders, decorations and medals. Regular lists issued. Subscription (four issues): £3.00 UK, £4.00 Europe, £6.00 elsewhere. Send SAE for sample copy.

Spink & Son Ltd
5 King Street, St James's, London SW1Y 6QS

Contact: David Erskine-Hill
Tel: 0171 930 7888
Fax: 0171 839 4853

Spink buy and sell Orders, Decorations and Medals of the World. A retail list, 'The Medal Circular' is published three times a year and available by subscription. Also three specialist auction catalogues per annum. Open Monday to Friday, 9am to 5.30pm.

Steve Johnson
PO Box 1SP, Newcastle upon Tyne, Tyne & Wear NE99 1SP

T L O Militaria
Longclose House, Common Road, Eton Wick, nr Windsor, Berkshire SL4 6QY

Contact: Tony Oliver
Tel: 01753 862637
Fax: 01753 841998

Specialists in DDR collectables at all levels of scarcity and price. Mail order; callers welcome seven days a week, BY PRIOR APPOINTMENT.

Toad Hall Medals
Toad Hall, Newton Ferrers, Devon PL8 1DH

Tel: 01752 872672
Fax: 01752 872723

Established mail order medal business producing 5 lists annually each with several hundred constantly changing items, including British Gallantry singles and Foreign sections. Also includes Nazi award documents and military miscellania. SAE for latest list.

Tony Radman
Denver House, 17 Witney Street, Burford, Oxfordshire OX18 4RU

Tel: 01993 822040
Fax: 01993 822769

Orders, medals, decorations, badges and militaria of all nations. We issue lists of different countries and different subjects. Callers welcome by appointment.

Toye Kenning & Spencer Ltd
Medal Chat, Newtown Road, Bedworth, Warwickshire CV12 8QR

Yeovil Collectors Centre
16 Hendford, Yeovil, Somerset BA20 1TE

Contact: Barry Scott
Tel/fax: 01935 33739

Quality selection of medals, badges, buttons, militaria always in stock. Plus models, thimbles, prints and many other collectables. Closed some Tuesdays / Wednesdays. Occasional lists.

MILITARY VEHICLES
(ACTUAL, NOT MODEL VEHICLES)

A C Miles (Insurance Consultants) Ltd
663 High Road, Leytonstone, London E11 4RD

Contact: Alan & Moira Cogden
Tel: 0181 539 6424
Fax: 0181 558 1226

Specialists in historic military vehicle owner's insurance.

A C V M A
79 Avenue de la Premiere Armee, 63300 Thiers, France

Contact: Joel Dosjoub

A N A C A
La Gerbe de Ble, Place J.Guihard, 44130 Blain, France

Anchor Supplies Ltd
Peasehill Road, Ripley, Derbyshire DE5 3JG

Contact: Barbara Merrett
Tel: 01773 570139
Fax: 01773 570537

Anchor Supplies, one of Europe's largest genuine government surplus dealers. Specialising in clothing, tools, electronics, domestic ware, furniture, watches, military vehicles, you name it! Goods are available mail order, or visit our Derbyshire or Nottingham depots. Please ring for directions.

Ardennes-Club 44
17 Rue Baudelaire, 08000 Charleville Mezieres, France

Contact: Jean Claude Avril

Association de Véhicules Militaires
Automobile Club du Rhone, 7 Rue Grolek, 69002 Lyon, France

C L V M A
51 Rue Daubree, 57245 Peltre, France

Contact: Isabelle Mann

C V M A
246 Route de Clisson, 44120 Vertou, France

Contact: J C Mallet

Champ Spares (UK) Ltd
PO Box 87, Lichfield, Staffordshire WS13 8YW

Tel/fax: 01543 253562

World's most complete stock of spare parts for Austin Champ FV1801 4x4; full postal spares service; selection of vehicles available.

Club d'Amateurs de Matériel Militaire
B.P.508, 77304 Fontainebleau Cedex, France

Club de l'Est de Véhicules Militaires
Musee de l'Automobile, Bainville aux Miroirs, 54290 Bayon, France

Club Europeen de Matériel Mil Hist
65 Avenue de Versailles, 75016 Paris, France

Contact: Christian Verrier

Club Lorrain de Véhicules Mil Allies
Impasse Didier Leroy, 57159 Morenchy Marauges, France

Contact: Alain Bertelotti

Cobbaton Combat Collection
Chittlehampton, Umberleigh, Devon EX37 9RZ

Contact: Preston Isaac
Tel/fax: 01769 540740

Private collection of about 50 vehicles, tanks and artillery, WWII British, Canadian, some Warsaw Pact, fully equipped. Home Front section. Militaria shop. Open daily April – November: Winter Mon - Fri.

Colls des Véhicules Militaires / Ouest
Le Quebec, 72330 Cerans-Foulletourte, France

Contact: Jean Royer

Cord Video
2B Cleveland Street, Bedford, Bedfordshire MK42 8DN

Tel/fax: 01234 840122

Dallas Auto Parts
Cold Ash Farm, Long Lane, Hermitage, Newbury, Berkshire RG18 9LT

Contact: Stephen Rivers
Tel/fax: 01635 201124

Military vehicles for sale; full workshop restoration and servicing facilities; engines and gearboxes rebuilt. Spares & accessories, inc. mail order. WWII Dodge and Jeep specialist.

Escadron de l'Histoire
35 Avenue des Gobelins, 75013 Paris, France

Contact: Andre Lecocq

Escadron Historique
190 Rue Edmond Rostand, 94310 Orly, France

Contact: Claude Daout

Fallingbostel WWII & 2RTR Regt.Museum
MBB3 Lumsden Barracks, Fallingbostel BFPO 38, Germany

Contact: Kevin Greenhalgh

Museum of WWII relics/items recovered from the actions around Fallingbostel, including the POW

camps 11B and 326. 2RTR Regimental Museum covers regimental history 1916 to the Gulf War; uniforms, large collection of WWII RTR items, photos, books, etc.

G C V M Ile de France
6 Allee de Milleportuis, 78450 Chavenay, France

Contact: Eric Hermann

G L K Video Arts
20 Butte Furlong, Haddenham, Buckinghamshire HP17 8JF

Tel/fax: 01844 291083

G M V C
Le Pierane, Rue Lamartine, 70000 Vesoul, France

Contact: J C Maguet

History on Wheels Motor Museum
Longclose House, Little Common Road, Eaton Wick, Nr.Windsor, Berkshire

Contact: Tony L Oliver
Tel/fax: 01753 862637

I N T Brokers Ltd
Insurance House, 301 Wakefield Road, Denby Dale, Huddersfield, Yorkshire (W.) HD8 8RX

Tel: 01484 865252
Fax: 01484 865231

Historic, veteran and vintage military vehicle insurance scheme, designed by an enthusiast to suit all your needs from pedal cycles to tanks.

Imperial War Museum / Duxford
Duxford Airfield, Cambridge, Cambridgeshire CB2 4QR

Contact: Frank Crosby
Tel: 01223 835000
Fax: 01223 837267

Houses the largest collection of military and civil aircraft in the country, totaling over 140. Also exhibiting over 100 military vehicles, artillery, and much more.

Invicta Military Vehicle Preservation Society
c/o 58 Laddds Way, Swanley, Kent BR8 8HW

Contact: V Burford
Tel/fax: 01322 408738

Society with over 750 members involved in the preservation of military vehicles of all types and ages. Monthly meetings and newsletter with quarterly magazine.

Jeeparts UK
Rose Villa, Dorrington, Shrewsbury, Shropshire SY5 7JD

Contact: Graham Lycett
Tel/fax: 01831 440831

Comprehensive range of parts for restorations and repairs; also sales and servicing. Fast, reliable mail order.

John & Mary Worthing
Spout House, Orleton, Ludlow, Shropshire SY8 4JG

Tel: 0158 474239
Fax: 0158 474554

High quality canvas authentic reproduction hoods and seats for restored US and British military vehicles. Ring or write for price list, photos and samples.

Lincolnshire Military Preservation Soc
Memories, 20 Market Place, Alford, Lincolnshire LN13 9EB

Contact: Trevor Budworth
Tel/fax: 01507 462541

Established 1981, and dedicated to WWII social, dances, military vehicles, battle re-enactment. British, American, German units with own field HQ at former RAF station. New recruits, ex-service, and associated groups welcome. Charity events undertaken.

M V C G Aquitaine
3 Quai Hubert Prom, 333000 Bordeaux, France

Contact: Jean Gomis

M V C G Artois
La Charite, 62400 Bethune, France

Contact: Dr Frederic Rey

M V C G Centre
40 Rue de la Gare, 36120 Ardentes, France

Contact: Eric Boussardon

M V C G Charente
La Tenaille, 17240 St Denis de Saintonge 17240, France

Contact: Francois Begouin

M V C G Cote D'Azur
Imm. Les Marguerites, 65 Avenue Raoul Dufy, 06200 Nice, France

Contact: Michel Caligaris

M V C G Est
5 Rue du Roerthal, 68530 Buhl, France

Contact: Bernard Rost

M V C G France
BP 24, 79201 Parthenay, France

Contact: J C L Fillon
Tel/fax: (0033) 5 49943945

French federation of military vehicles collectors' clubs, 700 members, 1200 vehicles.

M V C G Gascogne
Chemin de Gaouere, 32000 Auch, France

Contact: Alfred Algeri

M V C G Grand Rhone-Alpes
16 Place du Commerce, 71250 Cluny, France

Contact: M Noel

M V C G Ile de France
78 Rue de la Roquette, 75011 Paris, France

Contact: Jean Pisapia

M V C G Midi-Pyrenees
Locadour, RN 113, 47420 Bon Encontre, France

Contact: Thierry Jacques

M V C G Normandie
25 Rue de l'Observatoire, 76600 Le Havre, France

Contact: M Sautreuil

M V C G Pays de Loire
Les Esnauderies, Bouille Menard, 49520 Combrecy, France

Contact: Alain Quemener

M V C G Sixteens Club
42 Rue Victor Hugo, 16600 Magnac sur Tourve, France

Contact: Gilles Ducouret

M V C G Vallee du Rhone
Quartier Saint Laurent, 84350 Courthezon, France

Contact: Alainb Gomez

Memorial de Verdun
Fleury-devant-Douaumont, 55100 Verdun, France

Tel/fax: (0033) 29843534

Important WWI collection including uniforms, equipment, artillery, dioramas, etc.

Military Club de Normandie
Chateau de l'Abbaye, 27230 Le Theil Molent, France

Contact: Michel de Montrion

Military Vehicle Museum
Exhibition Park Pavillion, Newcastle-upon-Tyne, Tyne & Wear NE2 4PZ

Contact: Walter Tearse
Tel/fax: 0191 281 7222

The museum houses up to 50 military vehicles, mainly WWII, and 60 cabinets displaying WWI, WWII and post war memorabilia plus WWI mock up trench.

Military Vehicle Trust
PO Box 6, Fleet, Hampshire GU13 9PE

Contact: Nigel Godfrey
Tel/fax: 01264 392951
Email: nigelgodfrey@mut.org.uk

The Trust is an international group which supports the military vehicle/militaria enthusiast. There is a bi-monthly magazine, 'Windscreen', and newsletter. Club shop etc. http://www.mvt.org.uk or mailbox@mvt.org.uk.

Military Vehicle Trust (Avon)
16 Broncksea Road, Filton Park, Bristol, Avon BS7 0SE

Contact: Andrew John
Tel/fax: 0117 9793263

Meets last Thursday each month at Portcullis Inn, Tormarton, nr.Bristol.

Military Vehicle Trust (B'ham & W Mids)
133 Millfield Road, Handsworth Wood, Birmingham, Midlands (W.) B20 1EA

Contact: Andrew Jones
Tel/fax: 0121 357 2512

Meets first Tuesday each month at British United Services Club, Gough St., Birmingham.

Military Vehicle Trust (Berks & Oxon)
Higate, 150 Binfield Road, Bracknell, Berkshire RG12 1AY

Contact: Neil Randle
Tel/fax: 01344 425315

Meets second Wednesday each month at London Road Inn on B4009 at Benson, Oxon, between RAF Benson and Maines Scrapyard.

Military Vehicle Trust (Central & Southern Scotland)
5 Muirmont Crescent, Bridge of Earn, Perthshire PH2 9RG, Scotland

Contact: Jimmy Wood
Tel/fax: 0173 8812718

No area meetings held due to distance between members. For up to date information telephone 01738 812718/850673 regarding any military vehicle matter, or display.

Military Vehicle Trust (Chesh & M'side)
6 Grasmere Road, Ellesmere Port, South Wirral, Merseyside L65 9BR

Contact: B Alex Jenkins
Tel/fax: 0151 356 3330

Meets second Tuesday each month at Shell Club, Dunkirk Lane, Ellesmere Port.

Military Vehicle Trust (Cumbria)
7 Bowlands Drive, Kendal, Cumbria LA9 6LT

Contact: Jimmy Miller
Tel/fax: 01539 732931

Meets second Thurs of each month at Crooklands Hotel, Crooklands, Near Kendal.

Military Vehicle Trust (Devon/Cornwall)
8 Barnsfield Close, Glampton, nr Brixham, Devon

Contact: Peter Kay
Tel/fax: 01803 844602

Meets third Wednesday each month at Smugglers Inn, Steamer Quay, Totnes.

Military Vehicle Trust (Dorset)
35 Purbeck Close, Wyke Road, Weymouth, Dorset DT4 9QU

Contact: John Butcher
Tel/fax: 01305 778444

Meets third Tuesday each month at Parley Sports Club, Christchurch Rd., Parley Cross, Bournemouth.

Military Vehicle Trust (E Midlands)
Pegasus, Main Road, Nocton, Lincolnshire LN4 2BH

Contact: Nick Penistan
Tel/fax: 01526 322757

Meets second Monday each month at The Blue Tit, Forest Rd., New Ollerton, Notts.

Military Vehicle Trust (Essex)
24 Skylark Walk, Chelmsford, Essex CM2 8BB

Contact: Colin Tebb
Tel/fax: 01245 251857

Meets fourth Thursday each month at The Crown, Sandon, Chelmsford.

Military Vehicle Trust (Guernsey)
Sans Souci, Le Vilocq Castle, Guernsey, Channel Islands

Contact: Dave Malledent

Meets second Tuesday each month at St.Saviour's Hotel.

**Military Vehicle Trust
(Hertfordshire)**
I Colne Way, Garston,
Hertfordshire WD2 4BN

Contact: Andy Camp
Tel/fax: 01923 671056

Meets second Thursday each
month at Pinks Hotel, Rectory
Lane, Shenley, Herts.

Military Vehicle Trust (Jersey)
Champion Lodge, La Grande
Route de St Jean, St.Helier,
Jersey, Channel Islands JE2 3FN

Contact: Graeme Sty
Tel/fax: 01534 589999

**Military Vehicle Trust
(North West)**
II Balcary Grove, Bolton,
Lancashire BL1 4PY

Contact: M Cowcill
Tel/fax: 01204 496166

Meets first Thursday each month
at British Rail Staff Assoc. Club,
Edgeley Road, Stockport.

**Military Vehicle Trust
(S Midlands)**
12 The Ridgeway, Astwood
Bank, Redditch, Hereford &
Worcester B96 6LT

Contact: Neil Wedgbury
Tel/fax: 01527 892282

Meets on second Monday each
month at The Beckford Inn,
Beckford, on A435 between
Cheltenham and Evesham.

**Military Vehicle Trust
(S Sussex)**
27 Westminster Crescent,
Hastings, Sussex (E.) TN34 2AW

Contact: Greg Garland
Tel/fax: 01424 421761

Meets on the second Thursday of
each month at The Red Lion,
Magham Down, on A271 north of
Hailsham.

Military Vehicle Trust (S Wales)
46 Tydraw Street, Port Talbot,
Glamorgan (W.) SA13 1BT, Wales

Contact: Ernie Williams
Tel/fax: 01639 886140

Meets on the second Tuesday each
month at the BP Sports Club,
Llandarcy, nr Neath.

**Military Vehicle Trust
(Salop & Border)**
47A High Street, Church
Stretton, Shropshire SY6 6BX

Contact: Philip Robinson

Meets second Wednesday each
month at The Cound Lodge,
Cound, Shropshire.

**Military Vehicle Trust
(SE Midlands)**
110 Bush Hill, Northampton,
Northamptonshire

Contact: Brian Shoebridge
Tel: 01604 409442
Fax: 01604 580033

Meets on first Thursday each
month at TA Drill Hall, Clare
Street, Northampton.

Military Vehicle Trust (Solent)
7 Hart Plain Avenue, Cowplian,
Waterlooville, Hampshire PO8
8RP

Contact: Mrs L Taylor-Cram
Tel/fax: 01705 250463

Meets on third Wednesday each
month at Royal British Legion
Hall, Forest End, Waterlooville.

**Military Vehicle Trust
(Suffolk)**
Woodstock Barn, Chapel Lane,
Botesdale, Suffolk IP22 1DT

Contact: Mike Warner
Tel/fax: 01379 898085

Meets on the second Tuesday each
month at the Bacton Bull, just
north of Stowmarket.

**Military Vehicle Trust
(Surrey/Hants)**
19 Moulshamcopse Lane,
Yateley, Hampshire GU46 7RF

Contact: Robert Blundon
Tel/fax: 01252 873015

Meets on first Wednesday each
month at Normandy Cricket Club,
Hunts Hill Rd., Normandy,
Guildford, Surrey.

**Military Vehicle Trust
(W London)**
30 Common Close, Horsell,
Woking, Surrey GU21 4DB

Contact: Alan Meaghan
Tel/fax: 01483 724743

Meets third Tuesday each month
at The Fulwell Park Association,
1a Fortescue Avenue, Fulwell,
Middlesex.

Military Vehicle Trust (Wessex)
10 Easthams Road, Crewkerne,
Somerset TA18 7AQ

Contact: Richard Fryer
Tel/fax: 01460 74250

Meets on third Saturday each
month at The Windwhistle Inn on
A30 between Crewkerne and
Chard.

Military Vehicle Trust (Wilts)
8 Upper Isbury, Marlborough,
Wiltshire SN8 4AY

Contact: Neil Stevens
Tel/fax: 01672 515030

Meetings third Wednesday each
month at The Patriot's Arms,
Chiseldon, nr.Swindon.

**Military Vehicle Trust
(Yorkshire)**
38 Fountain Drive, Robertown,
Liversedge, Yorkshire (W.) WF15
7NY

Contact: Mike Lousada
Tel/fax: 01924 407482

Meetings on first Wednesday each
month at The Masons Arms,
Allerton Park, Knaresborough.

**Musée de la Bataille des
Ardennes**
08270 Novion-Porcien, France

Tel/fax: (0033) 24232013

1870, World War I, but particularly
important World War II vehicle,
uniform, and weapon collection.

Musée de la Cavalerie
Ecole d'AABC, 49409 Saumur,
France

Tel/fax: (0033) 41510543

Open by appointment, afternoons
except Fridays; closed August. All
periods, but important World War
II AFV collection, housed at French
Army Armoured Cavalry School.

Museum of Army Transport
Flemingate, Beverley,
Humberside (N.) HU17 0NG

Tel: 01482 860445
Fax: 01482 866459

History of Army transport from
the Boer War to the present day.
Over 110 vehicles; archives,
workshops, restoration, Sir Patrick
Wall model collection, Blackburn
Beverley aircraft, book/gift shop.
Open daily 10am-5pm; free
parking; cafeteria.

**National Museum of Military
History**
10 Bamertal, PO Box 104,
L-9209 Diekirch, Luxembourg

Contact: Roland Gaul
Tel: (00352) 808908
Fax: (00352) 804719
Email: mnhmdiek@pt.lu

Important collections from Battle
of the Bulge 1944-45, life-size
dioramas, uniforms, vehicles,
weapons, equipment. Open from
January 1 – March 31: daily
14–18h hrs; April 1 – November 1:
daily 10–18 hrs; November 2 –
December 31: daily 14–18 hrs; last
ticket sold 17.15 hrs.

Northern Allied Axis Society
39 Rosendale Grove, Spring Bank
West, Hull, Humberside (N.)

Contact: B Nuttall
Tel/fax: 01482 566144

NAAS are battle re-enactors and
also offer a mobile military
museum for events.

R E M E Museum
Isaac Newton Road, Arborfield,
nr Reading, Berkshire RG2 9LN

Tel/fax: 01734 760421

R R Motor Services Ltd
Bethersden, Ashford, Kent TN26
3DN

Contact: Mike Stallwood
Tel: 01233 820219
Fax: 01233 820494

Suppliers of all military vehicles,
spares, accessories – from tanks to
jeeps. Wide range of World War II
and postwar vehicles in stock, and
international contacts: if you are
looking to buy or sell, let us help.
We also specialise in military
vehicles restorations and repairs;
and can supply paint in many
authentic shades. Send SAE for
details.

Tank Museum
Bovington Camp, Wareham,
Dorset BH20 6JG

Tel: 01929 405096
Fax: 01929 405360

The world's most comprehensive
collection of AFVs. Free
'Firepower & Mobility' displays
Thursdays 12 noon, July, August,
September. Open daily 10am-5pm.

Tank TV
PO Box 9724, Wellington 6001,
New Zealand

Contact: Peter Cooke

A4 format newsletter covering
historical and current armoured
vehicles.

**Texas Oklahoma Vehicle
Group**
I Rue des Erables, 57330
Entrange Cite 57330, France

Contact: Eric Klamerec

U S Tank Corps
5bis Rue des Iris, 52000
Chaumont, France

V M H
7 Rue Montesquieu, 49000
Angers, France

Contact: Jacques Paul Callerot

MODELLING CLUBS & SOCIETIES

Amis d'Historex & Figurines Historiques
75 Rue Henri Barbusse, 77290 Mitry-Mory, France

Contact: M Hanin

Arms & Armour World War I (I P M S)
33 Keats Avenue, Cannock, Staffordshire WS11 2JY

Contact: Roy Carson
Tel/fax: 01543 503528

Special interest group of IPMS (UK) on arms, armour of WWI; seeks members interested in WWI land warfare.

B M S S Bristol Branch
The Old Coach House, 18 York Place, Clifton, Bristol, Avon BS8 1AH

Contact: Peter Sturgeon
Tel/fax: 0117 9732067

British Model Soldier Society. Meetings held at 19.30 hrs on the first Wednesday of each month at the Drill Hall Horfield Common Bristol. Visitors welcome.

B M S S Cumbria Branch
5 Collinfield, Kendal, Cumbria LA9 5JD

Contact: Michael Bunn
Tel/fax: 01539 721608

Meetings held at The Duke of Cumberland, Kendal, Cumbria, at 19.30 hrs, last Friday of each month.

B M S S Devon Branch
102 Warleigh Avenue, Keyham, Plymouth, Devon PL2 1NP

Contact: Harry Miller
Tel/fax: 01752 556811

Meetings held at St John House, North Hill, Plymouth, on the first Friday of each month.

B M S S Dorset (Mid-Wessex) Branch
21 Whitecliff, Mill Street, Blandford Forum, Dorset DT11 7BQ

Contact: Peter Ladyman
Tel/fax: 01258 454626

Meetings held regularly. Contact the Branch Organiser for details of dates and venues.

B M S S Ealing/West London Branch
93 Leighton Road, Ealing, London W13 9DR

Contact: Bruce Harron
Tel/fax: 0181 840 2284

Meetings held at the Parlour of Ealing Green Church Hall, Ealing Green, London, W5, at 19.30 hrs on the second Friday of each month.

B M S S East Midlands Branch
Twyford House, 194 Meanor Road, Smalley, Derbyshire

Contact: Tony Kettle
Tel/fax: 01773 762777

Contact the Representative for details of meetings.

B M S S Grampian Branch
Findon Croft, Findon, Portlethen, Aberdeen AB1 4RN, Scotland

Contact: Mitchell Davidson
Tel/fax: 01224 780606

Contact Representative for dates and venues of meetings.

B M S S Leicestershire Branch
Birch View, Kimcote, Lutterworth, Leicestershire LE17 5RU

Contact: Maria Cooper
Tel/fax: 01455 553430

Meetings held at the Red Lion, Gilmorton, near Lutterworth, at 19.30 hrs on the third Monday of each month.

B M S S London Branch, North
25 Bittacy Road, Mill Hill, London NW7 1BP

Contact: Eric Sanger
Tel/fax: 0181 349 0710

Meetings held at Millfield House Arts Centre, Silver Street, Edmonton, N18, at 19.00 hrs on the first Friday of each month.

B M S S London Branch, South
125 Lethbridge Close, Lewisham Road, London SE13 7QW

Contact: Malcolm Mayes
Tel/fax: 0181 691 1746

Meetings held at Becorp, Randlesdown Road, Bellingham, London SE6, mainly on second and fourth Mondays each month (except August) at 19.30 hrs.

B M S S Norfolk Branch
1 Mill Road, Loddon, Norfolk NR14 6DR

Contact: Roy Porter
Tel/fax: 01508 520327

Meetings held at Norwich City Supporters Club, Thorpe Road, Norwich, at 20.00 hrs on the fourth Monday of each month.

B M S S North Hertfordshire Branch
96 High Avenue, Letchworth, Hertfordshire SG6 3RR

Contact: Alan Jones
Tel/fax: 01462 676020

Meetings held at Plinston Hall, Broadway, Letchworth at 19.30 hrs on the third Wednesday of each month.

B M S S North Kent Branch
52 Imperial Drive, Gravesend, Kent DA12 4LN

Contact: Colin Bowen
Tel/fax: 01474 327003

Meetings held at Holy Trinity School, Trinity Road, Gravesend, at 14.00 hrs on the third Saturday of each month.

B M S S Northants Branch
35 St Barnabas Street, Wellingborough, Northamptonshire NN8 3HA

Contact: George Hanger
Tel/fax: 01933 383018

Meetings held at respective members homes, at 19.30 hrs on the first Wednesday of each month.

B M S S Northern Branch
24 Broad Oak Lane,, Penwortham, Preston, Lancashire

Contact: Harry Middleton
Tel/fax: 01772 745952

Meetings held at the LMSR Club, Store Street, Manchester (Piccadilly Station). Apply to Harry Middleton for monthly details and dates.

B M S S Oxfordshire Branch
1 Norman Avenue, Abingdon, Oxfordshire OX14 2HQ

Contact: Duncan Buller-West
Tel/fax: 01235 520324

Contact Representative for details of meetings.

B M S S Sunderland/Tyne & Wear Branch
31 Stansfield Street, Roker, Sunderland, Tyne & Wear SR6 0JY

Contact: Peter Watson
Tel/fax: 0191 564 1938

Meetings held at Brinkburn CA, Brinkburn School, South Shields, at 18.30 hrs on each Tuesday.

B M S S Wales Branch
305 Chepstow Road, Newport, Gwent NP9 8HJ, Wales

Contact: Mike Thomas
Tel/fax: 01633 272579
Email: MDTCymru@aol.com

Meetings held at the above address at 19.30 hrs on the last Wednesday of each month.

B M S S Waveney Branch
10 Green Drive, Lowestoft, Suffolk NR33 7SR

Contact: David Jarvis
Tel/fax: 01502 512245

To promote all forms of military modelling except wargames; minimum age 12; monthly meetings at members' homes on a rota basis; contact organiser for details.

B M S S Wiltshire Branch
Apple Garden, Corsley, Warminster, Wiltshire BA12 7QL

Contact: J A (Tony) Young
Tel/fax: 01373 832323

Contact representative for details of meetings which are held monthly in Salisbury and Mid-Wessex.

British Flat Figure Society
9 Church Gardens, Barningham, Bury St Edmunds, Suffolk IP31 1DE

Contact: Michael Creese
Tel/fax: 01359 221628

The Society publishes four journals a year, and catalogue reprints; and arranges bulk orders from Continental makers. Branch meetings held across the country.

British Model Soldier Society
44 Dane Mead, Hoddesdon,
Hertfordshire EN11 9LU

Contact: Mark Gilbert
Tel/fax: 01992 441078

Founded in 1935 to cater for the interests of the modeller and collector. Over 20 branches (see separate listings under 'BMSS') and members in 20 countries are kept in touch by magazine, newsletters, and branch open days. Competitions and auctions take place regularly.

City & East London Model Making Club
St John Ambulance Hall, East Avenue, Walthamstow, London E17

Contact: Paul Melton
Tel/fax: 0181 559 0189 (pm)

Cornwall Military Modelling Society
24 Central Avenue, St.Austell, Cornwall PL25 4JG

Contact: T Grainger-Allen

Darlington Military Modelling Society
127 Dinsdale Crescent, Darlington, Co.Durham DL1 1EZ

Contact: Colin Holmes
Tel/fax: 01325 489801

Meetings on second Friday of each month at 7.30pm in the Arts Centre, Vane Terrace, Darlington. Figures, vehicles, aircraft etc. Subscription £10 per year.

De Tinnen Tafelronde
Jan van Nassau Str 101, 2596 BR Den Haag, Netherlands

Contact: J R Mengarduque

East Midlands Model Club
65 Shilton Road, Barwell, Leicestershire LE9 8HB

Contact: Gordon Upton
Tel: 01455 848772
Fax: 01455 230952

Scale model club for AFV, figure, marine and aircraft modellers. Meets first Monday of each month at the RAFA Club, Lancaster Road, Hinckley, Leics. at 7.30pm. Subscription £5.00.

Essex Scale Model Society
50 Putney Road, Enfield, Middlesex EN3 6NN

Contact: Mark Gilbert
Tel/fax: 01992 715621

Figure, vehicle and aircraft modelling. Meetings at Star & Garter, Moulsham Street, Chelmsford, Essex, at 19.30 hrs on the first Wednesday of each month.

Figurina Helvetica
Postfach 649, CH-8025 Zurich, Switzerland

Glasgow Miniature Armour Group
15 Mill Street, Bridgeton, Glasgow, Strathclyde G40 1LT, Scotland

Contact: Robert Burns

Groupement des Clubs de Figurines
75 Rue Henri Barbusse, 77290 Mitry Mory, France

Contact: M Hanin

Halesowen Military Modelling Society
45 Comberton Avenue, Kidderminster, Hereford & Worcester DY10 3EQ

Tel/fax: 01562 823829

Holts' Tours (Battlefields & History)
15 Market Street, Sandwich, Kent CT13 9DA

Contact: John Hughes-Wilson
Tel: 01304 612248
Fax: 01304 614930
Email: www.battletours.co.uk

Europe's leading military historical tour operator, offering annual world-wide programme spanning history from the Romans to the Falklands War. Holts' provides tours for both the Royal Armouries Leeds and the IWM. Every tour accompanied by specialist guide-lecturer. Send for free brochure.

I P M S (UK)
2 St Augustine's Road, Edgbaston, Birmingham, Midlands (W.) B16 9JU

Contact: Paul Regan

International Plastic Modellers' Society central liaison. See separate regional branches for contact addresses.

I P M S (UK) Aberdeen Branch
Girdleness Lighthouse, Grey Hope Road, Aberdeen, Grampian AB1 3QX, Scotland

Contact: Charlie Reed

I P M S (UK) Abingdon Branch
1 Argentan Close, Sutton Fields, Abingdon, Oxfordshire OX14 5QW

Contact: Tony Clements

I P M S (UK) Avon Branch
8 Greenswood Drive, Alveston, Bristol, Avon BS12 2RH

Contact: Kevin Wells

I P M S (UK) Clacton Branch
42 Valley Road, Clacton-on-Sea, Essex CO15 4AJ

Contact: Peter Terry

I P M S (UK) Coventry Branch
12 Telford Avenue, Leamington Spa, Warwickshire CV32 7HL

Contact: Dave Eales

I P M S (UK) Cumbria Branch
c/o 26 School Knott Drive, Windermere, Cumbria LA23 2DY

Contact: Arthur Jackson

I P M S (UK) Edinburgh Branch
15 Mauricewood Rise, Penicuik, Lothian EH26 0BJ, Scotland

Contact: Jim Fraser

I P M S (UK) Essex Branch
37 Manor Close, Dagenham, Essex RM10 8BH

Contact: Gary McCrudden

I P M S (UK) Farnborough Branch
22 Sidlaws Rad, Cove, Farnborough, Hampshire GU14 9JL

Contact: Mark Worboys

I P M S (UK) Glasgow Branch
12 Woodfield Avenue, Bishopriggs, Glasgow, Strathclyde G64 1TT, Scotland

Contact: Bruce Smith

I P M S (UK) Gloucester Branch
123 Pheasant Way, Beeches Park, Cirencester, Gloucestershire GL7 1BJ

Contact: Jeffrey Brown

I P M S (UK) Kent Branch
13 Cecil Way, Bromley, Kent BR2 7JU

Contact: Norman Brice
Tel/fax: 0181 462 5177

Meet monthly in Sevenoaks; all scales and subjects; phone for details.

I P M S (UK) Lancashire/Cheshire Branch
77 Cheetham Hill Road, Dukinfield, Cheshire SK16 5DL

Contact: Charles Hanvy

I P M S (UK) Leeds Branch
107 Harrogate Road, Yeadon, Leeds, Yorkshire (W.) LS19 7BP

Contact: Tony McGannity

I P M S (UK) Magazine Editor
16 Green Street, Greasbrough, Rotherham, Yorkshire (S.) S61 4EF

Contact: Neil Robinson

I P M S (UK) Mid-Sussex Branch
130 Wiston Road, Whitehawk, Brighton, Sussex (E.) BN2 5PR

Contact: Keith Soutter

I P M S (UK) Newark Branch
51 Richmond Road, Lincoln, Lincolnshire LN1 1 LH

Contact: Ian Crawford

I P M S (UK) Plymouth Branch
Flat 10, Dunlewery House, 4b Seymour Road, Plymouth, Devon PL3 5AX

Contact: Ian Chanter

I P M S (UK) Portsmouth Branch
17 Victory Avenue, Horndean, Hampshire PO8 9PH

Contact: Richard Parkhurst

I P M S (UK) Publicity Officer
8 Oakwood Close, Stenson Fields, Derby DE24 3ET

Contact: Nick Allen

I P M S (UK) Rotherham Branch
42 Thickett Drive, Maltby, Yorkshire (S.) S66 7LB

Contact: Dave Neale

I P M S (UK) SE Essex Branch
4 Wheatley Road, Corringham, Essex SS17 9EQ

Contact: John Bryce

I P M S (UK) SE London Branch
13 Farley Road, Catford, London SE6 2AA

Contact: Alan Partington

I P M S (UK) Sheffield Branch
3 Regents Way, Aston, Sheffield, Yorkshire (S.) S31 0FP

Contact: Mike Prime

I P M S (UK) Shetland Branch
8 Dalsetter Wynd, Dunroseness, Shetland ZE2 9JQ, Scotland

Contact: Martin Smith

I P M S (UK) Southport/Liverpool Branch
97 Barrington Road, Liverpool, Merseyside L15 3HR

Contact: Trevor Green

I P M S (UK) Stafford Branch
29 Sidmouth Avenue, Stafford, Staffordshire ST17 0HG

Contact: John Tapsell
Tel/fax: 01785 603157

Meet at 8pm on the second Thursday of the month. The White Eagle Club (The Polish Club), Riverway, Stafford.

I P M S (UK) Stirling Branch
30A Adamson Place, Cornton, Stirling, Grampian FK9 5BS, Scotland

Contact: David Barbara

I P M S (UK) Suffolk Branch
17 Edward Fitzgerald Court, Woodbridge, Suffolk IP12 4LA

Contact: Mike Grzobien

I P M S (UK) Swindon & District
8 Goldsborough Close, Eastleaze, Swindon, Wiltshire SN5 7EP

Contact: Jim Totman

I P M S (UK) Tayside Branch
Western Millbank, 8 Upper Mill Street, Blairgowrie, Perthshire PH10 6AG, Scotland

Contact: Peter Mackinnon

I P M S (UK) Telford Branch
44 Rowan Road, Market Drayton, Shropshire TF9 1RP

Contact: Geoff Arnold

I P M S (UK) Tyneside Branch
7 West Avenue, Rowlands Gill, Tyne & Wear NE39 1EB

Contact: Rob Sullivan

Irish Model Soldier Society
12 Kingsland Parade, Portobello, Dublin 8, Ireland

Contact: Shane McElhatton
Tel/fax: (00353) 1453917

IMSS welcomes modellers from all disciplines – figures, vehicles, aircraft and ships. Workshops, trade stands, monthly and annual competitions. Annual subscription IR£15.00.

K L I O – A G V
Papiererstrasse 34, 84034 Landshut, Germany

Contact: Fritz Neureuther

K L I O Arbeitsgruppe Shogun
Schweidnitzer Strasse 4B, D-10709 Berlin, Germany

Contact: Till Weber
Tel/fax: (0049) 308913347

Society of Japanese military history enthusiasts, manufacturing 30mm flat tin figures, organizing exhibitions in museums and annnual sessions in Goslar (Germany). English spoken, enquiries welcome.

K L I O Berlin
Taunusstrasse 78, 12309 Berlin, Germany

Contact: Peter Rein

The Keep Military Museum
The Keep, Bridport Road, Dorchester, Dorset DT1 1RN

Contact: Len Brown
Tel: 01305 264066
Fax: 01305 250373

Covers Devonshire Regiment, Dorset Regiment, The Devonshire and Dorset Regiment, Queens Own Dorset Yeomanry, The Royal Devonshire Yeomanry, Royal North Devonshire Hussars, Dorset Yeomanry and 94th Field Regiment, RA.

L I H M C S
PO Box 118, Wantagh, New York 11793, USA

Contact: Kevin W Dunne
Tel/fax: (001) 516 764 7104

Long Island Historical Miniature Collectors Society – the major figure modelling club in the North-East USA – holds an annual exhibition and competition in November.

The Leicester Modellers
Thurmaston Working Men's Club, Old Melton Road, Thurmaston, Leicestershire

Tel/fax: 0116 2671057

Letchworth & N Herts Military Model Club
6 Caslon Way, Letchworth, Hertfordshire SG6 4QL

Contact: Frank Henson
Tel/fax: 01462 685039

Incorporating North Herts BMSS branch. Meetings held at Plinston Hall, Letchworth, on third Wednesday each month at 7.30pm.

Military Miniature Society of Illinois
PO Box 394, Skokie, Illinois 60077, USA

Miniature AFV Association
15 Berwick Avenue, Heaton Mersey, Stockport, Cheshire SK4 3AA

Contact: G E G Williams
Tel: 0161 432 7574
Fax: 0161 283 0283
Email: mafvahq@aol.com

Promotes interest in AFVs, military vehicles and artillery. Information is disseminated through 'TANKETTE', our bi-monthly magazine containing articles, photographs and plans of interest to both historian and model maker. Provides information service.

Miniature AFV Association (Cambridge)
39 The Leas, Baldock, Hertfordshire SG7 6HZ

Contact: Paul Middleton
Tel: 01462 896101
Fax: 01223 374101

Cambridge branch of MAFVA meets on the first Tuesday of each month at The Earl of Derby, Hills Road, Cambridge, from 7.30pm.

Miniature AFV Association (Gloucester)
63 Redswell Road, Matson, Gloucestershire GL4 6JJ

Contact: Bob Griffin
Tel/fax: 01452 534944

Gloucestershire MAFVA area secretary and information on modern British AFVs.

Miniature AFV Association (Herts)
Ashbourne, 9 Belswains Lane, Hemel Hempstead, Hertfordshire HP3 9PN

Contact: Chris Lloyd-Staples
Tel/fax: 01442 403273

Meetings 8pm at above address, second Wednesday of each month. New members are very welcome, both experts and complete novices – MAFVA membership not essential.

Miniature AFV Association (London)
102 Clockhouse Lane, Collier Row, Romford, Essex RM5 3QT

Contact: John Baumann

Promotes and operates the MAFVA's information service. Gathers & assists editor assemble the club journal 'Tankette' for this non-profit making volunteer organisation.

Miniature AFV Association (London)
84 Lillie Road, London SW6 1TN

Contact: Danny Taylor

Formed in the late 1960s, London MAFVA comprises both researchers and military modellers. Meetings are every two months, at the National Army Museum in Chelsea, London.

Miniature AFV Association (Notts)
63 Manvers Road, Westbridgeford, Nottinghamshire NG2 6DJ

Contact: Ian Sadler

Miniature AFV Association (Oxford)
39b Rockery Close, Shippon, Abingdon, Oxfordshire OX13 6LZ

Contact: P R May

Miniature AFV Association (Paris/France)
7 Square F Jammes, Elancourt 78990, France

Contact: Vincent Berko

Miniature AFV Association (S Wales)
3 St James Park, Brackla, Bridgend, Glamorgan, Mid- CF31 2NP, Wales

Contact: Paul Gandy
Tel/fax: 01656 660870

South Wales MAFVA meets on the first Monday of each month, details from Paul Gandy. We will be holding the 1998 MAFVA competitions in Cardiff.

Miniature AFV Association (Suffolk)
10 Clements Way, Beck Row, Bury St Edmunds, Suffolk IP28 8AB

Contact: Gary Wenko

The Miniature Armoured Fighting Vehicle Association is an enthusiast's group whose interests lie in the scale modelling of all types of military vehicles.

Miniature AFV Association (Sussex)
2 Green Close, Southwick Green, Southwick, Sussex (E.) BN42 4GR

Contact: Richard Allebone

Miniature AFV Association (W Australia)
33 Boronia Avenue, Nedlands, Western Australia 6009, Australia

Contact: Jon Baley

Miniature AFV Association (York)
29 Rosedale Avenue, Acomb, Yorkshire (N.) YO2 5LH

Contact: Mike Welsh

Miniature Figure Collectors of America
19 Mars Road, North Star, Newark, Delaware 19711, USA

Contact: Alban P Shaw

North East Modellers
6 Jude Place, Peterlee, Co.Durham SR8 5JW

Contact: Len Swaisland
Tel/fax: 0191 5867139

North London Military Modelling Society
4 College Court, Cheshunt, Hertfordshire EN8 9NJ

Contact: Jack Snary
Tel/fax: 01992 638046

Members' interests include figures, AFVs, aircraft, naval and fantasy. Meetings 1900hrs first Friday of the month at Millfield House, Silver Street, Edmonton, London N18.

North Surrey Military Group
5 Flimwell Close, Bromley, Kent BR1 4NB

Tel/fax: 0181 698 0890

Osterreichischer Zinn-Club
Postfach 59, A-1101 Wien, Austria

Peterborough Model Makers Society
58 Station Road, Nassington, Peterborough, Cambridgeshire PE8 6QB

Contact: Nick Tebbs
Tel/fax: 01780 782274

Plymouth Model Soldier Society
102 Warleigh Avenue, Keyham, Plymouth, Devon PL2 1NP
Contact: Harry Miller
Tel/fax: 01752 556811

Poole Viking Model Club
21 Abbotsbury Road, Broadstone, Dorset BH18 9DA
Contact: Ian Groves
Tel/fax: 01202 699504

Meetings held at the Civic Centre Social Club, Municipal Buildings, Poole, at 19.30 hrs on the first Wednesday of each month. Covers all aspects of modelling. (Includes BMSS and IPMS branches.)

S C A M M S
11456 Broadmead South, El Monte, California 91733, USA

Southern California Area Military Miniature Society hosts major exhibition / competition annually in March for figure and associated models. With seminars, auction, etc. International entries and attendance.

Shard End Modelling Club
91 Moorfield Road, Shard End, Birmingham, Midlands (W.) B34 6QY
Contact: Gareth Davis
Tel/fax: 0121 747 3370

South Hants Military Modelling Society
38 The Spinney, Fareham, Hampshire PO16 8QB
Contact: Terry Nappin
Tel/fax: 01329 236365

Meetings held at Staff Club House, University of Southampton, University Road, Highfield, Southampton at 20.00 hrs. on the first Wednesday each month.

Sutton Coldfield Model Makers Society
Ashgrove, Didgley Lane, Fillongley, Coventry, Warwickshire CV7 8DQ
Contact: Robert Day
Tel/fax: 01676 540469

The club caters for modellers of all kinds and ages. We hold competitions, workshops, social evenings and trips; entertain guest speakers; and display at shows.

Swedish Model Figure Society (Carolinen)
Alvkarleovagen 25, S-115 43 Stockholm, Sweden
Contact: Henrik Moberg
Tel/fax: (0046) 6644024

Association for collectors of old toy soldiers, plastic figures, flats and model figures. We publish a magazine twice a year. Meetings are held regularly.

Swindon Military Modellers
20 Bryanston Way, Nythe, Swindon, Wiltshire SN3 3PG
Contact: Michael Copland
Tel/fax: 01793 642020

Modelling group meets fortnightly (Sun). All standards and age groups.

Universal Modelling Society
Midland Adult School Union Ctr, Gaywood Croft, Cregoe Street, Lee Bank, Midlands (W.) B15 2ED
Contact: Den Karsey
Tel/fax: 0121 705 4085

Over fifty members whose interests cover a wide range of modelling including military figures, ships, planes, armoured fighting vehicles, space and science fiction. Participates in major shows and holds own annual show. Meetings weekly, Mon 7.30pm-10.00pm.

West Scotland Military Modelling Club
Koorydoon, Hollybush by Ayr, Ayrshire KA6 7EA, Scotland
Tel/fax: 0129 256 763

Scotland's largest figure modelling club covers all aspects and periods. Based in Glasgow; monthly meeting on Sundays; and hosts annual two-day show 'Soldiers'. Contact Secretary at above address for details.

Wirral Modelling Club
10 Beaumaris Road, Wallasey, Merseyside L45 8NH
Contact: Ian Weir
Tel/fax: 0151 639 3636

A club of widespread interests and abilities, meeting on the evening of the last Wednesday each month.

MODELLING EQUIPMENT & SERVICES

Alec Tiranti Ltd
70 High Street, Theale, Reading, Berkshire RG7 5AR
Contact: J Lyons
Tel: 0118 930 2775
Fax: 0118 932 3487
Email: tiranti@cyberweb.co.uk

Sculptors' tools, materials and equipment, including silicone rubbers, white metals, centricast machines, fine modelling tools and materials, Milliput. London branch: 27 Warren Street, W1P 5DG (071 636 8565), open Mon-Fri 9am-5.30pm, Sat 9.30am-1.00pm. Theale Mon-Fri 8.30am-5.00pm, Sat 9.30am-1.00pm (mail order). Access and Visa.

Applied Polymer Systems
Westburn House, Parish Ghyll Road, Ilkley, Yorkshire (W.) LS29 9NG

Art of War
5 Gilson Way, Kingshurst, Birmingham, Midlands (W.) B37 6BG
Contact: N C Mullis
Tel/fax: 0121 749 5676

Assembleur Multifonction
5 Ruelle du Colombier, 78410 Nezel, France
Contact: M Dussieux

Awful Dragon Management
3 Ransome's Dock, 35-37 Parkgate Road, London SW11 4NP

Bases and Cases
308 Church Road, Kessingland Beach, Lowestoft, Suffolk NR33 7SB
Contact: Len Burrell
Tel/fax: 01502 740791

Hardwood bases and diorama/flats display boxes made to order. Figure clamps and painting aids. Sculpture tools. Button punch. 32nd Street building components. Free brochure.

Battle Art
13 Cameron Drive, Auchinleck, Ayrshire KA18 2JE, Scotland

Bryce
52 Caradoc Street, Taibach, Port Talbot, Glamorgan (W.), Wales

Bugle & Musket
148 Valley Crescent, Wrenthorpe, Wakefield, Yorkshire (W.) WF2 0ND

C R Clarke & Co (UK) Ltd
Unit 3, Betws Industrial Park, Ammanford, Carmathenshire SA18 2LS, Wales
Contact: Christopher Clarke
Tel: 01269 593860
Fax: 01269 591890

Vacuum Forming equipment for model making starting at 9 X 10in cut sheet size. Able to form acrylic, polystyrene, polycarbonate, etc to high definition.

Capital Model Supplies
42 Anerley Hill, London SE19 2AE

Carn Metals Ltd
8 Carn View Terrace, Pendeen, Cornwall TR19 7DU
Tel/fax: 01736 787343

Manufacturer of high quality pewters, whitemetal casting alloys and soldiers from 250gms to 250kg.

Ceremonial Colours
A J Collier, 3 Coppin Close, Belmont, Hereford & Worcester

Charles Frank Ltd
Ronald Lane, Carlton Park, Saxmundham, Suffolk IP17 2NL
Tel/fax: 01728 603506

Le Cimier
38 Rue Ginoux, 75015 Paris, France
Contact: Jacques Vuyet

Coker Bros Ltd
Unit 5, Upper Brents Industrial Estate, Faversham, Kent ME13 7DZ
Tel: 01795 535008
Fax: 01795 532146

Collector Cabinets
Eastburn House, Green Lane, Eastburn,Keighly, Yorkshire (W.) BD20 8UT
Tel/fax: 01535 656030

Craft Supplies Ltd
The Mill, Miller's Dale, nr. Buxton, Derbyshire SK17 8SN
Tel: 01298 871636
Fax: 01298 872263

Supplier of glass domes and bases to display and protect models. Catalogue available on request. Shop and mail order.

D K L Metals Ltd
Avontoun Works, Linlithgow, Lothian EH49 6QD, Scotland
Tel/fax: 01506 847710

D R Models
47 Glebe Road, Ampthill, Bedfordshire MK45 2TJ
Tel/fax: 01525 840140

David Proops Sales
21 Masons Avenue, Wealdstone, Harrow, London HA3 5AH
Tel: 0181 861 5258
Fax: 0181 861 5404

Full range of miniature and precision tools. For a full illustrated catalogue please send £1.50 (UK) or £3.00 (overseas).

Daylight Studios
223a Portobello Road, London W11 1LU
Contact: Matthew Briggs
Tel: 0171 229 7812
Fax: 0171 727 1773

Specialists in daylight lighting, magnification and visual accessories,including two very popular modellers' tools – a special rimless 5" flexiarm magnifier, and clip-on spectacle magnifiers.

Dean Forest Figures
62 Grove Road, Berry Hill, Coleford, Gloucestershire GL16 8QX
Contact: Philip & Mark Beveridge
Tel/fax: 01594 836130

Wargame figure painting, scratch-built trees, buildings and terrain features. Large scale figure painting and scratch-building to any scale. SAE for full lists.

Diostar
9 Grand Rue, 11400 Villeneuve la Comptal, France
Contact: Patrice Raynaud

The Dreamsmith
13 Church Lane, Killamarsh, Sheffield, Yorkshire (S.) S31 8AS
Contact: Nick Reynolds
Tel/fax: 0114 283124

E M A Model Supplies Ltd
58-60 The Centre, Feltham, Middlesex TW13 4B4
Tel: 0181 890 8404
Fax: 0181 890 5321

Europe's largest range of trees including etched brass palms, cacti, deciduous, etc. Also the home of 'Plastruct' plastic shapes – see our latest glowing rods, ideal for fantasy and Sci-Fi projects. Plus moulding materials: Gelflex, Silicone Rubbers, Latex, powders etc. Closed half day Wednesday; mail order catalogue (100 pages) available for £3.00.

The Fife and Drum
479 Heather Path, Collydean, Glenrothes, Fife KY7 6TX, Scotland
Tel/fax: 01592 745377

Fine Art Castings & Machinery
Unit 3, Weyhill Road, Andover, Hampshire
Tel: 01264 333470
Fax: 01264 334120

Formation Plastix
Unit E4, Hilton Main Business Park, Featherstone, Staffordshire WV10 7HP
Tel/fax: 01902 723999

Geo W Neale Ltd
Victoria Road, London NW10 6NG
Tel: 0181 965 1336
Fax: 0181 965 1725

Manufacturers of white metal casting alloys.

Graphic Air Systems
8 Cold Bath Road, Harrogate, Yorkshire (N.) HG2 0NA
Contact: Doug Taylor
Tel: 01423 522836
Fax: 01423 525397

Mail order suppliers of airbrush/compressor & spraybooth equipment. Most makes of airbrush available at discount prices. Send for catalogue. Technical and product advice available.

Historex Agents
Wellington House, 157 Snargate Street, Dover, Kent CT17 9BZ
Contact: Lynn Sangster
Tel: 01304 206720
Fax: 01304 204528

Supplier of pyrogravures for plastic work, and punch and die sets, together with the finest Kolinsky sable paint brushes, etc.

Humbrol Ltd
Marfleet, Hull, Humberside (N.) HU9 5NE
Tel: 01482 701191
Fax: 01482 712908

Ian Overton
133 Parklands Drive, Loughborough, Leicestershire LE11 2TA
Tel/fax: 01509 231489

Jena Enterprises
58 Wiston Avenue, Worthing, Sussex (W.) BN14 7PT

Knight Designs
95 Tyrrell Avenue, Welling, Kent DA16 2BT
Contact: Steve Chipperfield
Tel/fax: 0181 304 4408

Private commissions undertaken, large or small from single master

figures to ranges – figures, planes, ships, scenery. Mould-making, casting, and painting services also available.

M R Pictures Ltd.
Claire Road, Kirby Cross, Essex CO13 0LY
Tel: 01255 850974
Fax: 01255 678534

Mac Warren
50 Sunnybank, Hull, Humberside (N.) HU3 1LQ
Tel/fax: 01482 48704

Historex Agents

Verlinden Productions	Model Cellar
Elite Miniatures Spain	Nuts & Bolts
Fonderie Miniatures	Shenandoah
Soldiers Lucchetti	Grandt Line
Andrea Miniatures	Wild West
Fort Royal Review	Le Cimier
G & M Miniatures	Sovereign
Mini Art Studios	Historex
Windrow & Greene	Puchala
Poste Militaire	Almond
Friulmodelismo	Nemrod
Wolf Miniatures	Ironside
New Connection	Mil–Art
Metal Models	Pegaso
Hecker & Goros	Pili Pili
Mascot Models	Starlight
Trophy Models	Tomker
Quardi Concept	Beneito
Show Modelling	Azimut
Tiny Troopers	Decima
White Models	Glory
Fort Dusquesne	Aber
Hornet Models	

Historex Agents
Wellington House, 157 Snargate Street, Dover, Kent CT17 9BZ Tel: 01304 206720 Fax: 01304 204528

Maquette 21
9 Rue du Chapeau Rouge,
Passage Bossuet, 21000 Dijon,
France

Metal Modeles
Quartier Engaspaty, Seillans,
83440 Fayence, France

Contact: Bruno Leibovitz

Milites Miniatures
Wild Acre, Minchinhampton,
Gloucestershire GL6 9AJ

Design and manufacture 25mm
and 40mm figures; Seven Years
War, Medieval, Plains Indians.
Mail order only.

Milliput Company
Unit 5, Marian Mawr Industrial
Estate, Dolgellau, Gwynedd LL40
1UU, Wales

Milton Keynes Metals
(Dpt WG) Unit A2, Ridge Hill,
Farm, Little Horwood Rd, Nash,
Milton Keynes, Buckinghamshire
MK17 0GH

Contact: Vivian Wilson
Tel: 01296 713631
Fax: 01296 714155
Email: 106333.667
@compuserve.com

White metal, silicone rubber,
vulcanised rubber, centrifugal
casting machines and equipment,
model metals and engineers' tools.
Mon-Fri 9am-5pm. Access & Visa
accepted. Consultancy service.

Minicraft
Westpoint, The Grove, Slough,
Berkshire SL1 1QQ

Mount Star Metals
Rail Works, Biggleswade,
Bedfordshire SG18 8BD

Contact: C Burton
Tel: 01767 318999
Fax: 01767 318912
Email: mountstar
@compuserve.com

British manufacturers of high
quality tin-lead and lead free
pewter casting alloys – suppliers
to the military modelling
industry, est. 1949.

Oakwood Studios
396 Ring Road, Beeston Park,
Middleton, Leeds LS10 4NX

Contact: Richard Wharton
Tel/fax: 0113 2719595

Oakwood Studios produce a huge
variety of bases in different
shapes, sizes and woods. Our
speciality is to taylor-make to
customers' requirements.

Optum Hobby Aids
PO Box 262, Haywards Heath,
Sussex (W.) RH16 3FR

Contact: David Bedding
Tel/fax: 01444 416795

Distributor of modelling
materials: Sylmasta A+B,
Kneadatite-Duro, Artificial Water,
Superglues, pewter sheet,
moulding rubbers, casting resins,
No 5 wax carver, modelling tools,
Milliput, Swiss Files.

The Paint Box
Southford Engineering Services,
31 Ely Close, Southminster,
Essex CM0 7AQ

Paintworks
99-101 Kingsland Road, London
E2 8AG

Contact: Dorothy Wood
Tel: 0171 729 7451
Fax: 0171 739 0439

Art materials shop and mail order
selling top quality oil and acrylic
paints and a comprehensive range
of brushes. Opening hours
9.30am-5.30pm Mon-Sat.

Picture Pride Displays
17 Willow Court, Crystal Drive,
Sandwell Park, Warley, Midlands
(W.) B66 1RD

Contact: Eileen Bourn
Tel: 0121 544 4946
Fax: 0121 552 9959

Solid wood display cabinets in all
finishes for your collectables. Wall
mounting, complete with polished
edged glass shelves. Totally dust
proof. Free colour brochure
available.

Picturesque
25/27 Tufton Street, Ashford,
Kent TM23 1QN

Tel: 01233 641682
Fax: 01233 630293

Display cases made to order;
enquiries to address above.

Proops Brothers Ltd
Technology House, 34
Saddington Road, Fleckney,
Leicestershire LE8 0AW

Rose Paints
Ceremonial Studios, 88 Orchard
Road, Southsea, Hampshire
PO4 0AB

Tel/fax: 01705 732753

**Rotring/Conopois Airbrush &
Spray Centre**
39 Littlehampton Road,
Worthing, Sussex (W.) BN13 1QJ

Contact: Kenneth Medwell
Tel: 01903 66991
Fax: 01903 830045

Airbrush and compressor
equipment and materials
suppliers. Main agents for Iwata
and Paasche, including own
manufactured products.
Extensive range of equipment
backed up with airbrush advice
line and own workshops.
Servicing, repairs and spares
available on most major makes of
equipment by factory trained
staff. Industrial spray equipment
also available.

S V F
Unit 14d, Whitebridge Industrial
Estate, Stone, Staffordshire

Tel/fax: 01785 814654

Scale Model Accessories Ltd
160 Green Street, Enfield,
London EN3 7LB

Contact: David Tweedle
Tel: 0181 804 5100
Fax: 0181 805 1751

1/48th scale airfield and military
products consisting of figures,
accessories and vehicles in resin,
metal and etched brass. WWII
period Axis and Allied airforces.
Contract resin casting service
facilities now available.

Scale Models International
Boundary Way, Hemel
Hempstead, Hertfordshire
HP2 7ST

Set Scenes
PO Box 63, Crawley, Sussex
(W.) RH11 8YR

Soldats de Plomb
Route de Bourg, 38490 Fittilou,
France

Contact: Joanny Jabouley

Stauve Galleries
34 Lower Street, Dartmouth,
Devon TQ6 9AN

Tel/fax: 0180 3835020

Sungro-lite Ltd
118 Chatsworth Road, London
NW2 5QU

Tel: 0181 459 2636
Fax: 0181 459 4130

'Natural' light bulbs – true colour
rendition, cool, glare-free diffused
light. Similar to indirect sunlight
for model making and painting.
Available in 75/100/150 watt.
Mail order service available. No
callers without appointments.
Phone or write for details or order
form.

Trains & Things
170/172 Chorley New Road,
Horwich, Bolton, Lancashire BL6
5QW

Contact: David Moss
Tel/fax: 01204 669203

Virgin Soldiers
81 Coleridge Road, Weston
super Mare, Avon BS23 3UJ

Contact: C Ormsby
Tel/fax: 01934 614153

Vitrine Center
67 Rue de la Haie-Coq, 93300
Aubervilliers, France

Vitrines Collectionneurs
4 Route de Seur, 41120
Cellettes, France

Whittlesey Miniatures
75 Mayfield Road, Eastrea,
Whittlesey, Peterborough,
Cambridgeshire PE7 2AY

Contact: Keith Over
Tel/fax: 01733 205131

Manufacturers and suppliers of
high quality, painted 'toy'
soldiers. Currently specialising in
Medieval and Ancient subjects in
54mm. Master and mould making
services are available.

MODEL MANUFACTURERS

A & F Neumeister (Flats)
Hauptstrasse 98, 98553
Hirschbach, Germany

A B Figures/Wargames South
24 Cricketers Close, Ockley,
Surrey RH5 5BA

Contact: Mike Hickling
Tel/fax: 01306 627796

Manufacturers of the 15mm AB
Figure range for wargamers and
collectors; plus 10mm WWII and
19th century wargames figures.
Mail order only; send SAE for lists.

**A Ochel / Kieler Zinnfiguren
(Flats)**
Schulperbaum 23, 24103 Kiel,
Germany

Contact: E Kroschewski

Accurate Armour Limited
Units 15-16, Kingston Industrial
Estate, Port Glasgow, Inverclyde
PA14 5DG, Scotland

Contact: David Farrell
Tel: 01475 743 955
Fax: 01475 743 746
Email: 101363.74
@compuserve.com

Largest UK producer of specialist
1/35 scale military models of
AFV's and accessories /
conversions / decals. All products
available by direct mail order,
catalogue-2 £4.95 + postage. Also
distributors for Collectors Brass
(USA), MR Models (Germany),
FCM Models (Germany) and
Kendall Model Co. (USA).

AL. BY Miniatures
BP 34, Valence d'Agen 82400,
France

Contact: Yannick Laffargue
Tel: (0033) 563 291 122
Fax: (0033) 563 393 090

We produce a range of about 200
resin kits and conversion sets to
1:35, 1:72 and 1:87 scales. We also
cast for other manufacturers.

Alban Miniatures
28 Glade Side, St Albans,
Hertfordshire AL4 9JA

Albion Miniature Figures
58 Elmcroft Avenue, Sidcup,
Kent DA15 8NN

Alexander Wilken (Flats)
Carl-Spitzweg-Platz 3, 85586
Poing, Germany

**Alexander Windisch-Sachs
(Flats)**
Postfach 2625, 26122 Offenburg,
Germany

Almond Sculptures
The Nook, Faussett Hill, Street
End, Lower Hardress,
Canterbury, Kent CT4 7AJ

Contact: J Almond
Tel/fax: 01227 464424

Manufacturer of cast white metal
military figurines/model soldiers.

Ancient Artillery Models
6 Arlington Close, Attleborough,
Norfolk NR17 2NF

Contact: Mathew Shearn
Tel/fax: 019053 452905

Manufacturer of working scale
models of Roman and Medieval
catapults, including the mangonel,
onager and ballistae; educational
kit of the 13th century trebuchet
siege engine.

Anglian Miniatures (Flats)
28 St Leonards Close, Scole,
Diss, Norfolk IP21 4DW

Contact: Ken Bright
Tel/fax: 01379 740358

Manufacturer of 30mm flat figures
which are predominantly, though
not exclusively, based on British
and American themes. The
company also offers a design
service.

Anne-Marie Amblard
4 Chemin du Luquet, 80120
Lannoy-les-Rues, France

Tel/fax: (0033) 22250654

Large scale figure models in wax,
fabrics, etc.; enquire for details of
one-off and limited edition
subjects.

Armour Models
12 Sycamore Terrace, Haswell
Village, Co.Durham DH6 2AG

Contact: Paul Wade
Tel/fax: 0191 526 0485

Importers for Tonda Vac-forms,
Des Kit etc. Limited supply of
East European armour kits and
figures. Producers of 1:35th scale
resin tank crew figures. Lists
available. Mail order only.

Art of War
5 Gilson Way, Kingshurst,
Birmingham, Midlands (W.)
B37 6BG

Contact: N C Mullis
Tel/fax: 0121 749 5676

Arthur Speyer (Flats)
Taunusstrasse 23, 65345
Eltville-Rauenthal, Germany

Artisan Miniatures
6 Overstone Court, Westwood,
Peterborough, Cambridgeshire
PE3 7JE

Arwed Ulrich Koch (Flats)
Lerchenstrasse 33, 73441
Bopfingen, Germany

Azimut Productions
8 Rue Baulant, 75012 Paris,
France

Tel: (0033) 143070616
Fax: (0033) 143471193

Dealer and manufacturer of
1/35th scale military models; we
offer over 60 ranges of
vehicles,figures, accessories and
books. Trade enquiries welcome.

B & B Military
1 Kings Avenue, Ashford, Kent
TN23 1LU

Contact: Len Buller
Tel/fax: 01233 632923

Makers of military model vehicles
and figures in Dinky 1/60 inch
scale. All white metal. Mail order
only: send SAE for list of kits and
finished models available. Colour
catalogue £3.50.

**Bahr & Schmidtchen
Zinnfiguren (Flats)**
Bennigsenstrasse 3, 04315
Leipzig, Germany

Ballantynes of Walkerburn Ltd
Tweedale Mills (East),
Walkerburn, Borders EH43 6AS,
Scotland

Contact: Neil Ballantyne
Tel: 0189687 697
Fax: 0189687 409

Manufacturers of quality
statuettes. Available in hand
painted, bronze resin, silver plate
and pewter. Full catalogue
available. Visa/Access accepted.
Trade enquiries welcome.

Berliner Zinnfiguren (Flats)
Knesebeckstrasse 88, 10623
Berlin, Germany
Contact: Hans-Gunther Scholz
Tel: (0049) 303130802
Fax: (0049) 303131180

Germany's largest manufacturer
and shop for flat tin figures.
Illustrated mail order catalogue
available. Shipping anywhere.

Bombardier Models
Barton Lane Workshops, Pound
Lane, Bradford on Avon,
Wiltshire

Borbur Enterprises
Unit 220, 62 Tritton Road,
London SE21 8DE
Contact: M Borrowman
Tel/fax: 0181 766 7227
Email: borburent@btinternet.com

Manufacturers of Steadfast
Soldiers and white metal casters.

Border Miniatures
Fernlea, Penrith Road, Keswick,
Cumbria CA12 4LJ
Contact: Peter Armstrong
Tel/fax: 017687 71302

White metal kits, 80mm, medieval
subjects, designed by Pete
Armstrong. Direct mail order or
through specialist shops
worldwide. Visitors welcome by
appointment. Illustrated catalogue
£2.

**Braunschweiger Zinnfiguren
(Flats)**
Schäfertiech 4a, 38302
Wolfenbuttel, Germany
Contact: Dr Peter Dangschat

Especially Georgian period,
Regency, maritime figures, ships,
French Revolution, Vendée,
Bonapartes, Napoleons, Prussia,
USA, Erotica.

Bremer Zinnfiguren (Flats)
Enscheder Strasse 13, 27753
Delmenhorst, Germany
Contact: Gunter Scharlowsky

C K Supplies
28 Hillbrow Road, Bromley, Kent
BR1 4JL
Contact: Carl Miller
Tel/fax: 0181 466 5543

Exclusive European distributors
of 'Soldiers of Russia', 54mm
lead/pewter models from St
Petersburg, Russia. Also exclusive
manufacturers of 280mm bronze
resin military figurines. Mail
order only.

C M S
3 Square Valerie, 95260 Parmain,
France

C P Golberg (Flats)
Christian-Rohlfs-Weg 11, 24568
Kaltenkirchen, Germany
Tel: (0049) 41911484
Fax: (0049) 419160746

Flats 30mm from Punic wars to
Napoleonic, soldiers and civilians.
Dealers for Gottstein-Terana-
Sivhed-Kovar-Schramm. 6800
different items, catalogue
available, worldwide delivery.

Caberfeidh Miniatures
Caberfeidh House, 15 Church
Road, Duffus, Moray IV30 2QQ,
Scotland
Tel/fax: 01667 56017

Cadre Miniatures
435 Auckland Drive, Chelmsley
Wood, Birmingham, Midlands
(W.) B36 0RG
Tel/fax: 0121 788 8429

Campaign Figures
377 Hainton Avenue, Grimsby,
Humberside (S.) DN12 9QP

Cavendish Miniatures
1 Little Buntings, Windsor,
Berkshire
Contact: Tony Kite
Tel/fax: 01753 855474

Metal handpainted 54mm models
of Marines, Guards, Scottish and
Irish Pipers, Germans, Knights of
Garter, Mounted Lifeguards,
Drum Horse, Falkland Party
artillery gun carriage.

Ceremonial Studios
Unit 215, Victory Business Ctr,
Somers Road North,
Portsmouth, Hampshire PO1 1PJ

Chas C Stadden Studios Ltd
Edwin Road, Twickenham,
London TW2 6ST

Cheshire Volunteer
2 Brookside Road, Fordsham,
Cheshire WA6 7BL
Contact: Alex Williams
Tel/fax: 0151 495 1254

Chosen Men
74 Rotherham Road, Holbrooks,
Coventry, Midlands (W.)
CV6 4FE
Tel/fax: 01203 666376

Manufacturer of resin figures in
120mm scale. British, North
American covering Victorian and
ACW periods. Mail order service.
Can supply to the trade.

Chota Sahib
124 Springfield Road, Brighton,
Sussex (E.) BN1 6DE
Contact: Sid Horton
Tel/fax: 01273 884170

Sid Horton's classic range of
(mainly) 54mm white metal
figures – all periods, but British
Victorian and Edwardian
specialist.

Classic Miniatures (Scotland)
Unit 12, Currie Road Industrial
Estate, Galashiels, Borders TD1
2BP, Scotland
Tel/fax: 01896 58377

Clydecast Products
97 Fereneze Avenue, Clarkston,
Glasgow, Strathclyde G76 7RT,
Scotland
Contact: Thomas Park
Tel/fax: 0141 638 1904

Manufacturer of model soldiers in
75, 90 and 110mm. Over 100 items
produced for collectors. Mail
order speciality. Also design and
contract casting work undertaken.
Suppliers to military and
regimental museums.

Collectair
32 West Hemming Street,
Letham, Angus DD8 2PU,
Scotland
Contact: Peter Fergusson
Tel/fax: 01307 818494

Makers of cast pewter scale model
aircraft (140+ models), 1/300 and
1/200 scales, for collectors and
wargamers. List available; trade
enquiries welcome. Available in
USA from Simtac Inc, 15G Colton
Road, East Lyme, CT 06333.

Conflict Miniatures
27 Leighton Road, Hartley Vale,
Plymouth, Devon PL3 5RT
Contact: Tim Reader
Tel/fax: 01752 770761

Manufacturer of wargaming
figures/models. Worldwide mail
order. Stockist of rules,
Kriegsspiel, paints, brushes etc.
Painting service available. Trade
enquiries welcome.

Continental Model Supply Co
36 Gray Gardens, Rainham,
Essex RM13 7NH

Manufacturers of AFVs and
trucks in resin and metal, 1/87th
scale; mail order only for Roco
Minitanks, Trident Models; send
stamped SAE for list.

Cortum Figuren (Flats)
Auf dem Kluschenberg 5, 23879
Molln, Germany
Contact: Christian Carl

Corvus Miniatures
Bards Hall Cottage, Chignal
Smealy, Chelmsford, Essex CM1
4TL

Cromwell Models
Regency House, 22 Hayburn
Street, Glasgow, Strathclyde G11
6DG, Scotland

Current Issue Miniatures
147 Faulds Gate, Aberdeen,
Grampian AB1 5RB, Scotland

D F Grieve Models
Westwood Road, Betsham, Nr
Gravesend, Kent DA13 9LZ
Contact: David Grieve

D M H Models
3 Greensand Road, Bearsted,
Maidstone, Kent ME15 8NY
Contact: Derek Hansen
Tel/fax: 01622 39186

Miniature figure designer and
manufacturer.

**D P & G
(Miniature Headdress)**
PO Box 186, Doncaster,
Yorkshire (S.) DN4 0HN

Miniature headdress, 1/5th scale –
British Household Cavalry,
Dragoon, Dragoon Guard,
helmets; British and French
Lancer caps, Hussar busbies;
Infantry blue cloth helmets;
German Pickelhaubes. Special
commissions undertaken; send 3
first class stamps for catalogue.

Daniel Horath (Flats)
Berkheimerstrasse 50, 73734
Esslingen am Neckar, Germany

Dartmoor Military Models
Woodmanswell House, Brentor,
Tavistock, Devon PL19 0NE
Tel/fax: 01822 82 250

Davco Productions
28 Brook Street, Wymeswold,
Loughborough, Leicestershire
LE12 6TU

Produces an extensive range of
1/3000th scale ship models and
harbour accessories for WWI,
WWII and modern periods. Also
produce a range of 1/300th
vehicle and aircraft models for
WWII and modern periods. All
ranges, including new Starguard
space fleet model kits, are retailed
through Skytrex Ltd.

Des Kits
27 Rue des Hauts de Bonne Eau,
94500 Champigny sur Marne,
France

Dieter Schubert (Flats)
Ritterstrasse 71, 79639
Grenzach-Wyhlen, Germany

Dieter Schulz (Flats)
Weinmeisterhornweg 169,
13593 Berlin, Germany

Dolp Zinnfiguren
Im Paradies 20, 8940
Memmingen, Germany

Dr Peter Hoch (Flats)
Meistersingerstrasse 8, 14471
Potsdam, Germany
Tel/fax: (0049) 331 974388

Artistically crafted models of
historical personalities, cast in
metal, in every scale; construction
of historically based dioramas,
models for exhibition purposes,
replicas of military monuments.

Dr Wolfgang Vollrath (Flats)
Am Ruhrstein 11, 45133 Essen,
Germany

Drew's Militia
1 Mosslea Road, Bromley, Kent
BR2 9PS
Contact: Andrew Steven
Tel/fax: 0181 460 0728

20mm 1944 British & US Airborne
figures, their equipment and their
German opponents, inc.light
vehicles, artillery etc. Also
exclusive supplier of Helmet
Miniatures 1/200th scale aircraft
for wargames. SAE for catalogue
and sample.

Drumbeat Miniatures
4 Approach Road, Ramsey, Isle of
Man IM8 1EB
Contact: Peter Rogerson
Tel/fax: 01624 816667

A large and rapidly expanding
range of toy soldiers covering
many periods and unusual
subjects. Sets can be varied on
request and a sculpting service is
offered.

E & G Tobinnus (Flats)
Matthaikirchstrasse 58, 30519
Hannover, Germany

E J Schulze (Flats)
Haustenbeck 2, 44319
Dortmund, Germany

E Kastner (Flats)
Eichenhain 6a, 90571 Schwaig,
Germany

Eagle Miniatures
Wild Acre, Minchinhampton,
Gloucestershire GL6 9AJ
Contact: David Atkins
Tel/fax: 01453 835782

Design, manufacture and
distribute 15mm, 25mm and
54mm figures from Medieval,
Seven Years War, Napoleonic and
ACW periods. Design and cast
service also available. Callers by
appointment.

Eastern Import & Export
35 Kennedy Court, Stonehouse
Drive, St Leonards-on-Sea,
Sussex (E.) TN38 9DH
Tel/fax: 01424 720958

Edith Fohler (Flats)
Isoppgasse 6, A-1238
Wien-Mauer, Austria

Egon Krannich (Flats)
Bergstrasse 6, 04828 Schmolen,
Germany

Elite Collection
Bank House, 127 High Street,
Crediton, Devon EX17 3LQ
Tel/fax: 01395 273126

Ensign Miniatures
Littlebury Hall, Station Road,
Kirton, Boston, Lincolnshire
PE20 LQ
Contact: Robert & Ann Rowe
Tel/fax: 01205 722101

Specialises in high quality 54mm
metal figures. Illustrated
catalogue of 250 figures, contains
mess dress, naval, undress,
Napoleonic, European, ACW and
knights. World wide mail order
service. Callers welcome.

Eric Austen
36 Mill Hill, Deal, Kent
CT14 9EW

Ernst Vogel (Flats)
Schwabelweiserweg 18, 93059
Regensburg, Germany

Ettlinger Zinnfiguren (Flats)
Bachstrasse 32, 76275 Ettlingen,
Germany
Contact: Dieter Schwarz

F M
36 Rue Charron, 93300
Aubervilliers, France

F R A SARL
125 Rue Fauburg Poissoniere,
75009 Paris, France

Fianna Minatures
78 Bridgefoot Street, Dublin 8,
Ireland
Contact: Richard Keane
Tel/fax: (01) 679 1697

F.M. offer a range of busts/figures
dealing with the Irish Army and
WWII subjects. All figures in the
range can be supplied painted and
finished.

Fire Force Products
Unit 26, Supa Shopper, 27-28
High Street, Kings Heath,
Birmingham, Midlands (W.)
B14 7LB

Fort Royal Review
25 Woolhope Road, Worcester,
Hereford & Worcester WR5 2AR
Contact: David Gerrett
Tel: 01905 356379
Fax: 01905 764265

Manufacturer and supplier of
finecast military figurine kits –
format resin and white metal.
Suppliers world wide. All
enquiries welcome.

Fortress Models
87 Yew Tree Road,
Southborough, Tunbridge Wells,
Kent TN4 0BJ

**Frankfurter Zinnfiguren
(Flats)**
Quellenweg 7, 61250 Usingen,
Germany
Contact: Helga Lampert

Frapel
6 Chemoin des Salines, 17590
Ars en Re, France

**Fusilier Miniatures (Mail
Order)**
The Command Post, 23 Ashcott
Close, Burnham-on-Sea,
Somerset TA8 1HW
Contact: Tony Moore
Tel/fax: 01278 786858

A wide range of hand-painted,
well-detailed figures and guns in
1:32 scale. Sets and single figures
available. SAE and £1.50 for
current catalogue.

G N M Miniatures
Micawber, Yewbank, Skipton
Road, Utley, Keighley, Yorkshire
(W.) BD20 6HJ
Contact: Graham Mollard
Tel/fax: 01535 691587

65mm Napoleonic miniatures
created by Graham Mollard,
strongly influenced by the work of
Fernande Metayer and Roger
Berdou.

Gavin Haselup
Kennel House, 150 Capel Street,
Capel-le-Ferne, Folkestone, Kent
CT18 7HA

Gerhard Fetzer (Flats)
Marktstrasse 76, 89537 Giengen,
Germany

**Gerlinde & Ulrich Lehnart
(Flats)**
Eurener Strasse 59, 54294 Trier,
Germany

Gert Grosse (Flats)
Sulzburgstrasse 8, 73268
Erkenbrechtsweiler, Germany

Glorious Empires (Flats)
1 Vicarage Walk, Bray, Berkshire
SL6 2AE
Contact: Jacques Vullinghs

Gunter Fuhrbach (Flats)
Im Himmelreich 4, 59320
Ennigerloh, Germany

H D Oldhafer (Flats)
Huttfleth 48, 21720
Grunendeich, Germany

Hans G Lecke (Flats)
Finkenstrasse 19, 31543
Rehburg, Germany

Hans Joachim Reibold (Flats)
Treuenbrietzener Strasse 14,
13439 Berlin, Germany

Hart Models Ltd
Cricket Green, Hartley Wintney,
Hampshire RG27 8QB
Tel/fax: 01252 842637

Manufacturers of the world's
largest range of high quality 1/48
scale military models and kits.
Send £1 or 2 I.R.C. for full list.

Hecker & Goros Zinnfiguren
Romerhofweg 51c, 8046
Garching, Germany

Heco
23 Addison Road, Brockenhurst,
Hampshire SO4 27SD

Heidrum Urich (Flats)
Guardini-Strasse 139, 81375
Munchen, Germany

Heinrich Jeismann (Flats)
Munsterstrasse 49, 44145
Dortmund, Germany

Heinz Tetzel (Flats)
Wespenstieg 3, 03090
Magdeburg, Germany

Helmet Soldiers
Mundy's Farm, Middle Road,
Aylesbury, Buckinghamshire
HP21 7AD
Contact: George Hill
Tel/fax: 01296 415710

Manufacturer of 54mm cavalry
and infantry kits. Mainly
Napoleonic. Illustrated spare
parts catalogues on request.
Suitable for dioramas, single
display figures or wargaming.

Herbu Zinnfiguren (Flats)
Weidenbruch 15, 21147
Hamburg, Germany

Heroics & Ros Figures
Unit 12, Semington Turnpike,
Semington, Trowbridge,
Wiltshire BA14 6LB
Tel: 01380 870228
Fax: 01380 871045

Highlander Miniatures
47 Spruce Armington,
Tamworth, Staffordshire

Hinchcliffe Models
28 Brook Street, Wymeswold,
Loughborough, Leicestershire
LE12 6TU
Tel: 01509 213789
Fax: 01509 230874

Cast metal model kits. Extensive
range 54mm 1/35th scale figures
and equipment, Medieval to
WW2. 20mm 1/76th scale WW2
vehicle kits. Mail order available.

Historia (Flats)
Lindenstrasse 39, 40789
Monheim, Germany
Contact: Marie-Luise Muller

Horst Tylinski (Flats)
Achtermannstrasse 53, 13187
Berlin, Germany

Howes
85 Gloucester Road, Bishopston,
Bristol, Avon
Tel/fax: 0117 9247505

Hussar Military Miniatures
3 Third Avenue, Hayes,
Middlesex UB3 2EF
Contact: Michael Hearn
Tel: 0181 573 4597
Fax: 0181 573 1414

We manufacture a large range of
white metal figures, mainly
Medieval and Samurai in 90mm
and 54mm scales. Colour
catalogue available at £4.00.

I/R Miniatures Inc
PO Box 89, Burnt Hills, New
York 12027, USA
Contact: William Imrie
Tel: (001) 518 885 6054
Fax: (001) 518 885 0100

Largest manufacturer of 54mm
kits supplying the trade in the
USA; over 1,000 subjects from
ancient to present military, plus
literary subjects.

Ingrid Trothe (Flats)
Narzissenweg 13, 06118 Halle,
Germany

**Irene & Michael Braune
(Flats)**
Dammweg 1, 01662 Miessen,
Germany

Irene Menz-Wille (Flats)
Im Holken 7, 58675 Hemer,
Germany

Irregular Miniatures Ltd
69a Acomb Road, Holgate,
Yorkshire (N.) YO2 4EP
Tel/fax: 01904 790597

J F Benassi
55 St Mungo Avenue, Glasgow,
Strathclyde G4 0PL, Scotland
Contact: Julian Benassi

Designer of white metal model
soldiers, produced in small
quantity and available by mail
order. Send for catalogue.
Painting service also available.

J M P
78 Rue de Paris, 94190
Villeneuve St Georges, France

J Peddinghaus Zinnfiguren
Beethovenstrasse 20, 5870
Henner 5870, Germany

Jacobite Miniatures Ltd
65 Byfield Road, Woodfield
Halse, Daventry,
Northamptonshire

Jurgen Schmittdiel (Flats)
Postfach 1220, 35276
Neustadt/Hessen, Germany

Karl Romund (Flats)
Lavesstrasse 19, 30159
Hannover, Germany

Karl-Werner Reiger (Flats)
Melsdorferstrasse 77, 24109 Kiel,
Germany

Kingscast
Unit 9, Industrial Estate,
Presteigne, Powys LD8 2UG,
Wales
Tel/fax: 01544 260130

Kjeld Lulloff (Flats)
Lauravej 14/60, Copenhagen
2000, Denmark
Contact: Kjeld Lulloff

Private collection figures of
Danish kings 8 centimetre.

Klaus Hofrichter (Flats)
Gebeschusstrasse 48, 65929
Frankfurt, Germany

Klaus Karbach (Flats)
Talstrasse 83, 41516
Grevenbroich, Germany

Klaus Kittlemann (Flats)
Halleschestrasse 62a, 06366
Kothen, Germany

Knight Designs
95 Tyrrell Avenue, Welling, Kent
DA16 2BT
Contact: Steve Chipperfield
Tel/fax: 0181 304 4408

Private commissions undertaken,
large or small from single master
figures to ranges – figures, planes,
ships, scenery. Mould-making,
casting, and painting services also
available.

Kuhn & Zimmerer (Flats)
Hellenstall 10a, Hinterwossen
Allgau, Germany

The Lancashire Collection
20 Clarendon Road, Tonse Fold,
Bolton, Lancashire BL2 6BT

Lancer
Forest View, Holt Pound,
Farnham, Surrey GU10 4JZ
Contact: Margaret Bracey
Tel/fax: 01420 22354

Manufacturer and retailer of high
quality military and civilian
figures for the collector and
modeller.

Lead Sled Models
Unit 3, Round House Craft Ctr,
Buckland in the Moor,
Ashburton, Devon TQ13 7HN
Tel/fax: 01364 52971

Liselotte Maier (Flats)
Steinstrasse 8, 91086 Aurachtal,
Germany

Lobauer Zinnfiguren (Flats)
Daimlerstrasse 16, 02708 Lobau,
Germany
Contact: Arnfried Muller

Lubecker Zinnfiguren (Flats)
Schaarweg 3, D-23683
Scharbeutz, Germany
Contact: Marleen Worbs
Tel/fax: (0049) 4503 74488

Prussian 18th century, 'Menzel'
figures; War of American
Independence 1775-1783; Lubeck
State & Militia troops 1750-1900;
Hanoverian 1820; complete
showcase sets.

M & E Models
62 Periwinkle Close,
Sittingbourne, Kent ME10 2JU

M C Export & Enterprises
The Flat, Shadowlawn, Kimcote,
Lutterworth, Leicestershire LR17
5RU
Tel/fax: 01455 53430

**M D M Les Grandes
Collections**
9 Rue Villedo, 75001 Paris,
France
Tel/fax: (0033) 142617601

Manufacturer of tin soldiers,
miniatures for collectors;
Napoleonics; French Regimental
standards of US War of
Independence; mail order service.

M L R Figures
17 Oakfield Drive, Upton Heath,
Chester, Cheshire CH2 1LG

M M S Models
26 Crescent Rise, Luton,
Bedfordshire LU2 0AU

MMS models represent the
current state-of-the-art in white
metal casting. Three expanding
ranges feature the vehicles, guns
and troops of World War Two, all
in constant 1:76 (20mm) scale. The
'CLASSIC' vehicle and 'GUNpak'
towed gun kits contain full
instructions; including general
tips, if you've never worked with
white metal before. Robust and
easy to assemble, they are ideally
suited to the Wargamer and yet,
have the accuracy and a wealth of
detail that gives satisfaction to the
serious collector.

M P Studios
23 Peregrine Close, Winshall,
Burton-on-Trent, Staffordshire
DE15 OEB
Tel/fax: 01283 37104

Professional model maker and
figure painter – the finest quality
painted figures available, by
national competition winner.
Commissions undertaken to your
own specifications; quality
guaranteed – a sure investment.
Character figures also available –
all we need is a photograph. Write
or call for more information.

Macs Models
133-135 Canongate, The Royal
Mile, Edinburgh, Lothian
EH8 8BP, Scotland
Tel/fax: 0131 557 5551

Marksmen Models
(Dept MDS), 7 Goldsmith
Avenue, London W3 6HR
Contact: Michael Ellis
Tel: 0181 992 0132
Fax: 0181 992 5980

High quality, low cost plastic
figures from 25mm to 120mm,
most periods from ancient to
Korean War. Mostly cast from
original Marx and Ideal moulds.
Also sculpting and visualising for
many leading manufacturers.

Marlborough Military Models
The Duchy, Pontycymmer,
Brigend, Glamorgan, Mid- CF32
8DU, Wales
Tel/fax: 01656 871774

Martin Andra (Flats)
Narzissenweg 13, 06118 Halle,
Germany

Martina Vogel (Flats)
Finkenstrasse 7, 72294
Grombach, Germany

Mellita von Droste (Flats)
Ulrichsrain 32, 71729
Erdmannshausen, Germany

Metal Modeles
Quartier Engaspaty, Seillans,
83440 Fayence, France
Contact: Bruno Leibovitz

Mick Lunn Metal Masters
46 Staindale, Guisborough,
Middlesbrough, Cleveland TS14
8JU
Tel/fax: 01287 652760

Mike French Models
19 Langton Road, Boscombe,
Bournemouth, Dorset BH7 6HS
Tel/fax: 01202 394514

65mm and 90mm white metal
figure kits; 120mm resin figure
kits. Price list available free of
charge.

Mike Papworth Miniatures
36 Rosedale Road, Kingsthorpe,
Northamptonshire NN2 7QF
Tel/fax: 01604 458916

WWI and WWII 1/76 white metal
figures, vehicles and artillery.

Mil-Art
41 Larksfield Crescent,
Dovercourt, Harwich, Essex
CO12 4BL
Tel/fax: 01255 507440

Producers of metal military figure
kits in 54mm, 80mm and 100mm
scales. Also retailer of new and
out of print military books.

Milestone Sculpture Design
10 Portland Street, Blyth,
Northumberland NE24 1NP
Contact: Peter Miles
Tel/fax: 01670 560866

Milicast
PO Box 711, Glasgow,
Strathclyde G41 2HX, Scotland
Contact: Tom Welsh
Tel/fax: 0141 649 0319

Manufacturers of authentic scale
model military kits in 1:76 scale in
polyurethane resin; over 100 kits in
the range. Also specialise in all
WWII AFVs, soft-skins, tanks,
armoured landing vehicles etc.
Send SAE for free catalogue. Mail
order.

Military Motors
16 Coolhurst Lane, Horsham,
Sussex (W.) RH13 6DH

British World War II vehicles in
white metal, 1/7600 scale, hand
made and painted for display or
wargaming; SAE for catalogue.

Military Pageant
45 Silverston Way, Stanmore,
Middlesex HA7 4HS

Militia Models
Rosedean, Gorsty Knoll,
Coleford, Gloucestershire
GL16 7LR
Contact: Esme Walker

Cottage industry still producing –
after sudden death of founder Ken
Walker – small number of new
54mm figures in limited edition
action sets, 1870-1902 period.

Mini Art Studio
PO Box 88215, Shamshuipo Post
Office, Shamshuipo, Kowloon,
Hong Kong
Contact: David C L Kan

AFV and figure resin models in
1/35th scale. Mail order
welcomed world-wide; trade
enquiries welcomed; free
catalogue on request.

Mitrecap Miniatures
Manorfield House, 46 Main
Street, Sheffield, Yorkshire (S.)
S31 0XJ

Mopicom
104 Avenue Pierre Semard,
95400 Villiers le Bel, France

Navwar Productions Ltd
11 Electric Parade, Seven Kings,
Ilford, Essex IG3 8BY
Tel/fax: 0181 590 6731

Manufacturers of 1/3000 and
1/1200 scale ships, Roundway
and Naismith 15mm figures;
1/300 figures, tanks & aircraft.
Closed Thursdays.

Neckel Zinnfiguren (Flats)
Hattenhoferstrasse 11, 73066
Uhingen, Germany

New Connection Models
Dorfgutingen 40, 91555
Feuchtwangen, Germany
Tel: (0049) 9852 4329
Fax: (0049) 9852 3594

High quality conversions and
figures in 1/35 scale (see
advertisemnt).

New Hope Design
Rijksweg 42, 6269 AC
Margraten, Netherlands
Contact: Deborah Lake
Tel: (0031) 43458 2211
Fax: (0031) 43458 2626
*Email: info@nh-design.com or
soldier@cobweb.ul*

Produces more than one thousand
white metal figure kits in 54mm
size. A comprehensive
listing/catalogue is available on
request.

Nik Studios
49 The Meads, Edgeware,
Middlesex
Tel/fax: 0181 959 5289

North West Frontier
61 Trafalgar Road, Bowerham,
Lancaster, Lancashire LA1 4DB

Old Glory
Institute House, New Kyo,
Stanley, Co. Durham DH9 7TJ
Contact: Andrew Copestake
Tel: 01207 283332
Fax: 01207 281902

UK & European agent for Old
Glory Manufacturing, Box 20,
Calumet, PA15621, USA, tel: 001
412 4233580, fax: 001 412 4236898.
15mm, 20mm & 25mm model
figures, limited edition
Napoleonic prints, large scale
military busts.

Old Glory Corporation
Institute House, New Kyo,
Stanley, (Co.) Durham DH9 7TJ
Contact: Andrew Copestake
Tel: 01207 283332
Fax: 01207 281902

UK and European agents for Old
Glory Corp, PO Box 20, Calumet,
PA 15621 USA. Phone 412 423 3580,
fax 412 423 6898. Manufacturers of
a wide range of 25mm, 15mm and
20mm scale figures in a wide range
of periods. Trade enquiries
welcome worldwide.

On Guard Miniatures
4 Co-operative Terrace, Hetton
le Hole, Tyne & Wear DH5 9EH
Tel/fax: 0191 5269413

**Original Heinrichsen-Figuren
(Flats)**
Marktstrasse 14, 76883 Bad
Bergzabern, Germany
Contact: Kurt Wilms

Parade Figures
65 Shilton Road, Barwell,
Leicestershire LE9 8HB
Contact: Gordon Upton
Tel: 01455 848772
Fax: 01455 230952

Distributors of quality
Russian-made 54mm figures,
featuring Italian Carabinieri,

Russian Napoleonic,
Revolutionary, 17th century, and
personality series. SAE for full list;
trade/mail order welcome.

Peter Ewald Kover (Flats)
Liechtensteinstrasse 66/5,
A-1090 Wien, Austria

Phoenix Model Developments
Earls Barton, Northamptonshire
NN6 0NA
Tel/fax: 01604 810612

Pierre Bourrilly (Flats)
81 Avenue de Montolivet, 13004
Marseille, France

Pierre Bretegnier (Flats)
505 Rue des Moulins, 28260 La
Chaussee d'Ivry, France

Piper Craft
4 Hillside Cottages, Glenboig,
Lanarkshire ML5 2QY, Scotland
Contact: Thomas Moles
Tel: 01236 873801
Fax: 01236 873044

Manufacturer of white metal
military and non-military figures
designed to a general scale of
75mm. Suppliers to museums,
places of historic interest, shops
and collectors. Established 1985.
Send SAE or 2 x IRC's for a
complete illustrated list.

Poste Militaire
Station Road, Northiam, Rye,
Sussex (E.) TN31 6QT
Contact: Ray or Norma Lamb
Tel: 01797 252518
Fax: 01797 252905

Comprehensive range of white
metal kits – 70mm 75mm 90mm
110mm by Ray & Julian Lamb,
Stefano Cannone Julian Hullis,
Derek Hansen Mike Good & Keith
Durham. Resin/white metal busts
1/10th scale by Julian Lamb. Full
colour catalogue – £6.50.

Present Arms
Berrows Business Centre, Bath Street, Hereford, Hereford & Worcester HR1 2HE

Tel/fax: 01432 276510

Military miniatures – scale model weapons, statuettes, wall shields, special commissions, etc. Suppliers of presentation models to HM armed forces, defence industries, police and retail outlets worldwide.

Prince August UK Ltd
Dept DM, Small Dole, Henfield, Sussex (E.) BN5 9XH

Pro Mods
Suite 2, Wincombe Estate, Albert Road, Bristol, Avon BS2 0XW

Tel/fax: 0117 9724433

Pro-Models
5 Briscoe Road, Henlane, Coventry, Midlands (W.) CV6 4JL

Contact: Mr Bailey
Tel/fax: 01203 685372

R Boverat (Flats)
41 Boulevard de Batignolles, 75008 Paris 8, France

R G King & Son
32 London Road, Teynham, Kent ME9 9QN

Contact: Diana Budd
Tel: 01795 521372
Fax: 01795 522944

Mail order distributors of 'Roco' precision plastic semi-assembled H.O. scale (1:87) models of military and civilian models. See advertisement for further information.

Ray Brown Design
Laundry House, Bowhill, Selkirk, Borders Region TD7 5EJ

Contact: Ray Brown
Tel/fax: 01750 20430

Designers and manufacturers of precision sculptures manufactured in resin, with pewter accessories. Statuettes are 8 inches tall, supplied with turned wooden bases, and pewter nameplates.

Windrow & Greene's
Europa Militaria Series
24 titles providing superb colour photographic reference for models of 20th century military forces. For complete catalogue call 0171 287 4570.

Redoubt Enterprises
49 Channel View Road, Eastbourne, Sussex (E.) BN22 7LN

Contact: Peter Helm
Tel: 01323 738022
Fax: 01323 738032

Manufacturers and retailers of 25mm figures for Zulu Wars, Sudan, Napoleonic, ECW, Pirates, AWI, Foreign Legion, Three Musketeers and others. Send £2 for catalogue.

Regimental Statuette Manufacturers
Littlebury Hall, Station Road, Kirton, nr.Boston, Lincolnshire PE20 1LQ

Contact: Robert & Ann Rowe
Tel/fax: 01205 722101

Family business providing limited edition, large scale (11" approx), high quality statuettes primarily for military establishments. Figures are cast in resin with white metal parts and finished either in bronze or are hand painted. Callers welcomed to discuss requirements.

Reheat Models
1a Oak Drive, North Bradley, Trowbridge, Wiltshire BA14 0SW

Specialists in accurate aircrew figures, etc.

Replicast Record Models
153 Upper Aughton Road, Southport, Merseyside

Tel/fax: 01704 550488

Retter / Brenner-Maurle (Flats)
Kleinhohenheimerstr. 32, 70619 Stuttgart, Germany

Roger Saunders
290 Queen's Road, London SE14 5JN

Tel/fax: 0171 639 9409

Manufacturer of 'Hornet' 1:35 scale military figures and accessories, and 'Wild West' 1:32 scale characters of the American frontier. Send SAE or IMO for lists. Mail order welcome. Agents for Wolf Models.

The Roll Call Military Miniatures
Hoar Park Farm Craft Village, Ansley, Nuneaton, Warwickshire CV10 0QU

Tel: 01203 394848
Fax: 01203 394499

Manufacturers of a quality range of fifty plus 120mm resin figurines from ancient times to WWII, the majority from the Napoleonic and Crimean Campaigns. Send £4.00 for catalogue, price list and newsletters.

Rosedale Figurines
8 China Street, Lancaster, Lancashire LA1 1EX

Tel/fax: 01524 65129

Rudiger Engel (Flats)
Grundelbach 8, 56329 St Goar, Germany

Rudolf Grunevald (Flats)
Larchenweg 28, 30900 Wedemark-Elze, Germany

S H A Zinnfiguren (Flats)
Schonblick 3, 74544 Michelbach/Bilz, Germany

Contact: Werner Fechner

Samone Hobbies
PO Box 42, Darlington, Yorkshire (N.) DL1 2XP

Scale Models
c/o Blackwells of Hawkwell, 733-735 London Road, Westcliffe-on-Sea, Essex SS0 9ST

Tel/fax: 01702 72248

Scientific Models
Unit 3, Point Pleasant Estate, Hadrian Road, Wallsend, Tyne & Wear NE28 6HA

Contact: Jason & Martin Purns
Tel: 0191 2340455
Fax: 0191 295 4822

Scientific Models are model makers to the Defence industry, and our experience as such has allowed us to produce specialists' 1:35 scale modern armour kits.

Scotia Micro Models
32 West Hemming Street, Letham, Angus DD8 2PU, Scotland

Contact: Robert Fergusson
Tel: 01307 818707
Fax: 01307 818494

Makers of white metal castings for wargaming and collectors. Figures, AFVs, landing craft etc. in 1/300, 10mm, 15mm, 20mm, 25mm. Also Sci-Fi figures in 1/300 and 25mm (Acropolis series) and Kryomek fantasy figures. Contract casting and modelling undertaken. Trade enquiries welcome. Catalogues available. Also produced in USA by Simtac Inc., 15G Colton Road, East Lyme, CT 06333.

Segom Miniatures (Flats)
23 Rue Faraday, 75017 Paris, France

Siegfried Nonn (Flats)
Kranzbergstrasse 14, 86316 Friedberg, Germany

Skybirds 86 Model Engineers
Orchard House, Chetnole, Sherborne, Dorset DT9 6PE

Contact: Michael Eacock
Tel/fax: 01935 872182

Skytrex Ltd
Unit 3, Canal Bank, Loughborough, Leicestershire LE12 6TU

Tel: 015092 13789
Fax: 015092 30874

Cast metal kits. Extensive armour and vehicles ranges in 1/200th and 15mm (1/100th) scales. WWII and Modern Naval metal model kits in 1/1200th scales, all periods. Mail order available.

Regina Sonntag, (Engraver)
Eschenworther Weg 7, 29690 Gilten, Germany

Tel/fax: (0049) 5071 3455

Carl Spitzweg, Jugendstil, Fairies, Laokoon, Teddy Bear's Picnic, Der Abscheid, Zany Zoo, Apokalyptische Reiter, Rubens' fruit angel, Chinese Lady with Dragon.

Sovereign Miniatures
4 Hawbeck Road, Gillingham, Kent

Spartan Studio
26 Majors Close, Chedburgh, Bury St Edmunds, Suffolk IP29 4UN

Tel/fax: 01284 850326

Starlux
BP 36, 24021 Perigueux Cedex, France

Syntown Miniatures
Knellstone Lodge, Udimore, Rye, Sussex (E.) TN31 6AR

T V Models
147 Fauldsgate, Aberdeen, Grampian AB1 5RB, Scotland

Tappert Zinnfiguren (Flats)
Neckarstrasse 9, 47051 Duisburg, Germany

Thirtysecond Street
1 Mill Road, Loddon, Norwich, Norfolk NR14 6DR

Tel/fax: 01508 520327

1/32nd scale brickwork, stonework, housefronts, walls, columns, arches etc. Send SAE for lists and prices.

Thistle Miniatures
Findon Croft, Findon, Aberdeen, Grampian AB1 4RN, Scotland

Tel/fax: 01224 571831

Manufacturer of white metal/resin figures in 1/24th, 1/20th and 1/16th scales. They are mainly Scottish, in ranges: modern barrack dress, The Great War, personalities, and mess dress. Also produce prints depicting the kilts and trews of the British Army. Available by mail or telephone order.

Thorild Sivhed (Flats)
Engelbrektsgatan 83, S-23134 Trelleborg, Sweden

Till Krall (Flats)
Blissestrasse 49, 10713 Berlin, Germany

Tiny Troopers
19 Laughton Road, Boscombe, Bournemouth, Dorset BH7 6HS

Trux Models
156 High Street, Yeadon, Leeds, Yorkshire (W.) LS19 7AB

Contact: Mike Simpson
Tel/fax: 0113 2502051

Manufacturers of 1/76 scale kits in polyurethane resin specialising in wheeled vehicles used by the British Army in WWII. Also publish sets of factsheets on the organisation and vehicles used in various campaigns. Mail order only. Quarterly list on receipt of SAE.

Tumbling Dice Miniatures/Shire Design
96 Sandfield Road, Arnold, Nottingham, Nottinghamshire NG5 6QJ

Contact: Paul Sulley
Tel/fax: 0115 926 8800

Designers and manufacturers of miniature figurines and models, cast in pewter or white metals. Quality design, mould making, casting and packing services available on request.

Tyresmoke Products
21 Brampton Court, Bowershill, Melksham, Wiltshire SN12 6TH

Ulrich Puchala
Zollhausstrasse 14, 7906 Blaustein-Wippingen, Germany

Ursula Schiller-Rathke (Flats)
Blumenstrasse 56a, 47057 Duisburg, Germany

V L S Corporation
Lone Star Industrial Park, 811 Lone Star Drive, O'Fallon MO 63366, USA

Contact: Tom Gerringer
Tel: (001) 314 281 5700
Fax: (001) 314 281 5750
Email: vlsmo1@il.net

Manufacturer/distributor of Verlinden Productions. Importer/distributor for over 13,000 hobby products for scale modellers.

Viking Miniatures Ltd
Littlebury Hall, Station Road, Kirton, Nr Boston, Lincolnshire PE20 1LQ

Vintage Ltd
104 Stanwell Road, Penarth, Glamorgan (S.) CF64 3LP, Wales

Contact: David Vanner
Tel/fax: 01222 701030

International mail order metal waterline model ship & aircraft business established 15 years. For illustrated lists of stocks and shows attended send £1.00 plus SAE.

Vladimir Nushdin (Flats)
Taunsstrasse 78, 12309 Berlin, Germany

Contact: Peter Rein

Walter Onken (Flats)
Alsterkrugchaussee 312, 22297 Hamburg, Germany

Wargames Foundry
4 Victoria Avenue, Norton, Stockton on Tees, Cleveland TS20 2QB

Tel/fax: 01642 553787

Designers and manufacturers of high quality, 25mm white metal historically accurate miniatures for collecting, diorama building and wargaming. Catalogue £2.50 post paid.

Werner Klotzsche (Flats)
Wilhelm-Pieck-Strasse 88, 01445 Radebeul, Germany

Western Miniatures (Flats)
1 Henacre Road, Lawrence Weston, Bristol, Avon BS11 0HB

Wilhelm Schweizer (Flats)
Herrenstrasse 7, 86911 Diessen/Ammersee, Germany

MODEL PAINTING SERVICES

Wolf Models
PO Box 64, Rochester, Kent
ME1 3JR

Contact: Nicholas Adams
Tel/fax: 01634 842523

Range of 1/35 scale WWII resin
figures, white metal accessories.
American Civil War, Napoleonic
and British Empire 1/32 scale
(54mm) metal and resin figures.
1/9 scale resin busts. Available
direct from Historex Agents (see
Section 11) and Roger Saunders
(see above).

Wolf-Dieter Weirich (Flats)
Goldener-Au-Strasse 34, 66280
Sulzbach, Germany

Wolf-Peter Sander (Flats)
Brauacker 1, 94419
Griesbach/Reisbach, Germany

Wolfgang Hafer (Flats)
Schlangenweg 14, 34117 Kassel,
Germany

Tel/fax: (0049) 1561 773113

Figurines of antiquity, hunting,
civil, military. The catalogue
covers 148 pages (A5), price DM48.

Wolfgang Hodapp (Flats)
Sophienstrasse 60, 76133
Karlsruhe, Germany

Wolfgang Unger (Flats)
Fregestrasse 5a, 04105 Leipzig,
Germany

Tel/fax: (0049) 341 9800092

Flat 30mm figures: Middle Ages,
18/19th century military and
civilians.

Wolfgang Windisch (Flats)
Hauptstrasse 80, 77652
Offenburg, Germany

Wolfgang Wohlmann (Flats)
Invalidenstrasse 55a, 10557
Berlin, Germany

Yeomanry Miniatures
34 Vesey Close, Cove,
Farnborough, Hampshire
GU14 8UT

Contact: Brian Harrison
Tel: 01252 523943
Fax: 01420 563221

Yeomanry Miniatures produce
fully sculptured 'toy style' figures
for the connoisseur representing
the Yeomanry Cavalry and the
regular cavalry of the British
Army.

A J Dumelow
53 Stanton Road, Stapenhill,
Burton-on-Trent, Staffordshire
DE15 9RP

Tel/fax: 01283 30556

Dealer and painter of wargames
figures in all scales.
Comprehensive stock to supply
complete armies or single figures,
painted and unpainted. Mail
order welcome.

Ace Models
Fountain Arcade, Dudley,
Midlands (W.) DY1 1PG

Contact: Dixon
Tel/fax: 01384 257045

All branches of military modelling
stocked. Professional building
service by award winning
modeller.

Alan E Jones
96 High Avenue, Letchworth,
Hertfordshire SG6 3RR

Tel/fax: 01462 676020

Figures, toy soldiers and military
models assembled and
hand-painted to your individual
requirements. Personal service
guaranteed. All scales and
subjects considered. Mail order
welcome.

Art Militaire
6 Gypsy Lane, Oulton, Leeds,
Yorkshire (W.) LS26 8SA

Contact: A Buttery

**B J Harris Professional Figure
Painting**
123 Coverside Road, Great Glen,
Leicester, Leicestershire LE8 9EB

Contact: Barry Harris
Tel: 0116 2592004
Fax: 0116 2593355

Member of the Guild of Master
Craftsmen. World-wide mail
order – all types of figures
painted. Send 7 x 1st class stamps
for full colour brochure.

Charles Baldwin
4 Cooper Row, Southgate,
Crawley, Sussex (W.) RH10 6DJ

Contact: Charles Baldwin
Tel/fax: 01293 521288

Model soldier painting service,
54mm upwards. Napoleonic
figures a speciality; also
traditional toy soldiers painted.
For details please send SAE.

**Bath Miniatures /
Bonaparte's Ltd**
1 Queen Street, Bath, Avon BA1
1HE

Tel/fax: 01225 423873

Stockists of figures, s/h plastic
kits, military books, painted
figures incl. chess sets; also glass
domes, cases, bases, etc; buying
uniform books, figures, plastic
kits. Painting service offered by
winning artist. Mail order
available, post free in UK.

C B Philcox
35 Nettlecombe, Crown Wood,
Bracknell, Berkshire RG12 3UG

Tel/fax: 01344 422845

Cameo Miniatures
103 Manor Way, Mitcham,
Surrey CR4 1EJ

Tel/fax: 0181 764 3228

Conflict Miniatures
27 Leighton Road, Hartley Vale,
Plymouth, Devon PL3 5RT

Contact: Tim Reader
Tel/fax: 01752 770761

Manufacturer of wargaming
figures/models. Worldwide mail
order. Stockist of rules,
Kreigspiel, paints, brushes etc.
Painting service available. Trade
enquiries welcome.

Dean Forest Figures
62 Grove Road, Berry Hill,
Coleford, Gloucestershire
GL16 8QX

Contact: Philip & Mark Beveridge
Tel/fax: 01594 836130

Wargame figure painting,
scratch-built trees, buildings and
terrain features. Large scale figure
painting and scratch-building to
any scale. SAE for full lists.

**Debra Raymond
Hand-Painted Figures**
Catbells, 1 Hillhead Cottages,
Rectory Road, Staplegrove,
Taunton, Somerset TA2 6ER

Tel/fax: 01823 333431

Professional military figure
painting service, at sensible prices.
Any scale, period or subject
considered. Please ring with any
enquiry (evenings).

Fernando Enterprises
107 Galle Road, Walana,
Panadura, Sri Lanka

Contact: Sanath Fernando
Tel/fax: (0094) 34 32833

Professional wargaming flat and
round figure painting service.
Any scale, competitive prices and
quick delivery. Contact us with
details of your job.

Fortaddae Design
93 Sturla Road, Chatham, Kent
ME4 5QJ

Tel/fax: 01634 401210

G M Painting Services
Church Lodge, Hanly Swan,
Hereford & Worcester WR8 0DE

G Otty
50 Willbye Avenue, Diss, Norfolk
IP22 3NW

Mail order painting service – all
periods and scales undertaken to
a high standard. Please write for
details.

H Q Painting Services
114 Windmill Hill Lane, Derby,
Derbyshire DE3 3BP

Contact: Paul Spencer
Tel/fax: 01332 298519

Specialists in the hand painting of
wargames figures and accessories.
All scales from 5mm-120mm.
Send SAE for full details.

Helion & Company
26 Willow Walk, Solihull,
Midlands (W.) B91 1UE

Contact: Duncan Rogers
Tel: 0121 705 3393
Fax: 0121 711 1315

Your figures painted down to
their last gaiter button!
Competitive prices – quality and
accuracy assured. All scales and
periods catered for. Discounts
available. Vignettes produced to
order.

The Iron Duke
Edgehill Cottage, Ropeyard,
Wotton Bassett, Wiltshire SN4
7BW

Contact: Ian Barstow
Tel/fax: 01793 850805

Professional wargames figure
painter and dealer in second-hand
painted figures and armies, all
scales. Guaranteed no
sub-contracting. All major credit
cards accepted.

J F Benassi
55 St Mungo Avenue, Glasgow,
Strathclyde G4 0PL, Scotland

Contact: Julian Benassi

Designer of white metal model
soldiers, produced in small
quantity and available by mail
order. Send for catalogue.
Painting service also available.

Kemco Models
106-108 Delce Road, Rochester,
Kent ME1 2DH

Tel/fax: 01634 407080

Knight Designs
95 Tyrrell Avenue, Welling, Kent
DA16 2BT

Contact: Steve Chipperfield
Tel/fax: 0181 304 4408

Private commissions undertaken,
large or small from single master
figures to ranges – figures, planes,
ships, scenery. Mould-making,
casting, and painting services also
available.

The Last Detail
196 Parlaunt Road, Langley,
Slough, Berkshire SL3 8AZ

Contact: David Seagrove
Tel/fax: 01753 547900

Professional figure painting
service producing high quality
artwork on 15-30mm figures for
discerning wargamers and
collectors. Figures available from
stock. Established 1984. SAE for
details.

M Dean
75 Castle Mona Avenue,
Douglas, Isle of Man
Tel/fax: 01624 624120

M P Studios
23 Peregrine Close, Winshall,
Burton-on-Trent, Staffordshire
DE15 0EB

Tel/fax: 01283 37104

Professional model maker and
figure painter – the finest quality
painted figures available, by
national competition winner.
Commissions undertaken to your
own specifications; quality
guaranteed – a sure investment.
Character figures also available –
all we need is a photograph. Write
or call for more information.

Paint Box
April Cottage, 49 Stonely,
Huntingdon, Cambridgeshire
PE18 0EP

The Painted Soldier
138 Friern Road, East Dulwich,
London SE22 0AY

Contact: Bill Brewer
Tel/fax: 0181 693 2449

Painting service for wargamers
and collectors established for 25
years. Work on show at studio
and has yearly trade stand at
'Salute', Chelsea and Kensington
Town Hall.

**Richard Newth-Gibbs
Painting Services**
59 Victor Close, Hornchurch,
Essex RM12 4XH

Tel/fax: 01708 448785

Any scale from 15mm up, single
figures, groups, dioramas,
artillery pieces, mounted gun
teams; British and Indian Army
specialist. Factual military work
only. Callers by appointment.

Warpaint Figure Painting
20 Swaledale Crescent, Barnwell,
Houghton le Spring, Tyne &
Wear DH4 7NT

Tel/fax: 0191 385 7070

MODEL SHOPS & SUPPLIES

Le 11e Hussard
15 Rue Trousseau, 75011 Paris,
France

Contact: J L Ribot
Tel/fax: (0033) 147007433

A D V M
15 Rue Dunoise, 41240 Verdes,
France

Ace Models
Fountain Arcade, Dudley,
Midlands (W.) DY1 1PG

Contact: Dixon
Tel/fax: 01384 257045

All branches of military modelling
stocked. Professional building
service by award winning
modeller.

Actramac Diffusion
31/33 Rue Esquirol, 75013 Paris,
France

Adventure Worlds
13 Gillingham Street, London
SW1

Anni-Mini
22 Boulevard de Reuilly, 75012
Paris, France

Another World
23 Silver Street, Leicester,
Leicestershire

Tel/fax: 0116 2515266

Antics (Bristol)
8 Fairfax Street, Bristol, Avon
BS1 3DB

Tel/fax: 0117 9273744

Stockists of modelling accessories
and military items.

Antics (Gloucester)
79 Northgate St, Gloucester,
Gloucestershire GL1 2AG

Tel/fax: 01452 410693

Stockists of modelling accessories
and military items.

Antics (Guildford)
89E Woodbridge Rd,, Guildford,
Surrey GU1 4QD

Tel/fax: 01483 39115

Stockists of modelling accessories
and military items.

Antics (Plymouth)
30 Royal Parade, Plymouth,
Devon PL1 1DU

Tel/fax: 01752 221851

Stockists of modelling accessories
and military items.

Antics (Stroud)
49 High Street, Stroud,
Gloucestershire GL5 1AN

Contact: Alan Tyndall
Tel/fax: 01453 765920

Stockists of modelling accessories
and military items.

Antics (Swindon)
8 Regent Circus, Swindon,
Wiltshire SN1 1JQ

Tel/fax: 01793 430417

Stockists of modelling accessories
and military items.

Antics (Worcester)
16 St Swithin Street, Worcester,
Hereford & Worcester WR1 2PS

Tel/fax: 01905 22075

Stockists of modelling accessories
and military items.

Arhisto
p/a Office du Tourisme,
CH-1530 Payerne, Switzerland

Au Paradis des Enfants
3bis Grande Rue, 91260
Juvisy-sur-Orge, France

Au Sanctuaire de la Miniature
7 Boulevard de l'Observatoire,
34000 Montpellier, France

Au Tapis Vert
40 Rue Voltaire, 47000 Agen,
France

Autoroute Models
168 Bishopric, Horsham, Sussex
(W.) RH12 1QR

Awful Dragon Management
3 Ransome's Dock, 35-37
Parkgate Road, London SW11
4NP

Azimut Productions
8 Rue Baulant, 75012 Paris, France

Tel: (0033) 143070616
Fax: (0033) 143471193

Dealer and manufacturer of 1/35th scale military models; we offer over 60 ranges of vehicles,figures, accessories and books. Trade enquiries welcome.

B & B Military
1 Kings Avenue, Ashford, Kent TN23 1LU

Contact: Len Buller
Tel/fax: 01233 632923

Makers of military model vehicles and figures in Dinky 1/60 inch scale. All white metal. Mail order only: send SAE for list of kits and finished models available. Colour catalogue £3.50.

Bath Miniatures / Bonaparte's Ltd
1 Queen Street, Bath, Avon BA1 1HE

Tel/fax: 01225 423873

Stockists of figures, s/h plastic kits, military books, painted figures incl. chess sets; also glass domes, cases, bases, etc; buying uniform books, figures, plastic kits. Painting service offered by winning artist. Mail order available, post free in UK.

Battlefield Model Supplies
12 Delta Drive, Musselburgh, Lothian EH21 8HR, Scotland

Contact: Ian Hanratty
Tel/fax: 0131 665 4087

Mail order company specialising in AFV models in plastic, resin and metal. Stock includes: Roco, Trident, M.A.A.G., C.M.S.C., Model Transport, J.B. Models, M.M.S., Trux, S & S Models, Strong-Point, Custom Miniatures. Send large SAE for list.

Beatties (Aberdeen)
18/22 Market Street, Aberdeen, Grampian AB1 2PL, Scotland

Tel/fax: 01224 590956

Beatties (Altrincham)
92/94 Stamford New Road, Graftons Precinct, Altrincham, Manchester, Gt. WA14 1DG

Tel/fax: 0161 928 4228

Beatties (Andover)
20 Chantry Way, Andover, Hampshire SP10 1LX

Beatties (Aylesbury)
13/15 High Street, Aylesbury, Buckinghamshire HP20 1SH

Tel/fax: 01296 85752

Beatties (Ayr)
21/25 Newmarket Street, Ayr, Strathclyde KA7 1LL, Scotland

Tel/fax: 01292 282945

Beatties (Banbury)
28 Bridge Street, Cherwell Centre, Banbury, Oxfordshire OX16 8PN

Tel/fax: 01295 253131

Beatties (Basingstoke)
8 New Market Square, Basingstoke, Hampshire RG21 1JA

Tel/fax: 01256 59958

Beatties (Birmingham)
26 South Mall, The Pallisades, Birmingham, Midlands (W.) B2 4XD

Tel/fax: 0121 643 8604

Beatties (Blackpool)
34-35 Hounds Hill Centre, Victoria Street, Blackpool, Devon FY1 4HU

Tel/fax: 01253 26461

Beatties (Boston)
45 West Street, Boston, Lincolnshire PE21 8QN

Tel/fax: 01205 311688

Beatties (Bournemouth)
98 Poole Road, Westbourne, Bournemouth, Dorset BH4 9EG

Tel/fax: 01202 762811

Beatties (Brighton)
4/8 Dyke Road, Brighton, Sussex (E.) BN1 3FE

Tel/fax: 01273 776626

Beatties (Bristol)
17/19 Penn Street, Bristol, Avon BS1 3AW

Tel/fax: 0117 9260259

Beatties (Bury St Edmunds)
62 Cornhill, Bury St Edmunds, Suffolk IP33 1BE

Tel/fax: 01284 761646

Beatties (Cardiff)
Northgate House, Kingsway, Cardiff, Glamorgan (S.) CF1 4AD, Wales

Contact: B St John
Tel/fax: 01222 397645

Suppliers of models, trains, die-cast, boys' toys, remote control cars, Scalextric, hobby tools and paints. Repairers of trains, Scalextric cars and remote control cars.

Beatties (Clydebank)
11 Britannia Way, Clyde Regional Centre, Clydebank, Strathclyde G81 2RZ, Scotland

Tel/fax: 0141 952 7368

Beatties (Croydon)
135a North End, Croydon, London CR0 1TN

Tel/fax: 0181 688 1585

Beatties (Cumbernauld)
11c Forth Walk, Cumbernauld, Strathclyde G67 1BT, Scotland

Tel/fax: 01236 731297

Beatties (Dumfries)
7 Church Place, Dumfries, Dumfriesshire DG1 1BW, Scotland

Tel/fax: 01387 64884

Beatties (Eastleigh)
8/9 Fryern Arcade, Chandlers Ford, Eastleigh, Devon S05 2DP

Tel/fax: 01703 269986

Beatties (Glasgow)
8 Olympic Centre, Glasgow, Strathclyde G74 1PG, Scotland

Beatties (Glasgow)
30 St. Enoch Square, Glasgow, Strathclyde G1 4DF, Scotland

Tel/fax: 0141 248 6867

Beatties (Halifax)
24/26 Commercial Street, Halifax, Yorkshire (W.) HX1 1TA

Tel/fax: 01422 353986

Beatties (Harrogate)
47/49 James Street, Harrogate, Yorkshire (N.) HG1 1SJ

Tel/fax: 01423 564335

Beatties (Hemel Hempstead)
203 Marlowes, Hemel Hempstead, Hertfordshire HP1 1BL

Tel/fax: 01442 53691

Beatties (High Wycombe)
27 White Hart Street, High Wycombe, Buckinghamshire HP11 2HL

Tel/fax: 01494 25177

Beatties (Kettering)
19/21 Gold Street, Newlands Centre, Kettering, Northamptonshire NN16 8BX

Tel/fax: 01536 512507

Beatties (Kilmarnock)
3/5 Bank Street, Kilmarnock, Strathclyde KA1 1HA, Scotland

Tel/fax: 01563 20262

Beatties (Kingston)
30-32 Eden Street, Kingston, Surrey KT1 1EP

Tel/fax: 0181 549 5464

Beatties (Liverpool)
36/37 Dawson Way, St John's Centre, Liverpool, Merseyside L1 1LJ

Tel/fax: 0151 709 0799

Beatties (London N)
10 The Broadway, Southgate, London N14 6PN

Tel/fax: 0181 886 4258

Beatties (London SE)
210 Lewisham High Street, London SE13 6JP

Tel/fax: 0181 852 0449

Beatties (London W)
72a Broadway, West Ealing, London W13 0SY

Tel/fax: 0181 579 9959

Beatties (London WC)
202 High Holborn, London WC1V 7BD

Tel: 0171 405 6285
Fax: 0171 405 8592

Beatties (Manchester)
4/6 Brown Street, Market Street, Manchester, Manchester, Gt. M2 1EE

Tel/fax: 0161 834 7780

Beatties (Milton Keynes)
64 Midsummer Arcade, Secklow Gate West, Milton Keynes, Buckinghamshire MK9 3ES

Tel/fax: 01908 604464

Beatties (Newbury)
25/26 Cheap Street, Newbury, Berkshire RG14 5DB

Tel/fax: 01635 46004

Beatties (Newcastle)
43/47 Pilgrim Street, Newcastle upon Tyne, Tyne & Wear NE1 6QE

Tel/fax: 0191 232 4161

Beatties (Newcastle)
2 High Friars, Eldon Square, Newcastle upon Tyne, Tyne & Wear NE1 7XG

Tel/fax: 0191 261 6432

Beatties (Northampton)
41/43 Princes Walk, Grosvenor Centre, Northampton, Northamptonshire NN1 2EL

Tel/fax: 01604 27726

Beatties (Nottingham)
3 Mount Street, Nottingham, Nottinghamshire NG1 6JW

Tel/fax: 0115 9411693

Beatties (Perth)
15 Canal Street, Perth, Perthshire PH2 8LF, Scotland

Tel/fax: 01738 39450

Beatties (Peterborough)
Unit 8, Queensgate Centre, Peterborough, Cambridgeshire PE1 1NT

Tel/fax: 01733 313158

Beatties (Portsmouth)
28 Arundel Street, Portsmouth,
Hampshire PO1 1NL
Tel/fax: 01705 823681

Beatties (Reading)
51/52 Broad Street Mall,
Shopping Centre, Reading,
Berkshire RG1 7QE
Tel/fax: 01734 586899

Beatties (Reading)
16 Meadway Shopping Centre,
Tilehurst, Reading, Berkshire
RG3 4AA
Tel/fax: 01734 575571

Beatties (Reading)
176 Crockhamwell Road,
Woodley, Reading, Berkshire
RG5 3JH
Tel/fax: 01734 691730

Beatties (Romford)
7/11 High Street, Romford, Essex
RM1 1JU
Tel/fax: 01708 724283

Beatties (Sheffield)
38 Pinstone Street, Sheffield,
Yorkshire (S.) S1 2HN
Tel/fax: 0114 2757864

Beatties (Sheffield)
23 High Street, Meadowhall
Shopping Centre, Sheffield,
Yorkshire (S.) S9 1EN
Tel/fax: 0114 2568267

Beatties (Southampton)
6 Lords Hill Shopping Centre,
Southampton, Hampshire SO1
8HY
Tel/fax: 01703 737646

Beatties (Southampton)
114 East Street, Southampton,
Hampshire SO1 1HD
Tel/fax: 01703 224843

Beatties (Swindon)
25 Bridge Street, Swindon,
Gloucestershire SN1 1BP
Tel/fax: 01793 497213

Beatties (Watford)
70 The Parade, High Street,
Watford, Hertfordshire W1D
2AW
Tel/fax: 01923 227563

Beatties (Winchester)
46/47 High Street, Winchester,
Hampshire SO23 9BT
Tel/fax: 01962 860188

Beatties (Wishaw)
Unit 8, Caledonian Centre, New
Ashtree Street, Wishaw,
Strathclyde ML2 7UR, Scotland
Tel/fax: 01698 350784

Beau Geste
121 Ipswich Road, Woodbridge,
Suffolk IP12 4BY

Contact: Stefan Lidvall

UK agents of Argentine made
figurines in 65mm, mainly
Ancient and Medieval themes,
frequently reviewed/advertised
in 'Military Modelling'. Mail order
only. Colour catalogue and
pricelist £3.50.

C T Gascoigne Ltd
101-103 Tavistock Street,
Bedford, Bedfordshire
Tel/fax: 01234 52596

Carl Gruen (Flats)
PO Box 2019, Remsenburg,
New York 11960-2019, USA

Central Loisirs
83 Rue du President Wilson,
92300 Levallois Perret, France

Chatelet Miniatures
15 Place du Chatelet, 45000
Orleans, France

Cheltenham Model Centre
39 High Street, Cheltenham,
Gloucestershire GL50 1DY

**Christian Schmidt
Fachbuchhandlung**
Sauerbruchstrasse 10, 81377
Munchen, Germany
Tel: (0049) 89703227
Fax: (0049) 897005361

Military-shipping-aviation books;
1/1250 scale models; catalogues
available. Store hours Mon-Fri
9am-6pm, Saturday 9am-12am,
closed Sundays. Mail order; Visa,
Amex Euor/Mastercard accepted.

Christian Terana
31 Boulevard Kellerman, 75013
Paris, France

Le Cimier
38 Rue Ginoux, 75015 Paris,
France

Contact: Jacques Vuyet

City Models
6 Stanley Street, Liverpool,
Merseyside L1 6AF

Coldstream Guards
13 Market Street, Coldstream,
Borders, Scotland

Commando
4 The Arcade, Hoe Street,
London E17 4QG
Tel/fax: 0181 509 3153

Computer & Games Centre
34 St Nicholas Cliff,
Scarborough, Yorkshire (N.)
YO11 2ES

Concorde Models
179 Victoria Road, Aldershot,
Hampshire GU11 1JU

Contact: Brian Ballard
Tel/fax: 01252 26825

All prominent makes of figures,
military kits and toy soldiers
stocked. Osprey, Verlinden,
Kalmbach, Windrow & Greene
books. Mail order.

Conflict Miniatures
27 Leighton Road, Hartley Vale,
Plymouth, Devon PL3 5RT

Contact: Tim Reader
Tel/fax: 01752 770761

Manufacturer of wargaming
figures/models. Worldwide mail
order. Stockist of rules,
Kriegsspiel, paints, brushes etc.
Painting service available. Trade
enquiries welcome.

Conquest Models
11 Forresters Path, School
Aycliffe, Co.Durham DL5 6TA

Contact Modèlisme
183 Grande Rue, 86000 Poitiers,
France

D Hewins Models & Hobbies
7B East St Mary's Gate, Grimsby,
Humberside (S.) DN31 1LH

Tel/fax: 01472 347088

Stockists of 'Men at Arms' books,
plastic kits, fantasy figures,
general modelling accessories,
fantasy board/role playing
games. Open Mon-Wed Fri-Sat,
9.00am-5.30pm.

Dentons
71b Eastbourne Road,
Willingdon, Sussex (E.)

Dorking Models
12-13 West Street, Dorking,
Surrey RH4 1BL

Contact: Anthony Lawrence
Tel/fax: 01306 881747

Stockists of most ranges of
military kits as well as aircraft and
ship models. Mail order a
speciality with many special
imports from Eastern Europe.

The Dragon and George
39 Parnie Street, Glasgow,
Strathclyde G1 5RJ, Scotland

Les Drapeaux de France
34 Gallerie Montpensier, Palais
Royal, 75001 Paris, France

**The Dungeon / Modeller's
Nook**
15-17 Winetavern Street, Belfast,
Co.Antrim BT1 1JQ, N. Ireland

Contact: Joseph F Barlow
Tel/fax: 01232 233862

Non radio control. Specialising in
military, aircraft, fantasy
modelling, manufacturing fantasy
wargaming scenery & figures,
new range of 'Hairball' figures.
Send £1.50 for catalogue & figure.

E D Models
64 Stratford Road, Shirley,
Solihull, Midlands (W.) B90 3LP

Tel: 0121 744 7488
Fax: 0121 733 2591

Model kits and accessories –
extensive range. International
mail order; catalogue available.
Manufacturers/distributors of
Airwaves, K K Castings,
Scaleplanes. Shop open Mon-Sat
9.15am-6.00pm.

Ecole Modèlisme SA
70 Boulevard St Germain, 75005
Paris, France

Electro Jouets
18 Rue Gougeard, 72000 Le
Mans, France

Esdevium Games
2 Morley Road, Farnham, Surrey
GU9 8LY

Fanakits 37
109 Rue Colbert, 37000 Tours,
France

Fanatic Modele Réduit
17bis Rue Roger Collerye, 89000
Auxerre, France

Fantasy and Military World
10 Market Square Arcade,
Hanley, Stoke-on-Trent,
Staffordshire
Tel/fax: 01782 279294

**Figurines d'Art/ Editions
Vanot**
Rue des Ancres, Amfreville,
50480 Ste Mere Eglise, France

Five Easy Pieces
102 Grande Place, 38100
Grenoble, France

Flying High
127 George Street, Romford,
Essex RM1 2EB

Tel/fax: 01708 735983

Fonderie Miniature
36 Rue Charron, 93300
Aubervilliers, France

Forbes & Thomson
Burgate Antiques Centre, 10c
Burgate, Canterbury, Kent

Contact: Rowena Forbes

Specialist dealers in old toy
soldiers by Britains, Johillco etc.
and related lead and tinplate toys.
Model soldiers & kits by Rose
Models and Chota Sahib. Painted
military miniature figures. Mail
order and office address, P.O. Box
375, South Croydon, CR2 6ZG.
Open Mon-Sat 10.00am-5.00pm.

Fortress Models
87 Yew Tree Road,
Southborough, Tunbridge Wells,
Kent TN4 0BJ

G K's Model Centre
390 Holdenhurst Road,
Bournemouth, Dorset BH8 8BL

Contact: Frank Parsons
Tel: 01202 394007
Fax: 01202 251260
Email:
frank@gk-models.demon.co.uk

Stockists of all leading
manufacturers; Tamiya, Dragon,
etc. Specialist kits & accessories by
Verlinden, Accurate Armour etc;
also books and magazines.

Gaite Confort
63 Boulevard Philippe August,
75011 Paris, France

Gamers in Exile
283 Pentonville Road, London
N1 9NP

Games
63 Allerton Road, Liverpool,
Merseyside L18

The Games Room
29A Elm Hill, Norwich, Norfolk

Tel/fax: 01603 628140

Gaugemaster Controls Plc
Gaugemaster House, Ford Road,
Arundel, Sussex (W.) BN18 0BN

Tel: 01903 884488
Fax: 01903 884321

Distributors of Preiser and
Elastolin figures and accessories.
Catalogue/price lists available
and mail order/retail shop open
Mon-Sat 9am to 5.30pm.

Golden Gains Model Shop
7-11 Wardwick, Derby,
Derbyshire DE1 1HA

Tel/fax: 01332 44822

Grenadier Models
94 Pier Avenue, Clacton on Sea,
Essex

Tel/fax: 01255 421963

Halifax Modellers World
55 The Arcade, The Piece Hall,
Halifax, Yorkshire (W.) HX1 1RE

Contact: David Smith
Tel/fax: 01422 349157

All types of model kits, figures,
paints, accessories. Open 7 days a
week, Mon-Fri 10.30am-5pm, Sat
9am-5pm, Sun 10.30am-4.30pm.
Visa/Access accepted.

Hannant's Mailorder
Harbour Road, Oulton Broad,
Lowestoft, Suffolk NR32 3LZ

Tel: 01502 517444
Fax: 01502 500521

Retailer and mail order suppliers
of plastic and resin model figures,
tanks, aircraft, ships, cars etc.

Hannant's Model Shop
Colindale Station House,
157-159 Colindale Avenue,
London NW9 5HR

Tel/fax: 0181 205 6697

Retailers of plastic and resin
model figures, tanks, aircraft,
ships, cars etc.

Henley Model Miniatures
24 Reading Road, Henley on
Thames, Oxfordshire RG9 1AB

Contact: David Hazell
Tel/fax: 01491 572684
Email: enquiries@toysoldier.co.uk

We stock the widest range of
painted toy soldiers available in
the UK, together with castings,
paints and modelling materials.
Mail order a pleasure.

Heroes Miniatures
7 Waverley Place, Worksop,
Nottinghamshire S80 2SY

Historex Agents
Wellington House, 157 Snargate
Street, Dover, Kent CT17 9BZ

Contact: Lynn Sangster
Tel: 01304 206720
Fax: 01304 204528

Mail order and wholesale
distributor of model soldier and
AFV kits. We represent the
following UK and overseas
manufacturers: Historex,
Verlinden, Le Cimier, Modelavia,
Fonderie Miniature, Metal
Modeles, Puchala, Hecker-Goros,
Beneito, Andrea, Taxdir, Pegaso,
Soldiers Luchetti,
Friulmodellismo, Kalmbach,
Mini-Arts Studios, Wolf, Hornet,
Poste Militaire, Almond, Belgo,
Chota Sahib, Tiny Troopers,
Sovereign, Tomker, Fort Royal
Review, Azimut, Mil-Art, Mascot,
Starlight, Ridwulf, Trophy, Aquila.

Hoopers
105 Cornwall Street, Plymouth,
Devon PL1 1PA

Contact: Brian Mardon
Tel/fax: 01752 667840

A newsagent shop with a
difference, specialising in military
and transport books and
magazines. Also stocks a vast
range of plastic kits and Verlinden
products.

Hovels Ltd
18 Glebe Road, Scartho,
Grimsby, Humberside (S.) DN33
2HL

Contact: Dennis Coleman
Tel/fax: 01472 750552

Howes
85 Gloucester Road, Bishopston,
Bristol, Avon

Tel/fax: 0117 9247505

I W Models
27 Saywell Road, Luton,
Bedfordshire LU2 0QG

Tel/fax: 01582 34036

Ideal Models
67 Boulevard Carnot, 31000
Toulouse, France

Imperial Figures
55a Leopold Road, Wimbledon,
London SW19 7JG

J E Hancock
19 Sydenham Road South,
Cheltenham, Gloucestershire
GL52 6EF

Je M'Amuse
3 Rue Saint Michel, 84000
Avignon, France

Jeux de Guerre Diffusion
6 Rue Meissonier, 75017 Paris,
France

Joanny Jabouley
Route de Bourg, 38490 Fittilieu,
France

Le Kangarou
58 Grande Rue C de Gaulle,
92600 Asnieres, France

Keep Wargaming
The Keep, Le Marchant
Barracks, London Road, Devizes,
Wiltshire SN10 2ER

Contact: Paul & Teresa Bailey
Tel/fax: 01380 724558

Shop and mail order service;
stockists of wargames figures,
books and equipment; some
military models and plastic kits.
Shop open Tues-Sat 10am to 6pm.

Kingsgrand Miniatures
Prinny's Gallery, The Lanes,
Brighton, Sussex (E.) BN1 1HB

Kit Bits – The Model Store
7 Market House, Market Hill, St
Austell, Cornwall PL25 5QB

Contact: M L Squire
Tel: 01726 72818
Fax: 01726 65424

We stock a varied selection of
paints, glues, models, including
cars, Warhammer, kites, boats,
planes, Star Trek, fantasy
militaria. Open 9am-5pm Mon-Sat.

Kit'n'Doc
144 Rue Martre, 92110 Clichy,
France

L S A Models
151 Sackville Road, Hove, Sussex
BN3 3HD

Contact: J Lake
Tel/fax: 01273 592775

Suppliers/importers of a wide
range of military-related models
in plastic and resin. Figures, tanks,
soft-skins and accessories. Phone
for more details.

The Land of Gondal
76 Main Street, Haworth,
Yorkshire (W.)

Tel/fax: 01535 44924

Langley Models
166 Three Bridges Road,
Crawley, Sussex (W.) RH10 1LE

Contact: Ian McLellan
Tel: 01293 516329
Fax: 01293 403955

Toy soldiers – foot, mounted,
bands, gunteams – as castings or
hand-painted. Also knights,
Romans, mythological. Illustrated
catalogue £3.55 post free. Mail
order by return.

Lelchat Fils
35 Rue Porte aux Saints, 78200
Mantes la Jolie, France

Little Soldier
58 Gillygate, York, Yorkshire (N.)

Tel/fax: 01904 642568

London Wargames Depot
56 Beaumont Place, Mogden
Lane, Isleworth, Middlesex TW7
7LH

Les Lutins
78 Boulevard Marechal Joffre,
92340 Bourg-la-Reine, France

M J N Maquettes
9 Rue Boirot, 63000
Clermont-Ferrand, France

Macs Models
133-135 Canongate, The Royal
Mile, Edinburgh, Lothian EH8
8BP, Scotland

Tel/fax: 0131 557 5551

THE MODEL SHOP

190-194 STATION ROAD HARROW MIDDLESEX HA1 2RH

TELEPHONE: 0181-863 9788 0181-427 0387

FAX: 0181-863 3839

Imex ¹/₇₂ Figure Sets	American Civil War Union Infantry Set £3.45
American Chuck Wagon Set £5.50	Italeri Union Infantry ¹/₇₂ £3.00
American Munitions Ambulance Set £5.50	Italeri Russian Grenadiers £3.00
American Civil War Union Cavalry Set £3.45	Imex ¹/₇₂ scale American Civil War,
American Sioux Indians Set £3.45	Union £3.95 & £10.95

ADD 10% POST & PACKING

La Maison du Livre Aviation
75 Boulevard Malesherbes,
75008 Paris, France

Maltby Hobby Centre
19 Morrell Street, Maltby,
Rotherham, Yorkshire (S.)
S66 7LL
Contact: Richard Moss
Tel/fax: 01709 798287

A wide range of plastic kits and
accessories always in stock. Also
stocked, Fantasy and Sci-Fi games,
model railways and Scalextric.
Please phone for details.

Marc Tolosano
Val du Carei, 06500 Menton,
France

Maurice Verdeun
1 bis, Rue Piliers de Tutelle,
33000 Bordeaux, France

Medway Games Centre
294-6 High Street, Chatham,
Kent
Tel/fax: 01634 847809

Micro-Model
Rue Percheret, La Visitation,
83000 Toulon, France

Mike's Miniatures
29 Loane Road, Sholing,
Southampton, Hampshire S02
7PF
Tel/fax: 01703 421956

Mili-Art
18 Marine Park Mansions,
Wellington Road, New Brighton,
Merseyside

Miniature 2000
63 Avenue Philippe-Auguste,
75011 Paris, France

Minimodels 87
11 bis Rue Jean Jaures, 87000
Limoges, France

Minimodels Ternois
29-31 Rue du Quai, 27400
Louviers, France

Mitregames
77 Burntwood Grange,
Wandsworth Common, London
SW18

Model & Hobby
Fredriksborgade 23, DK-1360
Copenhagen, Denmark

Model 25 SARL
24 Rue des Febvres, 25200
Montbeliard, France

Model Aerodrome
Unit 223, Stoneborough Centre,
Maidstone, Kent
Tel/fax: 01622 691184

Model Aerodrome
36 The Boulevard, Crawley,
Sussex (W.) RH10 1XP
Tel/fax: 01293 540331

Model Aerodrome
37 West Street, Brighton, Sussex
(E.) BN1 2RE
Tel/fax: 01273 26790

Model Aerodrome
68 Seaside Road, Eastbourne,
Sussex (E.) BN21 3P9
Tel/fax: 01323 644001

Model Images Retail
56 Station Road, Letchworth,
Hertfordshire

The Model Shop
209 Deansgate, Manchester,
Manchester, Gt. M3 3NW
Tel: 0161 834 3972
Fax: 0161 831 7459

Stockist of Dragon, Kirin,
Academy, AFV Club, Tamiya,
Italeri, Wolf, Hornet, Revell, etc.
Comprehensive range of books,
paints, airbrushes and accessories.
Open 7 days.

The Model Shop
190-194 Station Road, Harrow,
Middlesex
Tel/fax: 0181 863 9788

The Model Shop
18 Blenheim Street,
Newcastle-upon-Tyne, Tyne &
Wear NE1 4AZ
Tel/fax: 0191 232 2016

The Model Shop
179 Ferensway, Hull,
Humberside (N.) HU1 3UA

Model-Time (Wholesale) Ltd
64-66 Windmill Road, Croydon,
Surrey CRO 2XP
Tel/fax: 0181 689 6622

Suppliers to the retail trade of
Victoria 1/43 scale precision
die-cast military vintage and
contemporary models.

Modelcraft
White Hart Mews, Southgate,
Sleaford, Hampshire NG34 7RY

Modelisme 50
35bis Blvd Robert Schumann,
50100 Cherbourg, France

Modelisme 92
1 Rue de Billancourt, 92100
Boulogne Billancourt, France

Modelkits
4 Rue Georges Clemenceau,
10000 Troyes, France

Modellbaustube
Doeblinger Hpstr. 87, A-1190
Wien, Austria
Tel: (0043) 1369 1768
Fax: (0043) 1369 2955
Email: mbs_austria
@compuserve.com

Specialise in imported AFV and
aircraft kits, conversions and
accessories. Shop and international
mail order. Open Mon-Fri
10am-6pm, Sat 10am–1pm.

Models Galore
56 London Road, Stone,
Dartford, Kent
Tel/fax: 01322 278984

Motor Books
10 Theatre Square, Swindon,
Wiltshire SN1 1QN
Contact: Janet Terry
Tel: 01793 523170
Fax: 01793 432070

Specialist bookshop for all
military publications. Large range
of military figures. Worldwide
mail order service.

Nantes Modelisme
3 Allee Jean Bart, 44000 Nantes,
France

National Army Museum Shop
National Army Museum, Royal
Hospital Road, Chelsea, London
SW3 4HT
Contact: Tim Errock
Tel: 0171 730 0717
Fax: 0171 823 6573
Email: nam@enterprise.net

The Museum Shop sells a wide
range of military books, postcards
and prints, plus videos, tapes,
models and toy soldiers. Open
seven days a week, 10.00am to
5.15pm.

Neilsons
76 Coburg Street, Edinburgh,
Lothian, Scotland
Tel/fax: 0131 554 4704

Oakmere Hobbies
161 High Street, Potters Bar,
Hertfordshire
Tel/fax: 01707 42462

Olivier Loisirs
4 Rue de Varennes, 77120
Coulommiers, France

Orion Models
54 Ebrington Street, Plymouth,
Devon
Tel/fax: 01752 265987

L'Ours Martin
107ter Rue du Point du Jour,
92100 Boulogne Billancourt,
France

Oxford Games
6 Harper Road, Summertown,
Oxford, Oxfordshire OX2 7LQ

Palais de la Maquette
19 Rue du Mont Desert, 54000
Nancy, France

La Panthere Rose
33 Rue A Danvers, 62510
Arques, France

Paquebot Normandie
247 Rue de Tolbiac, 75013 Paris,
France

Patricks Toys and Models
107-111 Lillie Road, Fulham,
London
Tel/fax: 0171 385 9864

Paul Dobson Wholesale
23 Leeswood, Skelmersdale,
Lancashire WN8 6TH
Contact: Paul Dobson
Tel/fax: 01695 21703

Distributor to trade, and retail by
mail order (ring/fax 9am -9pm)
The widest choice of fine scale
metal military miniatures from a
single source; our illustrated lists of
over 2000 figures are available for
only £7.50 – over 120 new items
will be added over the next year.

Pelta
16 Swietokrzyska Street, 00
Warsaw 050, Poland

Contact: Mark Machala
Tel: (0048) 2227661
Fax: (0048) 22269186

Leading distributor of Polish and
Russian books on aviation, vehicles,
AFVs, naval and militaria subjects
(e.g. uniforms), covers all historical
periods. ALSO military models,
plastic kits, vacuforms, figures
(30mm, 54mm) painted and
unpainted. The widest selection in
the world, available for trade and
mail order world wide. Free
catalogues and super prices.

Peter Brooke
Hay Lodge, Hay Street,
Braughing, nr Ware,
Hertfordshire SG11 2RQ

Tel/fax: 01920 821372

Le Petit Diable
584 Boulevard Poincare, 62400
Bethune, France

Philibert SA
12 Rue de la Grange, 67000
Strasbourg, France

Prince August
BP 29, 78770 Thoiry, France

Punctilio Model Spot
Waterloo Road, Ruby Road
Corner, Hinckley, Leicestershire
LE10 0QJ

Tel/fax: 01455 230952

Q T Models
17 Hilderthorpe Road, Bridlington,
Yorkshire (E.) YO15 3AY

Quartermasters
6 Hamilton Road, Sidcup, Kent
DA15 7HB

R A E Models
Unit 2, Service Road, (off Corrie
Road), Addlestone, Surrey
KT15 2LP

**Railway Book and Model
Centre**
The Roundway, Headington,
Oxford, Oxfordshire OX3 8DH

Tel/fax: 01865 66215

The Red Lancers
14 Broadway, Milton, PA 17847,
USA

Tel: (001) 717 742 8118
Fax: (001) 717 742 4814

Military miniatures for the
collector. Phone orders welcome –
Mastercard or Visa. (See
advertisement).

The Regiment
Baildon Craft Centre, Browgate,
Baildon, Yorkshire (W.)

Tel/fax: 01274 547671

Richard Kohnstam
13-15 High Street, Hemel
Hempstead, Hertfordshire

Tel/fax: 01442 61721

Royal Air Force Museum
Grahame Park Way, Hendon,
London NW9 5LL

Tel: 0181 205 2266
Fax: 0181 200 1751

Britain's National Museum of
Aviation, displays 70 full size
aircraft. Open daily. Extensive
library research facilities (weekdays
only). Large free car park. Licensed
restaurant. Souvenir shop with
extensive range of specialist books
and model kits.

Le Royaume de la Figurine
BP 3004, 24000 Perigeux, France

Le Régiment
195 Rue Beauvoisine, 76000
Rouen, France

S A Le Cercle
Art de Vivre, 78630 Orgeval,
France

S N C Bonnefoy
8 Avenue Emile Bouyssou, 46100
Figeac, France

La Sabretache
1 Rue Pargaminieres, 31000
Toulouse, France

Le Santa Fé
49 Rue Bersot, 25000 Besancon
25000, France

Saratoga Soldier Shop
5 Curtis Ind Pk Blvd, Ballston
Spa, New York 12020, USA

Contact: William F Imrie
Tel: (001) 518 885 1497
Fax: (001) 518 885 0100

We stock the complete line of I/R
Miniatures – 54mm metal soldier
kits with historical texts and
colouring guides, ancient world to
Vietnam. Illustrated catalogue
US$11.00 by airmail.

Second Chance Games
62 Earlston Road, Wallasey,
Merseyside

Sentinel Miniatures
4 Broadway, no.102, Valhalla,
New York 10595, USA

Tel/fax: (001) 914 682 3932

Military miniatures – figure kits,
resin armour, accessories,
publications. Painted figures on
display. Mail order catalogue
£8.50. Visa/MC/Access. Also
1/43 die cast vehicles.

Simon's Soldiers
14 Cae Ffynnon, Brackla,
Bridgend, Glamorgan, Mid-
CF31 2HG, Wales

Soldaten
76 Fortune Green Road, London
NW6 1DS

Soldats de Plomb
Route de Bourg, 38490 Fittilou,
France

Contact: Joanny Jabouley

Southsea Models
69 Albert Road, Southsea,
Portsmouth, Hampshire PO5 2SG

Space City Gifts
33 Marine Terrace, Margate,
Kent CT9 1XJ

Tel/fax: 01843 294906

Sud Modèles Diffusion
290 Chemin de Bertoire, BP 22,
83910 Pourrieres, France

Swansea Models & Hobbies
3 Shoppers Walk Arcade, Oxford
Street, Swansea SA1 3AY, Wales

Contact: Derek Matthews
Tel: 01792 652877
Fax: 01792 463933

Wales' premier model shop, mail
order a speciality. Open Tues-Sat
9.30am-5.30pm. Mastercard, Visa
and Switch accepted.

Temps Libre
22 Rue de Sevigne, 75004 Paris,
France

Torbay Model Supplies Ltd
59 Victoria Road, Ellacombe,
Torquay, Devon

Tel/fax: 01803 297764

The Toy Soldier
16 Magdalene Street, Cambridge,
Cambridgeshire CB3 0AF

Contact: Jason P Davis
Tel/fax: 01223 67372

We stock a wide variety of
soldiers, books, modelling
materials, and 'BB' guns. Five
minutes from the M11 in the city
centre; open Mon-Sat,
10am-5.30pm.

Toytub
100a Raeburn Place, Edinburgh,
Lothian EH4, Scotland

Tradition H Zorn
Bettenfeld 21, 91541
Rothenburg, Germany

Tel/fax: (0049) 98612611

Tradition of London Ltd
33 Curzon Street, Mayfair,
London W1V 7AE

Contact: Steve Hare
Tel: 0171 493 7452
Fax: 0171 355 1224

Largest range of painted and
unpainted figures in 54mm,
90mm, 110mm plus over 300
different 'Toy Soldier' style sets,
including 'Sharpe' range. Full
mail order, credit cards. Shop
open Mon–Fri 9.00-5.30, Sat
9.30-3.00.

Trafalgar Models
122 Lazy Hill Road, Aldridge,
Walsall, Midlands (W.) WS9 8RR

Trains & Things
170/172 Chorley New Road,
Horwich, Bolton, Lancashire BL6
5QW

Contact: David Moss
Tel/fax: 01204 669203

Under Two Flags
4 Saint Christopher's Place,
London W1M 5HB

Contact: Jock Coutts
Tel/fax: 0171 935 6934

Stockist of toy soldiers, model
kits, military books, painted
figures & dioramas. Open
Mon-Sat, 10.00am-5.00pm.

V C Miniatures
16 Dunraven Street, Aberkenfig,
Bridgend, Glamorgan, Mid- CF32
9AS, Wales

Contact: Lyn Thorne
Tel/fax: 01656 725006

Established 1987, manufacturer of
hand painted 54mm toy soldiers,
sculpted by Lyn Thorne and

specialising in the Victorian
period. In addition to sets of
traditional toy soldiers, military
and civilian vignettes and
individual figures are also
produced. Recently acquired are
Burlington Models range of
racehorses, jockeys and
showjumping figures.

**Van Nieuwenhuijzen
Modelshop**
Oude Binnenweg 91, 3012 JA
Rotterdam, Netherlands

Contact: Mike Lettinga
Tel: (0031) 10 4135923
Fax: (0031) 10 4141324

Waterloo Models
101 Eastbourne Road, Southport,
Merseyside PR8 4EH

Tel/fax: 01704 62423

Zinnfiguren Hofmann
Rathausplatz 7, D-90403
Nürnberg, Germany

Contact: Claudia Hofmann
Tel/fax: (0049) 911 204848

Tin figures, mainly 30mm flats by
Heinrichsen/Scheibert, period
1618–48 and 1805–15. Mailing
only within the EC.

MUSEUMS

101 Northumbrian Fd Regiment RA (V)
TA Centre, Knightsbridge, Gosforth, Newcastle upon Tyne, Tyne & Wear NE3 2JJ
Tel/fax: 0191 284 4789

13th/18th Royal Hussars Museum
Cannon Hall, Cawthorne, Barnsley, Yorkshire (S.)
Tel/fax: 01226 790270

Set within the beautiful country park and gardens, with shop, tea rooms etc. Open Tues-Sun inclusive; closed Mondays but open on Bank Holidays.

15th/19th King's Royal Hussars Museum
John George Joicey Museum, 1 City Road, Newcastle upon Tyne, Tyne & Wear NE1 2AS
Tel/fax: 0191 232 4562

1st The Queen's Dragoon Guards Museum
Regimental Museum, Cardiff Castle, Cardiff, Glamorgan (S.) CF1 2RB, Wales
Contact: C J Morris
Tel: 01222 222253
Fax: 01222 227611 x8384

The museum comprises a collection of medals, firearms, standards, guidons and accoutrements etc, dating from 1685 to the Gulf War.

21st Special Air Service Museum
B Block, Duke of York's HQ, King's Road, London SW3 4SE
Tel/fax: 0171 930 4466

4th (V) Bn Royal Green Jackets Museum
56 Davies Street, London W1
Tel/fax: 0171 748 3677

5th Royal Inniskilling DG Museum
The Castle, Chester, Cheshire CH1 2DN
Tel/fax: 01244 347203

9th/12th Royal Lancers Museum
Derby Museum and Art Gallery, The Strand, Derby, Derbyshire DE1 1BS
Contact: Angela Kelsall
Tel: 01332 716657
Fax: 01332 716670

Information panels, audio system, and displays relating the history of the Regiment and its predecessors from 1715 to the present. Enquiry service.

Abingdon Museum
County Hall, Market Place, Abingdon, Oxfordshire
Tel/fax: 01235 23703

Airborne Forces Museum
Browning Barracks, Aldershot, Hampshire GU11 2BU
Contact: Diana Andrews
Tel/fax: 01252 349619

The largest collection of airborne weapons and equipment ever assembled. A concise history of British Airborne Forces from their WWII formation to the present day.

Airborne Museum
Place du 6 Juin 1944, 50480 Ste Mere Eglise, France
Tel/fax: (0033) 33414135

Large private museum with important US Airborne collection.

Airborne Museum Hartenstein
Utrechtseweg 232, 6862 AZ Oosterbeek, Netherlands
Contact: W Boersma
Tel: (0031) 263337710
Fax: (0031) 263334035

The Airborne Museum commemorates the Battle of Arnhem, September 1944 – A Bridge Too Far. Open daily 11.00 till 17.00; Sunday 12.00 till 17.00.

Aldershot Military Museum
Queen's Avenue, Aldershot, Hampshire GU11 2LG
Tel/fax: 01252 314598

Depicts military life over 150 years as Aldershot grew up around the camp. Models, photographs, reconstruction barrack room, tailor's shop; Canadian WWII gallery. Open daily 10.00am – 4.30pm.

Anne Ford
Maytree House, Woodrow Lane, Bromsgrove, Hereford & Worcester B61 0PL
Contact: Alan Ford
Tel/fax: 0121 453 6329

Suppliers to museums, gift and surplus shops – inert ammunition, all calibres. Bullet keyrings, souvenirs, replica grenades, de-activated weapons, MG belts, etc. Trade and retail lists available.

Apsley House – The Wellington Museum
149 Piccadilly, Hyde Park Corner, London W1V 9FA
Contact: Leah Tobin
Tel: 0171 499 5676
Fax: 0171 493 6576

Apsley House, 'No. 1, London', described as "The most renowned mansion in the capital", houses the Duke's collection of paintings, silver, porcelain, sculpture and furniture.

Army Physical Training Corps Museum
Army School of Physical Trg, Queens Avenue, Aldershot, Hampshire GU11 2LB
Contact: A A Forbes
Tel: 01252 347168
Fax: 01252 340785

History, pictorial records, militaria of APTC since 1860. International sports achievements of APTC members. Free entry/parking, wheelchair access. M3/J4-A325-Lynchford Road-Queens Avenue.

Artists Rifles Museum
21 SAS Regt (Artists), Duke of Yorks HQ, Kings Road, Chelsea, London SW3 4RX
Tel/fax: 0171 414 5385

Auto + Technik Museum
D-74889 Sinsheim, Germany
Contact: Herr Boeckle
Tel: (0049) 7261 92990
Fax: (0049) 7261 13916

Ayrshire Yeomanry Museum
Rozelle House, Monument Road, Alloway by Ayr, Ayrshire KA9 4NO, Scotland
Tel/fax: 01292 264091

Bastogne Historical Center
B-6600 Bastogne, Belgium

Battle Historical Museum
Langton House, Abbey Green, Battle, Sussex (E.) TN33 0AQ

Bayerisches Armeemuseum
Paradeplatz 4, 85049 Ingolstadt, Germany

Bedford Yeomanry Collection
c/o Bedford Museum, Circle Lane, Bedford, Bedfordshire MK40 3XD
Tel/fax: 01234 353323

Bedfordshire & Hertfordshire Regiment Collection
c/o Luton Museum Service, Wardown Park, Luton, Bedfordshire LU2 7HA
Contact: Robin Holgate
Tel: 01582 546723
Fax: 01582 546763

Display of medals, weapons, equipment and memorabilia of the Bedfordshire and Hertfordshire Regiment, from the Marlborough campaigns to the end of the Second World War.

Bevrijdingsmuseum 1944
Wylerbaan 4, 6561 KR Groesbeek, Netherlands
Tel/fax: (0031) 889174404

Bjarni's Boots (Handsewn Footwear & Leathergoods)
The Craft Court, Royal Armouries Museum, Leeds, Yorkshire (W.) LS10 1LT
Contact: Mark Beabey
Tel/fax: 0113 245 8824

We offer hand-cut and hand-sewn work made up on a bespoke basis and constructed to exacting standards of historical accuracy. An extensive portfolio of work covering 2,000 years of footwear styles. Leather goods include: armour, vessels, scabbards, cases, holsters, saddlery and harness. Clients include: Historic Royal Palaces, Museum of London, Royal Armouries, National Army Museum, MOD (Household Cavalry).

Black Watch Museum
Balhousie Castle, Hay Street,
Perth, Perthshire PH1 5HR,
Scotland

Contact: S J Lindsay
Tel: 01738 621281
Fax: 01738 643245

250 years of military history of
Scotland's oldest Highland
regiment. Open May-Sept:
Mon-Sat. Oct-Apr: Mon-Fri. Entry
free.

Border Regt & K O R Border Regt Museum
Queen Mary's Tower, The
Castle, Carlisle, Cumbria
CA3 8UR

Contact: S A Eastwood
Tel: 01228 32774
Fax: 01228 21275

The museum relates the story of
Cumbria's County Infantry
Regiment, The Border Regiment
and its successor The King's Own
Royal Border Regiment from 1702
to date. Located in Carlisle Castle,
the Regiment's home since 1873,
the displays on two floors include
uniforms, weapons, equipment,
medals, silver, pictures,
memorabilia, dioramas, video
presentations, armoured car, field
and anti-tank guns. Museum
shop; archives by appointment;
enquiries welcome. Admission by
paid entry to Carlisle Castle; open
April–Sept, Mon–Sun
9.30am–6pm, Oct 9.30am to dusk,
Nov–March Mon–Sun 10am–4pm.
Closed 24–26 December.

Brabants Airborne Museum
Spoorstraat 1, Best, Netherlands
Tel/fax: (0031) 499897473

British In India Museum
Newtown Street, Colne,
Lancashire BB8 0JJ
Tel/fax: 01282 870215

Uniforms, medals, models,
photographs and paintings
covering all aspects, 17thC – 1947.
Open Mon, Wed-Sat 10am-4pm.
Closed December & January, all
Bank Holidays, two weeks in July,
one week in September.
Admission charge.

Buckinghamshire Military Museum
The Old Gaol, Market Hill,
Buckingham, Buckinghamshire
MK18 1EW

Contact: John Roberts

Military Museum Trust exhibits at
the Old Gaol Museum. Open 1
April to mid-September, and
November/December; Mon-Sat
10am-4pm, closed Thurs; Sun,
2pm-4pm; July/August open until
5pm & bank holidays.

Burrell Collection
Pollok Country Farm, 2060
Pollokshaws Road, Glasgow,
Strathclyde G43 1AT, Scotland
Tel/fax: 0141 649 7151

Cameronians (Scottish Rifles) Collection
Low Parks Museum, 129 Muir
Street, Hamilton, Lanarkshire
ML3 6BJ, Scotland

Contact: Liz Hancock
Tel: 01698 283981
Fax: 01698 283479

Housed in the historic Low Parks
Museum, this new exhibition tells
the story of the world-wide
service of the Regiment from the
covenanters to disbandment.

Cannon Hall Museum
Cannon Hall, Cawthorne,
Barnsley, Yorkshire (S.) S75 4AT
Tel/fax: 01226 790270

Regimental museum of 13th/18th
Hussars. 18th century house with
gardens and parkland. Open
Tues-Sat, 10.30-17.00hrs; Sun
14.30-17.00hrs. Parking:
refreshments: disabled facilities.

Carmarthen Museum
Abergwili, Carmarthen, Dyfed
SA31 2JG, Wales

Tel: 01267 231691
Fax: 01267 223830

Regional museum with collection
related to local militia and
volunteer units: Carmarthenshire
Militia, Carmarthen Yeomanry,
Cavalry.

Cheshire Military Museum
The Castle, Chester, Cheshire
CH1 2DN

Tel/fax: 01244 327617

Two regiments of Dragoon
Guards, Cheshire Yeomanry and
Cheshire Regiment combine
displays, medals, uniforms. Small
charge; open daily 10am-5pm
except Christmas.

Chesterholm Museum
Bardon Mill, Hexham,
Northumberland

Displays relating to this important
Roman fort site, where e.g. the
unique Vindolanda tablets were
recovered.

Cholmondeley Model Soldier Collection & Militaria
Houghton Hall, Kings Lynn,
Norfolk

Tel: 01485 528569
Fax: 01485 528167

Clayton Museum / Chesters Roman Fort
Chollerford, Hexham,
Northumberland

Cobbaton Combat Collection
Chittlehampton, Umberleigh,
Devon EX37 9RZ

Contact: Preston Isaac
Tel/fax: 01769 540740

Private collection of about 50
vehicles, tanks and artillery, WWII
British, Canadian, some Warsaw
Pact, fully equipped. Home Front
section. Militaria shop. Open daily
April – November: Winter Mon -
Fri.

Colchester & Essex Museum
The Hollytrees, The High Street,
Colchester, Essex

Tel/fax: 01206 712481/2

Combined Operations Museum
Argyll Estates Office, Cherry
Park, Inveray, Argyll PA32 8XE,
Scotland

Contact: James Jepson
Tel: 01499 500218
Fax: 01499 302421

The story of 250,000 Allied troops
who trained in Inveraray for the
early commando raids and major
landings of World War Two.
Open first Saturday in April to
second Sunday in October.

Cornwall Aircraft Park (Helston) Ltd
Flambards, Culdrose Manor,
Helston, Cornwall TR13 0GA

Contact: James Kingsford Hale
Tel: 01326 573404
Fax: 01326 573344
Email: flambards
@connexions.co.uk

Classic military aircraft, renowned
war galleries; WWII, Falklands,
Gulf memorabilia; acclaimed
recreation of wartime street in
'Britain in the Blitz'. Many other
collections and exhibitions.

D H Mosquito Aircraft Museum
Salisbury Hall, London Colney,
nr. St Albans, Hertfordshire

D-Day Museum & Overlord Embroidery
Clarence Esplanade, Southsea,
Hampshire PO5 3NT

Contact: David Evans
Tel: 01705 827261
Fax: 01705 875276
Email: pzlei130@hants.gov.uk

Tells the story of the Normandy
landings, 6th June 1944. Open
daily except 24th-26th December,
10.00am to 5.30pm, last admission
4.30pm.

Derbyshire Infantry Museum
City Museum & Art Gallery, The
Strand, Derby, Derbyshire DE1
1BS

Contact: Angela Kelsall
Tel: 01332 716657
Fax: 01332 716670

Displays relating to Derbyshire's
regular and irregular infantry
since 1689, including militia,
volunteer, 95th (Derbyshire)
Regiment and Sherwood Foresters
items. Enquiry service.

Derbyshire Yeomanry Museum
City Museum & Art Gallery, The
Strand, Derby, Derbyshire
DE1 1BS

Contact: Angela Kelsall
Tel: 01332 716657
Fax: 01332 716670

Display illustrating the history of
the Derbyshire Yeomanry &
Mounted Volunteers since 1794.
Some regimental records. Enquiry
service.

Doncaster Museum & Art Gallery
Chequer Road, Doncaster,
Yorkshire (S.) DN1 2AE

Contact: G Preece
Tel: 01302 734293
Fax: 01302 735409

Small displays relating to local
Yeomanry, Hussars and Militia,
and the KOYLI museum
collection. Open Mon-Sat,
10am-5pm; Sun 2pm-5pm.
Admission free.

Duke of Cornwall's Light Infantry Museum
The Keep, Bodmin, Cornwall
PL31 1EG

Contact: Hugo White
Tel/fax: 01208 72810

History of the Duke of Cornwall's
Light Infantry, and the local
Cornish forces. Excellent display
of uniforms, medals and weapons.
Very comprehensive reference
library.

Duke of Lancaster's Own Yeomanry Museum
Museum of Lancashire, Stanley
Street, Preston, Lancashire
PR1 4YP

Contact: Stephen Bull
Tel/fax: 01772 264075

The museum of the Duke of
Lancaster's Own tells the story of
the regiment from the eighteenth
century to the present.

Duke of Wellington's Regt Museum

Bankfield Museum, Akroyd Park, Halifax, Yorkshire (W.) HX3 6HG

Contact: J D Spencer
Tel: 01422 352334
Fax: 01422 349020

The museum displays the history of the 33rd and 76th Foot from 1702 to the present, including material relating to the 'Iron Duke'. Admission free.

Dumfries & Galloway Aviation Museum

Former Control Tower, Heathhall Industrial Estate, Dumfries, Dumfriesshire DG1 3PH, Scotland

Contact: Davie Reid
Tel/fax: 01387 259546 (eves.)

Museum situated in the former Control Tower, Heathhall, Dumfries. Open Easter – end Oct, weekends only.

Durham Light Infantry Museum

Aykley Heads, Durham, Co.Durham DH1 5TU

Tel/fax: 0191 384 2214

Houses exhibition on The Durham Light Infantry from 1758-1968 including uniforms, medals, weapons, photographs and Regimental treasures. Open Tues-Sat 10.00am-5.00pm, Sun 2.00pm-5.00pm. Shop and coffee bar. Access for disabled.

Edgehill Battle Museum

The Estate Yard, Farnborough Hall, Banbury, Oxfordshire OX17 1DU

Contact: Peter Dix
Tel: 01926 332213
Fax: 01926 336795

The museum commemorates the first Battle of the English Civil War with dioramas, maps and models. Open 2-6pm Wed and Sat April to Sept. Admission £1.00.

Elizabeth Castle

St.Aubin's Bay, Jersey, Channel Islands

Tel: 01534 23971
Fax: 01534 30511

Essex Regiment Museum

Oaklands Park, Moulsham Street, Chelmsford, Essex CM2 9AQ

Contact: Ian Hook
Tel: 01245 260614
Fax: 01245 350676
Email: chelmsfordbc
@dial.pipex.com

Displays of weapons, uniforms, medals, silver and relics of The Essex Regiment (now Royal Anglian) and its forebears, the 44th and 56th Foot, including the Eagle captured at Salamanca,

1812. Biographical index for family history research. Open Mon-Sat 10am-5pm, Sun 2pm-5pm; admission free.

Essex Yeomanry Museum Collection

Canterbury Lodge, Margaretting, Essex CM4 0EE

Tel: 01245 361258
Fax: 01245 360662

Fallingbostel WWII & 2RTR Regt.Museum

MBB3 Lumsden Barracks, Fallingbostel BFPO 38, Germany

Contact: Kevin Greenhalgh

Museum of WWII relics/items recovered from the actions around Fallingbostel, including the POW camps 11B and 326. 2RTR Regimental Museum covers regimental history 1916 to the Gulf War; uniforms, large collection of WWII RTR items, photos, books, etc.

Fife and Forfar Yeomanry Museum

Yeomanry House, Castlebank Road, Cupar, Fife KY14 4BL, Scotland

Tel/fax: 01534 52354

Fire Services National Museum Trust

9 Morland Way, Manton Heights, Bedford, Bedfordshire MK41 7NP

Contact: Maurice Cole
Tel/fax: 01234 355453

A project to create a National Museum of Firefighting devoted to the exciting story of the fire services and fire prevention in this country.

Flag Institute

44 Middleton Road, Acomb, Yorkshire (N.) YO2 3AS

Contact: Michael Faul
Tel/fax: 01904 781026

Flagship Portsmouth Trust

Porter's Lodge, College Road, HM Naval Base, Portsmouth, Hampshire PO1 3LR

Contact: Group Bookings Dept
Tel: 01705 839766
Fax: 01705 295252
Email: signals@flagship.
compulink.co.uk

HMS Victory, Mary Rose, HMS Warrior 1860, Royal Naval Museum. The world's greatest historic ships.

Fleet Air Arm Museum

RNAS Yeovilton, Ilchester, Yeovilton, Somerset BA22 8HT

Contact: Carol Rendell
Tel: 01935 840565
Fax: 01935 840181

World's leading naval aviation museum. History since 1908, exhibitions on World Wars, Kamikaze, WRNS, recent conflicts. New Carrier Experience – a flight deck on land – plus 11 aircraft.

Fort de Douaumont

55100 Verdun, France

Tel/fax: (0033) 29841885

Fort de Vaux

55100 Verdun, France

Tel/fax: (0033) 29841885

Fusiliers Museum of Northumberland

Abbot's Tower, Alnwick Castle, Alnwick, Northumberland NE66 1NG

Tel/fax: 01665 602152/510211

Fusiliers Museum, Lancashire

Wellington Barracks, Bolton Road, Bury, Lancashire BL8 2PL

Tel/fax: 0161 764 2208

The Fusiliers Museum, Lancashire at Wellington Barracks is open Mon, Tues, Fri, Sat 9.30am-4.30pm. Admission £1, OAPs and Children 50p.

Fusiliers Volunteer & Territorial Museum

213 Balham High Road, London SW17 7BQ

Tel/fax: 0181 672 1168

Genie Museum

Brederodekazerne, Lunettenlaan 102, 5263 NT Vught, Netherlands

Tel/fax: (0031) 73 881867

German Underground Hospital

St Lawrence, Jersey, Channel Islands

Tel/fax: 01534 63442

Large displays connected with German wartime occupation of Channel Islands, housed in underground complex.

Glamorgan Artillery Volunteers Museum

104 Air Defence Regt RA (V), Raglan Barracks, Newport, Gwent NP9 5XE, Wales

Tel/fax: 01633 840443

Glasgow Art Gallery & Museum

Kelvingrove, Glasgow, Strathclyde G3 8AG, Scotland

Contact: Robert Woosnam-Savage
Tel: 0141 305 2654
Fax: 0141 221 9600

European arms & armour (1000 AD – present); includes 'Avant' armour, c.1445 (part of the R.L. Scott Collection); Whitelaw Collection (Scottish arms); Martin Collection (firearms); Scottish W.Coast Volunteer Units (18-19thC). Library closed Dec 25, Jan 1.

Gordon Highlanders Museum

St Lukes, Viewfield Road, Aberdeen, Grampian AB15 7XH, Scotland

Contact: S W Allen
Tel: 01224 311200
Fax: 01224 319323

Recently refurbished museum with new exhibitions, audio-visual theatre, inter-active displays, gift shop, tea room, museum gardens. Opening times: please telephone 01224 311200.

The Green Howards Museum

Trinity Church Square, Richmond, Yorkshire (N.) DL10 4QN

Tel/fax: 01748 822133

Over 300 years of service to the country is displayed in three galleries in the attractively converted church in Richmond's historic market square. Includes important Crimean War items.

Grosvenor Museum

27 Grosvenor Street, Chester, Cheshire CH1 2DD

Tel/fax: 01244 21616

The Guards Museum

Wellington Barracks, Birdcage Walk, London SW1A 2AX

Tel/fax: 0171 930 4466

Guernsey Militia Museum

Guernsey Museums & Galleries, Candie Gardens, St Peter Port, Guernsey, Channel Islands GY1 1UG

Contact: Brian R Owen
Tel: 01481 720513
Fax: 01481 728671
Email: brian@museum.
guernsey.net

Uniforms, equipment and ephemera of the Guernsey Militia and Royal Guernsey Light Infantry. Open April to October – and in winter by appointment.

Gurkha Museum
Peninsular Barracks, Romsey Road, Winchester, Hampshire SO23 8TS

Contact: Christopher Bullock OBE MC
Tel: 01962 842832
Fax: 01962 877597

Open 10am-5pm Tues-Sat & Bank Holiday Mondays. Admission: adults £1.50; OAPs 75p; accompanied children free. Ten minutes walk from rail station. Shop and mail order service.

Hall of Aviation
Albert Road South, Southampton, Hampshire SO1 1FR

Tel/fax: 01703 227343

Hereford Regt & Light Infantry Museum
TA Centre, Harold Street, Hereford, Hereford & Worcester

Tel/fax: 01432 272914

Hertfordshire Regimental Collection
c/o Hertford Museum, 18 Bull Plain, Hertford, Hertfordshire SG14 1DT

Tel/fax: 01992 582686

Open Tuesday – Saturday, 10am-5pm, admission free.

Hertfordshire Yeomanry & Artillery Museum
Hitchin Museum & Art Gallery, Paynes Park, Hitchin, Hertfordshire SG5 1EQ

Contact: Isabel Wilson
Tel: 01462 434476
Fax: 01462 431315

A small display of medals and uniforms of the Hertfordshire Yeomanry.

History on Wheels Motor Museum
Longclose House, Little Common Road, Eaton Wick, Nr.Windsor, Berkshire

Contact: Tony L Oliver
Tel/fax: 01753 862637

HMS Belfast
Morgan's Lane, Tooley Street, London SE1 2JH

Contact: Sarah Hoben
Tel: 0171 407 6434
Fax: 0171 403 0719

Europe's largest preserved WWII warship, now a national museum moored close to London Bridge. Seven decks of exhibitions to explore. Open every day from 10am.

Holts' Tours (Battlefields & History)
15 Market Street, Sandwich, Kent CT13 9DA

Contact: John Hughes-Wilson
Tel: 01304 612248
Fax: 01304 614930
Email: www.battletours.co.uk

Europe's leading military historical tour operator, offering annual world-wide programme spanning history from the Romans to the Falklands War. Holts' provides tours for both the Royal Armouries Leeds and the IWM. Every tour accompanied by specialist guide-lecturer. Send for free brochure.

Home Front / Danelaw Village
Murton Park, York, Yorkshire (N.) YO1 3UF

Contact: Angela Hardman
Tel: 01904 489966
Fax: 01904 489159

Educational 'living history' site covering 10th to 20th century for schools. School visits, lectures, demonstrations all periods, nationwide. Site open all year.

Honourable Artillery Company Museum
Armoury House, City Road, London EC1Y 2BQ

Household Cavalry Museum
Combermere Barracks, St Leonards Road, Windsor, Berkshire SL4 3DN

Contact: Major A W Kersting
Tel/fax: 01753 5868222 X5203

Collection contains uniforms, weapons, horse furniture, standards and curios of The Regiment of Household Cavalry. Over 300 years of the history of The Sovereign's Mounted Bodyguard. Open Mon-Fri, 9.30am-12.30pm & 2.00pm-4.30pm. Free admission.

Hull City Museums
Wilberforce House, High Street, Hull, Humberside (N.)

Tel/fax: 01482 222737

Hunterian Museum
Glasgow University, Glasgow, Strathclyde G12 8QQ, Scotland

Tel/fax: 0141 339 8855

Huntly House Museum
142 Canongate, Edinburgh, Lothian EH8 8DD, Scotland

Contact: Elaine Finnie
Tel: 0131 529 4012
Fax: 0131 557 3346

Edinburgh's main museum of local history. Displays include collections relating to the life of Field-Marshal Earl Haig (1861-1928), with a reconstruction of his WWI HQ. Open Mon-Sat

10.00am-5.00pm, Sunday 2.00pm-5.00pm (during Edinburgh Festival). Admission free.

Imperial Press
Pantiles, Garth Lane, Knighton, Powys LD7 1HH, Wales

Contact: Terence Wise

Guide to Military Museums in the UK: full details of over 200 regimental and other military museums listed. 25th year of publication, 8th edition. £3.95 from Imperial Press at above address.

Imperial War Museum
Lambeth Road, London SE1 6HZ

Tel: 0171 416 5000
Fax: 0171 416 5374

Unique institution telling the story of 20th Century warfare. Exhibitions on the two World Wars, a 'large exhibits' hall, art galleries, library and photo archives, cafe and shop. Open Mon to Sun 10.00am to 6.00pm. Adults £4.10, concessions £3.10. Tube Lambeth North, Elephant & Castle; British Rail Waterloo, Elephant & Castle.

Imperial War Museum / Duxford
Duxford Airfield, Cambridge, Cambridgeshire CB2 4QR

Contact: Frank Crosby
Tel: 01223 835000
Fax: 01223 837267

Houses the largest collection of military and civil aircraft in the country, totaling over 140. Also exhibiting over 100 military vehicles, artillery, and much more.

Inns of Court & City Yeomanry Museum
10 Stone Buildings, Lincoln's Inn, London WC2A 3TG

Contact: Major M O'Beirne
Tel: 0171 405 8112
Fax: 0171 414 3496

Intelligence Corps Museum
Templer Barracks, Ashford, Kent TN23 3HH

Tel/fax: 01252 25251 X208

The Invasion Museum
The Wish Tower, King Edward's Parade, Eastbourne, Sussex (E.) BN21 4BY

Tel/fax: 01323 35809

John George Joicey Museum
1 City Road, Newcastle upon Tyne, Tyne & Wear NE1 2AS

Tel/fax: 0191 232 4562

Museum dedicated to The 15th and 19th King's Royal Hussars and Northumberland Hussars.

The Keep Military Museum
The Keep, Bridport Road, Dorchester, Dorset DT1 1RN

Contact: Len Brown
Tel: 01305 264066
Fax: 01305 250373

Covers Devonshire Regiment, Dorset Regiment, The Devonshire and Dorset Regiment, Queens Own Dorset Yeomanry, The Royal Devonshire Yeomanry, Royal North Devonshire Hussars, Dorset Yeomanry and 94th Field Regiment, RA.

Kent & Sharpshooters Yeomanry Museum
Hever Castle, Hever, Edenbridge, Kent TN8 7NG

Tel: 01732 865224
Fax: 01732 866796

Museum housed in Hever Castle. Castle admission rates apply – confirm museum opening times before arrival. Castle open daily 12noon-5pm, 1 March-30 November.

Kent Battle of Britain Museum
Aerodrome Road, Hawkinge Airfield, Folkestone, Kent CT18 7AG

Contact: Mike Llewelyn
Tel/fax: 01303 893140

Most important collection of Battle of Britain artefacts on show in the country. Aircraft, vehicles, weapons, flying equipment, prints, relics from nearly 600 crashed aircraft.

King's Own Royal Regiment Museum
City Museum, Market Square, Lancaster, Lancashire LA1 1HT

Contact: Peter Donnelly
Tel: 01524 64637
Fax: 01524 841692

Museum of the King's Own Royal Regiment (Lancaster) and its constituent militias, 1680 – 1959. Displays and regimental archive.

King's Own Scottish Borderers Museum
The Barracks, Berwick-on-Tweed, Northumberland TD15 1DG

Contact: C G O Hogg DL
Tel: 01289 307426
Fax: 01289 331928

This unique regimental museum tells the story of the Borderers from 1689 until today. From Killiecrankie to the Gulf. From the Broadsword to SA80 Rifle.

King's Own Yorkshire Light Infantry Museum
Doncaster Museum & Art Gallery, Chequer Road, Doncaster, Yorkshire (S.) DN1 2AE

Tel: 01302 734293
Fax: 01302 735409

History of the K.O.Y.L.I. and 51st and 105th regiments. Open Mon-Sat, 10am-5pm; Sun 2pm-5pm. Admission free.

King's Regiment Collection
Museum of Liverpool Life, Pier Head, Liverpool L3 1PZ

Contact: Simon Jones
Tel: 0151 478 4409
Fax: 0151 478 4090

Currently closed to public, re-opens in 1999. The Regimental collection of the King's Liverpool Regiment up to 1958, and the King's Regiment from 1958.

King's Regiment Museum
Manchester Collection, Old Town Hall, Ashton-under-Lyne, Lancashire OL6 6DL

Tel/fax: 0161 344 3078

Lancaster City Museum
Market Square, Lancaster, Lancashire LA1 1HT

Tel/fax: 01524 64637

Langford Lodge Wartime Centre
Station 597, Gortnagallon Road, Crumlin, Co.Antrim BT29 4QR, N. Ireland

Contact: Mrs Trudy Watson
Tel/fax: 01232 650451

'Station 597' – permanent collection of wartime memorabilia. Open: Sat & Sun, Easter-end November, 12noon-6pm. Weekday group visits by arrangement (concessions). Charges: adults £2.50, children & OAPs £1.50. Cafe/shop; free car park; facilities for disabled.

Leeds City Museum
Municipal Buildings, Leeds, Yorkshire (W.) LS1 3AA

Tel/fax: 0113 2462632

Leeds Rifles Museum Trust
1 Ledgate Lane, Burton Salmon, Leeds, Yorkshire (W.) LS25 5JY

Contact: R Addyman
Tel/fax: 01977 676835

Items of regimental silver, pictures, scrolls, medals and other effects accomodated exclusively in TA barracks. Access by prior arrangement only but enquiries welcome.

Leicestershire & Derbyshire Ymny Museum
Derby Museum & Art Gallery, The Strand, Derby, Derbyshire DB1 1BS

Tel/fax: 01332 255586

Lincolnshire Regiments Collection
Museum of Lincolnshire Life, Burton Road, Lincoln, Lincolnshire LN1 3LY

Tel: 01522 528448
Fax: 01522 521264

Royal Lincolnshire Regiment gallery and collection; Lincolnshire Yeomanry display; WWI tank. Open May-Sept every day 10am-5.30pm; Oct-April Mon-Sat 10am-5.30pm, Sun 2pm-5.30pm.

Liverpool Scottish Regiment Museum
TA Centre, Forbes House, Score Lane, Childwall, Liverpool, Merseyside L16 2NG

Tel/fax: 0151 647 4342

London Irish Rifles Museum
Duke of York's HQ, Kings Road, Chelsea, London SW3 4RX

London Scottish Regimental Museum
95 Horseferry Road, Westminster, London SW1P 2DX

Contact: J P Haynes
Tel/fax: 0171 630 1630

Private museum viewed by appointment only. Wednesday evening only, 6pm-8.30pm.

London War Museum
1-5 Crucifix Lane, London Bridge, London SE1

Loughborough War Memorial Museum
Carillon Tower, Queen's Park, Loughborough, Leicestershire LE11 2EW

Contact: L Butler

Three floors of displays of military interest; Leics Yeomanry, British & US Airborne, etc. Open Good Friday – 30 Sept, 2pm – 6pm. Museum run by volunteers, no phone or postal delivery; contact through Mr L. Butler, Chairman, 30 Glebe St, Loughborough, Leics.

Lunt Roman Fort
Coventry Road, Baginton, Coventry, Midlands (W.)

Contact: Susan Mileham

Partially reconstructed fort including gateway, ramparts, granary and unique gyrus. Interpretive exhibition and tape-tour guide. Good access for disabled. Address for

correspondence: Herbert Art Gallery & Museum, Jordan Well, Coventry CV32 6DY, tel:01203 832381, fax:01203 832410.

Luton Museum
Wardown Park, Luton, Bedfordshire LU2 7HA

Tel/fax: 01582 36941

Museum of the Bedfordshire and Hertfordshire Regiments.

Maidstone Museum
St Faith's Road, Maidstone, Kent ME14 1LH

Tel/fax: 01622 54497

The Queen's Own Royal West Kent Regimental Museum.

Maritime Museum
Albert Dock, Liverpool, Merseyside L3 4AA

Contact: Lorraine Knowles
Tel: 0151 207 0001
Fax: 0151 478 4590

Martello Tower 73
The Wish Tower, King Edward's Parade, Eastbourne, Sussex (E.) BN21 4JJ

Contact: Michael Moss
Tel/fax: 01323 410300

One of a chain of Martello Towers, it tells the story of the construction, manning and eventual fate of all the towers built.

Mary Rose Ship Hall & Exhibition
HM Naval Base, Portsmouth, Hampshire PO1 3LX

Contact: Andy Newman
Tel: 01705 750521
Fax: 01705 870588
Email: maryrose @cix.compulink.co.uk

Experience the breathtaking hull and exhibition of unique artefacts from King Henry VIII's favourite warship. Open all year, it is a moment captured in time.

Mary Rose Trading Company
No. 5 Boathouse, HM Naval Base, Portsmouth, Hampshire PO1 3PX

Contact: Sally Charleton
Tel: 01705 839938
Fax: 01705 870588
Email: maryrose @cix.compulink.co.uk

From replica artefacts to clothing and confectionary, the Mary Rose Gift Shop has something for everyone interested in Tudor life and times. Mail order available.

Mémorial de Verdun
Fleury-devant-Douaumont, 55100 Verdun, France

Tel/fax: (0033) 29843534

Important WWI collection including uniforms, equipment, artillery, dioramas, etc.

Mémorial Sud-Africain du Bois Delville
Longueval, 80360 Somme, France

Tel: (0033) 22850217
Fax: (0033) 22851360

Museum and memorial to South African troops in France, First World War. Open, free of charge, April to October 10am-5.45pm; November to March 10am-3.45pm; closed January, every Monday, and local holidays.

Middlesex Yeomanry History Trust
TA Centre, Elmgrove Road, Harrow, Middlesex HA1 2QA

Tel/fax: 01992 57018

Midland Air Museum
Coventry Airport, Baginton, Warwickshire CV8 3AZ

Tel/fax: 01203 301033

Militarhistorisches Museum Dresden
Olbrichtplatz 3, 01099 Dresden, Germany

Contact: Näser
Tel: (0049) 3518230
Fax: (0049) 3518232805

Military & Aero Museums Trust
Wavell House, Cavan's Road, Aldershot, Hampshire GU11 2LQ

Tel/fax: 01252 21048

Military Heritage Museum
West Street Auction Galleries, Lewes, Sussex (E.) BN7 2NJ

Contact: Roy Butler
Tel/fax: 01273 480208

Collection of Military History, 1660-1914, including uniforms, headdress, weapons and equipment. Admission by prior appointment only.

Montrose Air Station Museum
104F Castle Street, Montrose, Angus DD10 8AX, Scotland

Contact: G McIntosh
Tel/fax: 01674 673107

Montrose Air Station Museum. Open Sundays, 12-5pm. Outside usual hours Tel 01674 672035/ 675401/ 674210 or 673107 for general RFC/RAF and wartime artifacts etc.

Musée d'Art et d'Histoire
Chateau de Belfort, 90000 Belfort, France

Musée d'Art Militaire
I Rue Jeanne-d'Arc, 88440
Nomexyu, France
Tel/fax: (0033) 29674665

Musée Berrichon des Trois Guerres
Diors, 36130 Deols, France
Tel/fax: (0033) 54260163

Musée d'Hist Mil Champagne-Ardenne
68 Rue Leon Bourgeois, 51000
Chalons sur Marne, France

Musée de Guerre
Citadelle souterraine, 55100
Verdun, France
Tel/fax: (0033) 29841885

Musée de l'Armée
Hotel National des Invalides,
75007 Paris, France
Tel/fax: (0033) 145559230

The French national museum of
military history; not to be missed
by any Paris visitor.

Musée de l'Armistice
Carrefour de l'Armistice, 60200
Compiegne, France

Displays connected with WWI
Armistice railway carriage; WWI
photo-library.

Musée de l'Emperi
Chateau de l'Emperi, 13300
Salon de Provence, France
Tel: (0033) 490562236
Fax: (0033) 490560812

The former Brunon Collections,
now owned by the Musée de
l'Armée – large, superbly
presented, and of international
importance, particularly
Napoleonic and Second Empire
exhibits.

Musée de l'Infanterie
Quartier Guillaut, Avenue Lepic,
34057 Montpellier, France

Closed weekends.

Musée de la Bataille des Ardennes
08270 Novion-Porcien, France
Tel/fax: (0033) 24232013

1870, World War I, but
particularly important World War
II vehicle, uniform, and weapon
collection.

Musée de la Cavalerie
Ecole d'AABC, 49409 Saumur,
France
Tel/fax: (0033) 41510543

Open by appointment, afternoons
except Fridays; closed August. All
periods, but important World War
II AFV collection, housed at
French Army Armoured Cavalry
School.

Musée de la Caverne du Dragon
Chemin des Dames, 02160
Oulche la Vallee Foulon, France

Musée de la Cooperation Fr.-Americaine
Blerancourt, 02300 Chauny,
France
Contact: Philippe Grunchec
Tel/fax: (0033) 23396016

Open 15 April to 15 October
(except Tuesdays), 10am to 12.30
and 2pm to 5pm; 16 October to 14
April, 2pm to 5pm weekdays, also
10am to 12.30 Saturdays and
Sundays.

Musée de la Figurine Historique
28 Place de l'Hotel de Ville,
60200 Compiegne, France
Contact: Eric Blanchegorge
Tel/fax: (0033) 44407255

Closed Mondays. Large displays
and dioramas. 100,000 civil and
military figurines, individual or in
settings.

Musée de la Legion Etrangère
Quartier Vienot, 13400 Aubagne,
France
Tel/fax: (0033) 42030320

Major official museum of the
Foreign Legion, all periods,
housed at the Legion's depot.
Closed Mon, and Sat mornings,
June-Sept; open Wed, Sat and Sun
Oct to May.

Musée de la Libération de Normandie
Route Nationale 13, Surrain,
14710 Trevieres, France
Tel/fax: (0033) 31225756

Musée de la Poche de Royan
BP15, 17600 Le Gua, France

Musée de la Seconde Guerre Mondiale Message 'Verlaine'
4 bis Avenue de la Marne, 59200
Tourcoing, France

The XVth German Army
headquarters, where the D-Day
message was decoded. Located in
Generaloberst von Salmuth's
bunker.

Musée des Chasseurs a Pied
Chateau de Vincennes, 94300
Vincennes, France

Musée des Equipages & du Train
Quartier de Beaumont, Rue du
Plat-d'Etain, 37034 Tours, France

Musée des Parachutistes
Camp d'Astra, Route de
Bordeaux, 64023 Pau, France
Tel/fax: (0033) 59320597

Musée des Spahis
Musee du Haubergier, 2 Place
Notre Dame, 60300 Senlis,
France

Musée des Troupes de Marine
Quartier Colonel Lecocq, Route
de Bagnols en Foret, 83608
Frejus, France

French Colonial and Marine
Infantry collection. Open June to
Sept 10am-12am, 2pm-5.30pm;
Oct to May, pm only; closed
Tuesday.

Musée du Fort de La Pompelle
Route de Chalons-sur-Marne,
51100 Reims, France

Musée du Para
27 Rue de la Bienvenue, 45000
Orleans, France

Musée du Souvenir General Estienne
Route Nationale 44,
Berry-au-Bac, 02190
Guignicourt, France
Tel/fax: (0033) 23799525

Displays connected with WWI
founder of French tank arm. By
appointment only.

Musée Franco-Australien
1, Route de Corbie, 80380
Villers-Bretonneux, France
Contact: Robert Bled
Tel/fax: (0033) 22480765

Musée Lucien-Roy
Chemin de Maillot, 25720 Beure,
France
Tel/fax: (0033) 81526130

French army, 20th century; closed
Saturdays.

Musée Massey
Jardin Massey, 65000 Tarbes,
France

Major international collection of
Hussar uniforms.

Musée Militaire de Bordeaux
Caserne Boudet, 192 Rue de
Pessac, 33000 Bordeaux, France
Tel/fax: (0033) 56505633

Musée Militaire du Hackenberg
Veckring, 57920
Kedange-sur-Canner, France

Important Maginot Line museum;
open to large groups daily by
arrangement; individuals, Sat and
Sun afternoons.

Musée Militaire du Perigord
32 Rue des Farges, 24000
Perigueux, France

Musée Pierre Noël/Musée de la Vie dans les Hautes-Vosges
Place Georges Trimouille, 88107
St Die-des-Vosges, France
Contact: Daniel Grandidier
Tel: (0033) 329516035
Fax: (0033) 329526689

Important military collection,
France/Germany 1800-1960.
Open: Wednesdays
10am-noon/2pm-5pm; Thursday
to Sunday 2pm-5pm; 2pm-7pm in
summer (May to September).

Musée Regional d'Argonne
Chateau de Braux-Ste Cohiere,
51800 Sainte Menehould, France

Musée Serret
3 Rue Clemenceau, 68550
Saint-Amarin, France

Open afternoons, May to
September.

Museum of Army Flying
Army Air Corps Centre, Middle
Wallop, Hampshire SO20 8DY
Tel/fax: 01264 62121

Museum of Army Transport
Flemingate, Beverley,
Humberside (N.) HU17 0NG
Tel: 01482 860445
Fax: 01482 866459

History of Army transport from
the Boer War to the present day.
Over 110 vehicles; archives,
workshops, restoration, Sir Patrick
Wall model collection, Blackburn
Beverley aircraft, book/gift shop.
Open daily 10am-5pm; free
parking; cafeteria.

Museum of Artillery
The Rotunda, Woolwich, London
SE18 4BQ
Contact: Curator
Tel: 0181 781 3127
Fax: 0181 316 5402

Comprehensive museum of
artillery weapons spanning nearly
500 years. Free admission, open
Monday to Friday afternoons.
Closed weekends and public
holidays.

Museum of Badges and Battledress
The Green, Crakehall, Bedale,
Yorkshire (N.) DL8 1HP
Tel/fax: 01677 424444

Museum covers British
battledress, badges, and
equipment of the Army, HG,
Navy, RAF, RN, Civil Defence
and Land Army, 1900 to present
day (shop buys & sells militaria –
quarterly lists).

Museum of British Military Uniforms
Chapel Street, Billingborough, Lincolnshire

Contact: J Livermore

Museum of Lancashire
Stanley Street, Preston, Lancashire PR1 4YP

Contact: Stephen Bull
Tel/fax: 01772 264075

The museum covers the local regiments, especially the 14/20th King's Hussars, Queen's Lancashire Regiment, Duke of Lancaster's Own Yeomanry Volunteers. Open Mon-Sat except Thurs, 10.30am-5pm.

Museum of Lincolnshire Life
Old Barracks, Burton Road, Lincoln, Lincolnshire LN1 3LY

Tel/fax: 01522 528448

Features displays relating to The Lincolnshire Yeomanry, a WWI tank developed in Lincoln, and an entire gallery devoted to the 300 year history of The Lincolnshire Regiment. Open May-Sept everyday, 10.00am-5.30pm, Oct-April Mon-Sat 10.00am-5.30pm, Sun 2.00pm-5.30pm.

Museum of the Manchesters
Ashton Town Hall, Market Place, Ashton-under-Lyne, Manchester, Gt. OL6 6DL

Tel: 0161 344 3078
Fax: 0161 344 3070

A social & military history of The Manchester Regiment from 1758 to 1958. Open Mon-Sat, 10am-4pm. Admission free. Park on town centre car parks.

Museum of the Order of St. John
St. John's Gate, St. John's Lane, London EC1M 4DA

Contact: Piers de Salis
Tel: 0171 253 6644
Fax: 0171 336 0587

Sixteenth century gatehouse containing armour, silver, coins, medals and other objects of the Knights of St. John. Also uniforms, equipment, memorabilia and records of St. John Ambulance and wartime medical services. Mon-Fri 10.00am-5.00pm, Sat 10.00am-4.00pm. Tours of Gatehouse and Norman Crypt Tues, Fri & Sat 11.00am-2.30pm.

Museum of the Scottish Horse
The Cross, Dunkeld, Perthshire PH8 OAA, Scotland

Contact: M McInnes

This museum holds a unique collection of uniforms, arms, trophies, maps, photographs, and regimental records of this famous regiment, The Scottish Horse.

Nationaal Oorlogs- en Verzetsmuseum
Museumpark 1, 5825 AM Overloon, Netherlands

Tel/fax: (0031) 4788 1820

National Army Museum
Royal Hospital Road, Chelsea, London SW3 4HT

Contact: Julian Humphrys
Tel: 0171 730 0717
Fax: 0171 823 6573
Email: nam@enterprise.net

The story of the British soldier from Agincourt to the present. Videos, models, and reconstructions bring the soldier's story to life. Treasures on display include medals, paintings, weapons, silver and uniforms. Regular special exhibitions (see press for details). Normally open daily 10am to 5.30pm. Admission free.

National Maritime Museum
Park Row, Greenwich, London SE10 9NF

Tel/fax: 0181 858 4422

Britain's national museum of naval and maritime history, set in magnificent surroundings by the Thames at Greenwich. Historic exhibits, models, art collection, documentary and pictorial archives. Open September to April, Mon-Sat 10am-5pm, Sun 12am-5pm; May-October, closes 6pm.

National Museum of Antiquities/Scotland
Queen Street, Edinburgh, Lothian EH2 1JD, Scotland

Tel/fax: 0131 556 8921

National Museum of Military History
10 Bamertal, PO Box 104, L-9209 Diekirch, Luxembourg

Contact: Roland Gaul
Tel: (00352) 808908
Fax: (00352) 804719
Email: mnhmdiek@pt.lu

Important collections from Battle of the Bulge 1944-45, life-size dioramas, uniforms, vehicles, weapons, equipment. Open from January 1 – March 31: daily 14–18h hrs; April 1 – November 1: daily 10–18 hrs; November 2 – December 31: daily 14–18 hrs; last ticket sold 17.15 hrs.

Newark Air Museum
The Airfield, Winthorpe, Newark, Nottinghamshire NG24 2NY

Contact: Howard Heeley
Tel/fax: 01636 707170

Based on an original World War II bomber dispersal, nearly half of our 40 aircraft are displayed under cover; artefacts and engine display; book and souvenir shop.

Newark Museum
Appleton Gate, Newark, Nottinghamshire NG24 1JY

Contact: M J Hall
Tel: 01636 702358
Fax: 01636 610417

Collections relating to the 8th Battalion, Sherwood Foresters. Admission free. Open: Mon-Sat (closed Thurs), 10am-1pm and 2pm-5pm; Sun (April to September) 2pm-5pm.

Norfolk & Suffolk Aviation Museum
The Street, Flixton, Bungay, Suffolk

Tel/fax: 01986 896644 Sun/Tue

Over 20 historic aircraft. Unique indoor exhibitions. 446th BG museum & memorial, ROC museum. Admission free. Open Sun & Bank Holidays, Easter – October, 10am-5pm.

North Irish Horse Museum
c/o The Lord O'Neill,TD,DL, Shanes Castle, Co.Antrim BT41 4NE, N. Ireland

Northamptonshire Regt & Ymny Museum
Abington Park Museum, Abington, Northampton, Northamptonshire NN1 5LW

Tel/fax: 01604 31454

Oxfordshire Regimental Museum
TA Centre, Slade Park, Headington, Oxfordshire OX3 7JL

Tel/fax: 01865 778479

Regimental museum of The Oxfordshire and Buckinghamshire Light Infantry.

Panzermuseum Munster
Hans Kruger Strasse 33, 29633 Munster, Germany

Passmore Edwards Museum
Romford Road, Stratford, London E15 4LZ

Tel/fax: 0181 519 4296

Pembroke Yeomanry Trust
Castle Museum & Art Gallery, The Castle, Havorfordwest, Dyfed SA61 2EF, Wales

Tel/fax: 01437 3708

Polish Institute & Sikorski Museum
20 Princes Gate, London SW7

Tel/fax: 0171 589 9249

Access by arrangement.

Preston Hall Museum
Yarm Road, Stockton, Cleveland

Tel/fax: 01642 781184

Princess Louise's Kensington Regt Museum
c/o 31 Signal Regiment, 190 Hammersmith Road, Hammersmith, London W6 7DL

Tel/fax: 0181 748 6394

Q A R A N C Museum
Keogh Barracks, Ash Vale, Aldershot, Hampshire GU12 5RQ

Tel: 01252 340294
Fax: 01252 340224

The QARANC Museum relates the history of military nursing, Crimea to date, providing fascinating displays including uniforms, medals, memorabilia and photographs. Open Mon-Fri 0900-1530 hours.

Quebec House (National Trust)
Quebec Square, Westerham, Kent TN16 1TD

Tel/fax: 01959 562206

National Trust memorial to life of General James Wolfe. Open April to October, 2pm-6pm on Tuesday and Sunday. Quebec campaign model soldiers for sale.

Queen's Lancashire Regiment Collections
Fulwood Barracks, Preston, Lancashire PR2 8AA

Contact: Major M J Glover
Tel: 01772 260362
Fax: 01772 260583

Open Tues, Weds, Thurs, 9.30am-4.30pm; admission free. Weapons, uniforms, medals, silver, documentary archive, photographs, and memorabilia of the 30th, 40th, 47th, 59th, 81st and 82nd Regiments of Foot and the East Lancashire, South Lancashire, Loyal North Lancashire and Lancashire Regiments.

Queen's Own Highlanders Regt Museum
Fort George by Inverness, Inverness-shire, Scotland

Tel/fax: 01463 224380

Exhibits from Queen's Own Highlanders, Seaforth, Camerons, Lovat Scouts. Publishers of Regimental history. Pipe music, piping history. Shop and mail order. Books, postcards, prints, tapes etc. Open April-Sept Mon-Fri 10am-4pm, Sun 2pm-6pm; Oct-March Mon-Fri 10am-4pm. NB Administrative address, mail etc: RHQ The Highlanders, Cameron Barracks, Inverness, IV2 3XD. Tel: 01463 224 380.

Queen's Own Hussars Museum

Lord Leycester Hospital, High Street, Warwick, Warwickshire CV34 4BH

Contact: Major P J Timmons FISM
Tel: 01926 492035
Fax: 0171 414 8700

The regimental museum of The Queen's Own Hussars, the senior light cavalry regiment of the British Army. Open Tues-Sun incl., 10.00am-5.00pm. Museum shop and restaurant.

Queen's Own Royal West Kent Regt Museum

Maidstone Museum & Art Gallery, St Faith's Street, Maidstone, Kent ME14 1LH

Contact: The Curator
Tel: 01622 754497
Fax: 01622 602193

Uniforms, medals, weapons and general militaria of the regiment from 1756. Admission free. Museum shop and coffee shop. Open 10am-5.15pm.

Queen's Regiment Museum

Dover Castle, Dover, Kent CT16 1HH

Tel/fax: 01227 763434

Queen's Royal Irish Hussars Museum

The Redoubt Fortress, Royal Parade, Eastbourne, Sussex (E.) BN22 7AQ

Contact: Michael Moss
Tel: 01323 410300
Fax: 01323 732240

History of the 4th and 8th Hussars, including Charge of the Light Brigade and Gulf War exhibits and a Centurion tank from the Korean War.

ROYAL ARMY MEDICAL CORPS MUSEUM

The story of the Army Medical Services from 1660 to the present including the Gulf War and Bosnia. Over 3000 medical and military historical items on display plus ambulances and a superb medal collection. Open Mon to Fri 0830 – 4pm.

Keogh Barracks,
Ash Vale, Aldershot, Hants GU12 5RQ
Tel 01252 340212

The Queen's Royal Lancers Museum

c/o Home HQ, Prince William of Glos Bks, Grantham, Lincolnshire NG31 7TJ

Contact: Capt J M Holtby
Tel: 0115 9573295
Fax: 0115 95 3195

Traces the history of the 5th, 16th, 17th, 21st, 16th/5th and 17th/21st Lancers and the newly formed The Queen's Royal Lancers. Displays include uniforms, weapons, medals, and relics from 1759 to present day. Museum open April-September 11am-5pm except Fri & Mon.

The Queen's Royal Surrey Regt Museum

Clandon House, Clandon Park, W.Clandon, Guildford, Surrey GU4 7RQ

Contact: Penny James
Tel/fax: 01483 223419

Museum open Easter-end October, Tues, Wed, Thurs, Sun & Bank Hols 12-5pm. Research by arrangement: 01483 223419. Admission free.

R A F Regiment Museum

R A F Regiment Depot, R A F Catterick, Yorkshire (N.)

Tel/fax: 01748 811441 X202

R A M C Museum

Keogh Barracks, Ash Vale, Aldershot, Hampshire GU12 5RQ

Tel: 01252 340212
Fax: 01252 340224

Catering for all aspects of the history of the Royal Army Medical Corps and its predecessors, including uniforms, insignia, medals and rolls of honour. Open Mon-Fri 8.30am-3.30pm.

R E M E Museum

Isaac Newton Road, Arborfield, nr Reading, Berkshire RG2 9LN

Tel/fax: 01734 760421

Regiments of Gloucestershire Museum

Custom House, Gloucester Docks, Gloucester, Gloucestershire GL1 2HE

Tel/fax: 01452 522682

Award-winning museum tells story of Glorious Glosters and Gloucestershire Yeomanry. Open all year, closed Mondays. Admission charge. Shop. Send SAE for mail order list.

Roman Army Museum

Carvoran, Greenhead, Carlisle, Cumbria

Tel/fax: 0169 72485

Royal Air Force Museum

Grahame Park Way, Hendon, London NW9 5LL

Tel: 0181 205 2266
Fax: 0181 200 1751

Britain's National Museum of Aviation, displays 70 full size aircraft. Open daily. Extensive library research facilities (weekdays only). Large free car park. Licensed restaurant. Souvenir shop with extensive range of specialist books and model kits.

The Royal Armouries

Armouries Drive, Leeds LS10 1CT

Contact: Nicholas Boole
Tel: 0115 220 1948
Fax: 0113 220 1955

The Royal Armouries is the national museum of arms and armour and museum of the Tower of London. It has three sites each dealing with a different aspect of its subject. Royal Armouries HM Tower of London (Tower history), is scheduled to re-open fully to the public in summer 1998, Tel:0171 480 6358, Royal Armouries at Fort Nelson near Portsmouth(artillery collection), Tel:01329 233734, and Royal Armouries Museum in Leeds(arms and armour) 0990 1066 66. Internet: http://www.armouries.org.uk

Royal Army Chaplains' Department Museum

Netheravon House, Salisbury Road, Netheravon, Wiltshire SP4 9SY

Contact: M A Easey

The Museum's exhibits include photographs, POW items, church silver & memorabilia, uniforms & medals. Temporarily in storage prior to moving into Amprot House in 1998.

Royal Army Dental Corps Museum

HQ & Central Group RADC, Evelyn Woods Road, Aldershot, Hampshire GU11 2LS

Tel: 01252 347976
Fax: 01252 347726

Illustrates the connections between dentistry and the Army, from the restoration of the monarchy in 1660 to the present day.

Royal Army Education Corps Museum

R A E C Centre, Wilton Park, Beaconsfield, Buckinghamshire HP9 2RP

Tel/fax: 01494 683263

The museum of education in the British Army. Exhibits including uniforms, weapons & photographs. Extensive archives.

Royal Army Veterinary Museum

R A V C Support Group, Gallwey Road, Aldershot, Hampshire

Tel/fax: 01252 24431 X3527

Royal Devon Yeomanry Museum

40 Oakleigh Road, Barnstaple, Devon EX32 8JT

Contact: A J Hoddinott
Tel: 01271 45471
Fax: 01271 23687

Royal Engineers Museum

Prince Arthur Road, Gillingham, Kent ME4 4UG

Contact: J E Nowers
Tel: 01634 406397
Fax: 01634 822371

See 24 Victoria Crosses, Wellington's Waterloo battlemap, mementoes of Gordon of Khartoum and Field Marshal Lord Kitchener, Engineer tanks, Harrier aircraft and much more.

Royal Fusiliers Museum

HM Tower of London, Tower Hill, London EC3N 4AB

Tel/fax: 0171 488 5612

Royal Glos, Berks & Wilts Regt Museum

The Wardrobe, 58 The Close, Salisbury, Wiltshire SP1 2EX

Contact: Major J H Peters
Tel/fax: 01722 414536

The Royal Gloucestershire, Berkshire and Wiltshire Regiment (Salisbury) Museum offers excellent collections, a fine medieval house, riverside garden and renowned tearoom, all in Salisbury Cathedral Close. Open daily April to October, 10am-4.30pm.

Royal Green Jackets Museum

Peninsula Barracks, Romsey Road, Winchester, Hampshire SO23 8TS

Tel/fax: 01962 863846

Regimental museum for The Oxfordshire and Buckinghamshire Light Infantry, The King's Royal Rifle Corps and The Rifle Brigade. The Waterloo Diorama has twenty thousand model figures and a sound and light commentary. For further details contact the museum.

Royal Hampshire Regt Museum

Serle's House, Southgate St, Winchester, Hampshire SO23 9EG

Tel/fax: 01962 863658

Museum of the Royal Hampshire Regiment 1702-1992.

Royal Highland Fusiliers Museum
518 Sauchiehall Street, Glasgow, Strathclyde G2 3LW, Scotland

Tel/fax: 0141 332 0961

Royal Hospital Museum
The Royal Hospital, Royal Hospital Road, Chelsea, London SW3 4SR

Contact: Major R A G Courage
Tel/fax: 0171 730 0161

Open: 10am-12 noon & 2pm-4pm Mon-Sat daily; April-Sept only, Sun 2pm-4pm. Pictures, medals, uniforms connected with the Royal Hospital. Admission free; no parking.

Royal Hussars Museum
Peninsula Barracks, Romsey Road, Winchester, Hampshire SO23 8TS

Tel/fax: 01962 828539

Royal Inniskilling Fusiliers Museum
The Castle, Enniskillen, Co.Fermanagh BT74 7BB, N. Ireland

Contact: Hugh Forrester
Tel: 01365 323142
Fax: 01365 320359

Regimental museum of the Inniskillings, telling the story of the famous regiment from 1689 to 1968. Open all year, half day Mondays.

Royal Irish Fusiliers Museum
Sovereign's House, The Mall, Armagh, Co.Armagh BT61 9DL, N. Ireland

Contact: Amanda Moreno
Tel/fax: 01861 522911

Housed in a fine Georgian residence, this museum relates the history of this Regiment from 1793 to 1968. Exhibits include the Battle of Barossa and the local Militias. Open daily.

Royal Irish Regiment Museum
c/o RHQ Royal Irish Regiment, Ballymena, Co.Antrim BFPO 808, N. Ireland

Royal Leicestershire Regt Museum
The Newarke, Oxford Street, Leicester, Leicestershire

Tel/fax: 0116 2554100

Royal Logistic Corps Museum
Princess Royal Barracks, Deepcut, Nr Camberley, Surrey GU16 6RW

Contact: Frank O'Connell
Tel: 01252 340871
Fax: 01252 340875
Email: frank.oconnell @btinternet.com

Open Mon – Fri 10.00-16.00 hrs; Sat 10.00 – 15.00 hrs. Admission free. Guided tours, lectures, access to archives available by prior arrangement with Curator.

Royal Marines Museum
Southsea, Portsmouth, Hampshire PO4 9PX

Contact: Jorj Jarvie
Tel: 01705 819385
Fax: 01705 838420

The history of the Royal Marines from 1664 to the present. Open 7 days a week. Library and archives by arrangement.

Royal Military Police Museum
Roussillon Barracks, Chichester, Sussex (W.) PO19 4BL

Tel/fax: 01243 786311

Royal Naval Museum
HM Naval Base, Portsmouth, Hampshire PO1 3NH

Contact: Fiona Gordon
Tel: 01705 727562
Fax: 01705 727575
Email: rnmuseum.compulink.co.uk

The history, traditions and people of the Royal Navy are celebrated in the five extensive galleries of the Royal Naval Museum.

Royal Navy Submarine Museum
Haslar Jetty Road, Gosport, Hampshire PO12 2AS

Contact: Site Supervisor
Tel: 01705 529217/510354
Fax: 01705 511349
Email: rnsubs@submarine-museum.demon.co.uk

World's most comprehensive museum of underwater warfare with actual submarines, many models, memorabilia. Library, extensive photographic collection, research facilities by appointment. Open winter 10am-4.30pm, summer 10am-5.30pm, last tour 1 hour before closing; adult admission £3.75.

Royal Netherlands Army and Arms Museum
Armamentarium, Korte Geer 1, 2611 CA Delft, Netherlands

Contact: Jan A Buyse
Tel: (0031) 152 150500
Fax: (0031) 152 150544

Depicts military life from the Roman period onwards until recent army actions, worldwide (NATO, UN). Weapons, armour, uniforms, flags, models, transport, communications, paintings, photographs, books. Open all week, except Mondays. Admission charge. Tours on request.

Royal Norfolk Regiment Museum
Shirehall, Market Avenue, Norwich, Norfolk NR1 3JQ

Contact: Miss K Thaxton
Tel: 01603 223649
Fax: 01603 765651

Social and military history of the county regiment from 1685 to 1959. Open Mon-Sat 10am-5pm.

Museum of The Royal Scots
The Castle, Edinburgh, Lothian EH1 2YT, Scotland

Contact: R P Mason
Tel: 0131 310 5014
Fax: 0131 310 5019

Interesting and colourful story of The Royal Scots covering over 350 years service of the oldest Regiment of Infantry in the British Army. Excellent shop.

Royal Signals Museum
Blandford Camp, Blandford Forum, Dorset DT11 8RH

Contact: Dr Peter Thwaites
Tel: 01258 482267
Fax: 01258 482084

Displays items relating to the history of army signalling since the Crimean War and the history of Royal Signals. Open Mon-Fri 10.00am-5.00pm, Sat/Sun June-Sept only, 10.00am-4.00pm.

Royal Sussex Regiment Museum
The Redoubt Fortress, Royal Parade, Eastbourne, Sussex (E.) BN22 7AO

Contact: Michael Moss
Tel: 01323 410300
Fax: 01323 732240

History of the Royal Sussex Regiment from its formation in 1701. Exhibits include Afrika Korps General Von Arnim's staff car, and rare Napoleonic Volunteer uniforms.

Royal Ulster Rifles Museum
RHQ The Royal Irish Regt, 5 Waring Street, Belfast, Co.Antrim BT1 2EW, N. Ireland

Tel/fax: 01232 232086

Royal Warwickshire Regt Museum
St John's House, Warwick, Warwickshire

Tel/fax: 01926 491653

Royal Welch Fusiliers Museum
Queen's Tower, Caernarfon Castle, Caernarfon, Gwynedd LL55 2AY, Wales

Contact: Peter Crocker
Tel/fax: 01286 673362

History of Wales's oldest line regiment since 1689. Marvellous collection of uniforms, weapons, medals, pictures, etc. The Regiment of Sassoon, Graves, David Jones, Frank Richard, "Hedd Wyn", Thomas Atkins, and the 'Black Flash'.

Rutland County Museum
Catmos Street, Oakham, Rutland LE15 6HW

Tel: 01572 723654
Fax: 01572 757576

Museum of Rutland Life. Gallery on history of the volunteer soldier in Leicestershire and Rutland; museum housed in Napoleonic riding school; admission free.

Science Museum
Exhibition Road, South Kensington, London SW7 2DD

Tel: 0171 938 8000
Fax: 0171 938 8112

Scottish United Service Museum
The Castle, Edinburgh, Lothian EH1 2NG, Scotland

Contact: Stephen Wood
Tel: 0131 225 7534
Fax: 0131 225 3848

A large and comprehensive range of Scottish military antiquities and a library and archive, available for use by appointment (tel. ext. 204). Open 9.30am-5.30pm.

Sherwood Foresters Museum
The Castle, Nottingham, Nottinghamshire NG1 6AF

Tel/fax: 0115 9411881

Sherwood Rangers Museum
TA Centre, Carlton, Nottingham, Nottinghamshire NG4 3DX

Tel/fax: 0115 9618722

Shropshire Regimental Museum
The Castle, Shrewsbury, Shropshire SY1 2AT

Contact: John Taylor
Tel: 01743 358516
Fax: 01743 262542

New display of the collection of the King's Shropshire Light Infantry, Shropshire Yeomanry and Shropshire Royal Horse Artillery. Open Tuesday to Sunday, 10am-4.40pm.

The Shuttleworth Collection
Old Warden Aerodrome,
Biggleswade, Bedfordshire SG18
9EP

Tel: 01767 627288
Fax: 01767 627745

Large collection of historic
aeroplanes and vehicles on static
display – open daily; flying
displays each month. Adults
£5.00, children/OAPs £2.50; group
and school party rates.

Somerset Military Museum
County Museum, The Castle,
Taunton, Somerset TA1 4AA

Contact: Brig A I H Fyfe
Tel/fax: 01823 333434

Correspondence address: c/o
Light Infantry Office, 14 Mount
Street, Taunton TA1 3QE.

South Nottingham Hussars Ymnry Museum
TA Centre, Hucknall Lane,
Bulwell, Nottinghamshire
NG6 8AQ

Tel/fax: 0115 9272251

South Somerset District Council Museum
Hendford, Yeovil, Somerset

Tel/fax: 01935 24774

South Wales Borderers & Monmouthshire Museum of the Royal Regiment of Wales
The Barracks, Brecon, Powys
LD3 7EB, Wales

Contact: J M Grundy
Tel: 01874 613310
Fax: 01874 613275

The museum contains interesting
artefacts and archives relating to
the history of the South Wales
Borderers (24th Foot) from 1689 to
the present day as the Royal
Regiment of Wales. The collection
contains 16 Victoria Crosses and
many items relating to the battles
of Saratoga, Chillianwallah,
Isandhlwana and Rorke's Drift
and both World Wars.

St Peter's Bunker Museum
St Peter's Village, Jersey, Channel
Islands JE3 7AF

Contact: Anita Mayne
Tel: 01481 481048
Fax: 01481 481630

The largest collection of genuine
German militaria, occupation
relics and photographs, housed in
an actual German bunker. Open
daily, March to October.

Staff College Museum
Camberley, Surrey GU15 4NP

Contact: Col P S Newton
Tel/fax: 01276 412602

Deals with the history and dress
of the staff of the British Army
since the formation of the Staff
College in 1799. By appointment
only.

Staffordshire Regiment Museum
Whittington Barracks, Lichfield,
Staffordshire WS14 9PY

Contact: Major E Green
Tel: 0121 311 3229
Fax: 0121 311 3205

Open weekdays 9am-4pm,
admission free. Displays of the
Regiment's history, uniforms,
medals, badges, weapons,
vehicles, memorabilia. Car park;
shop; picnic area; school parties
welcome.

Stirling Regimental Museum
Stirling Castle, Stirling, Grampian
FK8 1EJ, Scotland

Tel/fax: 01786 75165

Stoke Museum & Art Gallery
Bethesda Street, Hanley, Stoke
on Trent, Staffordshire ST1 3DW

Tel/fax: 01782 202173

Suffolk Regiment Museum
The Keep, Gibraltar Barracks,
Out Risbygate, Bury St Edmunds,
Suffolk IP33 3RN

Sussex Combined Services Museum
The Redoubt Fortress, Royal
Parade, Eastbourne, Sussex (E.)
BN22 7AQ

Contact: Michael Moss
Tel: 01323 410300
Fax: 01323 732240

A restored Napoleonic fortress,
Museum illustrates the history of
three services in Sussex. Over
5,000 exhibits including uniforms,
medals and weapons.
Re-enactment weekend in July.

Sussex Yeomanry Museum
198 Dyke Road, Brighton, Sussex
(E.) BN1 5AS

Tel/fax: 01273 556041

Tank Museum
Bovington Camp, Wareham,
Dorset BH20 6JG

Tel: 01929 405096
Fax: 01929 405360

The world's most comprehensive
collection of AFVs. Free
'Firepower & Mobility' displays
Thursdays 12 noon, July, August,
September. Open daily 10am-5pm.

Tenby Museum
Castle Hill, Tenby,
Pembrokeshire, Dyfed, Wales

Tel/fax: 01834 2908

Towneley Hall Art Gallery
Towneley Hall, Burnley,
Lancashire BB11 3RQ

Contact: Miss J S Bourne
Tel: 01282 424213
Fax: 01282 36138

Exhibits relating to The East
Lancashire Regiment.

Trenchart
PO Box 3887, Bromsgrove,
Hereford & Worcester B61 0NL

Tel/fax: 0121 453 6329

Suppliers to museums, gift and
surplus shops. Inert ammunition,
all calibres; bullet keyrings,
dogtags, souvenirs, replica
grenades, de-activated weapons,
MG belts and clips, large cartridge
cases, etc. Trade only supplied.

Victoria & Albert Museum
South Kensington, London
SW7 2RL

Tel/fax: 0171 589 6371

Wallace Collection
Hertford House, Manchester
Square, London W1M 6BN

Tel: 0171 935 0687
Fax: 0171 224 2155

This museum contains a
magnificent collection of Medieval
and Renaissance arms and
armour, and also a superb array of
Oriental weaponry,
predominantly Indian and
Persian. Admission free. Location:
North from Oxford Street, behind
Selfridges. Opening Hours:
10am-5pm, Mon-Sat. 2pm-5pm,
Sun. Nearest tube stations: Bond
Street or Baker Street.

Walmer Castle
Deal, Kent

Warship Preservation Trust
HMS Plymouth, Dock Road,
Birkenhead, Merseyside PL41 1DJ

Tel: 0151 650 1573
Fax: 0151 650 1473

Historic warships at Birkenhead –
Falklands War veterans, the
former Royal Navy frigate
'Plymouth' and submarine 'Onyx'
are open to the public daily from
10am. New – German U Boat 534
now open.

Warwickshire Yeomanry Museum
The Court House, Jury Street,
Warwick, Warwickshire CV34
4EW

Tel/fax: 01926 492212

Open Good Friday to end
September, Fridays to Sundays
and Bank Holidays, 10am to 4pm.
Uniforms, medals and militaria of
the regiment since 1794. Entrance
fee.

Waterloo Museum
Crow Hill, Broadstairs, Kent
CT10 1HN

Contact: D P Saunders

Uniforms and weapons, artefacts,
paintings, prints; and over 22,000
model soldiers in dioramas and
singles. All related to Waterloo
campaign.

Weapons Museum
School of Infantry, Warminster,
Wiltshire BA12 0DJ

Tel/fax: 01985 214000

Welch Regiment Museum
Black & Barbican Towers, Cardiff
Castle, Cardiff, Glamorgan (S.)
CF1 2RB, Wales

Tel/fax: 01222 229367

History of the Welch Regiment
(41st and 69th Foot), 1719-1969;
Militia and Volunteers of South
Wales (41st Regimental District),
1794-1908; and the services of the
Royal Regiment of Wales, 1969 to
present day. Opening hours
Thurs/Fri/Sat 10.30am-5.30pm,
Wed 1pm-5.30pm (May 1 to Sept
30): Thurs/Fri/Sat
10.30am-4.30pm, Wed
1.30pm-4.30pm (Oct 1 to Apr 30).

West Highland Museum
Cameron Square, Fort William,
Inverness, Highland PH33 6AJ,
Scotland

Contact: Fiona C Marwick
Tel/fax: 01397 702169

Small collection of 18th and 19th
century weapons used in the
Highlands, some uniforms,
medals etc.

PAINTINGS, PRINTS & POSTCARDS

Abacus
Ukiyo-E, 6 Blandford Road, Reading, Berkshire RG2 8RE

Contact: Mike Brown
Tel/fax: 01734 874946

Importer of Japanese woodblock prints offers for sale a range of classical reprints plus limited edition prints. For copy catalogue send 4 x 1st class stamps.

Alix Baker
Exmoor House, Castle Hill, Brenchley, Kent TN12 7BL

Contact: Alix Baker
Tel/fax: 01892 722866

Postcards, prints, paintings. Many regiments in stock. Internationally collected artist working full-time for British Army units. See advertisement for details in Paintings, Prints & Postcards section.

Battle Scene Pictures
Sherwood House, 14 Norwood Drive, Brierley, Barnsley, Yorkshire (S.) S72 9EG

Contact: Andy Taylor
Tel/fax: 01226 717195

Burland Fine Art UK Ltd
Highgate Street, Hunslet, Leeds, Yorkshire (W.) LS10 1QR

Tel/fax: 0113 2760437

C F Seidler
Stand G12, Grays Antique Mkt, 1-7 Davies Mews, Davies Street, London W1V 1AR

Contact: Christopher Seidler
Tel/fax: 0171 629 2851

American, British, European, Oriental edged weapons, antique firearms, orders and decorations, uniform items; watercolours, prints; regimental histories, army lists; horse furniture; etc. We purchase at competitive prices and will sell on a consignment or commission basis. Valuations for probate and insurance. Does not issue a catalogue but will gladly receive clients' wants lists. Open Mon-Fri, 11.00am-6.00pm. Nearest tube station Bond Street (Central line).

Castita Designs
55a Castle Street, Truro, Cornwall TR1 3AF

Tel/fax: 01872 77413

Cranston Fine Arts
Torwood House, Torwoodhill Road, Helensburgh, Dunbartonshire G84 8LE, Scotland

Contact: David Higgins
Tel: 01436 820269
Fax: 01436 820473

Leading publishers of military prints, over 800 images, from Ancient period to IFOR Bosnia. 5 catalogues available, £7 each. See inside cover for our advert.

David Cartwright
Studio Cae Coch Bach, Rhosgoch, Anglesey, Gwynedd LL66 0AE, Wales

Contact: Sara Cartwright
Tel/fax: 01407 710801

Military artist specializing in Napoleonic and Crimean scenes on canvas. Commissions undertaken. Contact for further details of originals and limited edition prints.

E G Frames
7 Saffron, Amington, Tamworth, Staffordshire B77 4EP

Contact: Tony Cooper
Tel/fax: 01827 63900

All kinds of framing undertaken; we specialise in framing medals and regimental badges. Trade discounts given to regimental associations/ shops, etc. Phone for details.

Fine Detail Art
The Studio, 80 Westgate Street, Gloucester, Gloucestershire GL1 2NZ

Tel/fax: 01452 713057

Flintlock Publishing
10 Westbourne Road, Walsall WS4 2JA

Contact: Rob Chapman
Tel/fax: 01922 644078

Publishers of postcards, prints and exhibition materials from original artwork by leading artists in military and costume history. Commissions/collaborations undertaken.

Frontispiece
Concourse Level, Cabot Place East, Canary Wharf, London E14 4QS

Contact: Reginald Beer
Tel: 0171 363 6336
Fax: 0171 515 1424

Half a million antique prints. British Army campaigns; American Civil War; Napoleonic uniforms. Thousands of naval prints (and some airforce). 24 hour answerphone.

Gallery Militaire
1 Holstock Road, Ilford, Essex IG1 1LG

Contact: Rodney Gander
Tel: 0181 478 8383
Fax: 0181 533 4331

Fine and investment art dealers and publishers, supplying original paintings, limited edition prints, reproductions, plates and postcards. All types of framing and art commissions undertaken. European dealers and distributors for major military artists. Gallery viewing by appointment. Mail order. Large A4 illustrated catalogue £3.00 UK, £4.00 Europe $10.00 USA Airmail.

Geoff White Ltd.
Rushmoor Lane, Blackwell, Bristol, Avon BS19 3JA

Contact: Geoff White
Tel/fax: 01275 462346

Publishers of military postcards. All items are designed to supply reference material for collectors, modellers and other enthusiasts. Send SAE for illustrated catalogue.

George Opperman
Flat 12, 110/112 Bath Road, Cheltenham, Gloucestershire GL35 7JX

Contact: George Opperman

Private collector with large collection to sell. 50,000 lead soldiers, military books and magazines, post and cigarette cards. Send SAE and 3 x 1st class stamps for lists. Mail order only.

Historex Agents
Wellington House, 157 Snargate Street, Dover, Kent CT17 9BZ

Contact: Lynn Sangster
Tel: 01304 206720
Fax: 01304 204528

UK distributors of uniform plates and prints by Rousselot, Le Cimier, Le Hussard du Marais and R.J.Marrion. Napoleonic, Ancien Regime, and American War of Independence and Civil War.

Historic Art Company
Ashleigh House, 236 Wokingham Road, Reading, Berkshire RG6 1JS

Contact: H G Crabtree
Tel: 01734 261236
Fax: 01734 314432

Limited and open edition prints, cards and notelets exclusively featuring the work of Christopher Collingwood. Trade and private commission enquiries welcome. Please call for details.

I K Wren (Postcard Publisher)
14 Elmbridge Road, Cranleigh, Surrey GU6 8NH

Tel/fax: 01483 272551

Modern artist-painted postcards; prints produced by other publishers and regiments are also sold. A finders' service for pre WWI postcards also operates.

Inkpen Art Productions
12 Westdene Crescent, Caversham, Reading, Berkshire RG4 7HD

J C Mummery Military Art
16 Northumberland Crescent, Bedfont, Middlesex TW14 9SZ

Contact: Jack Mummery
Tel/fax: 0181 751 4599

Publisher and dealer in military prints, bronzes, hand coloured engravings, varnish texturing and custom framing. Mail order facility, SAE for lists on request. Personal visitors by arrangement.

John Brindley
29 Hambledon Road, Clanfield, Hampshire PO8 0QU

John Wayne Hopkins
Dept.MD, Gatooma, 58 Queen Victoria Road, Llanelli, Carms SA15 2TH, Wales
Tel/fax: 01554 750761

Artist specialising in 20th century military and aviation paintings in oils. Regularly commissioned by British Army. Rhodesian Fireforce and Army Air Corps limited edition prints available. Commissions accepted. Presently researching Dark Ages and Medieval periods and Battle of Bosworth for a series of Welsh history paintings. SAE for free details.

Librairie Uniformologique Internationale
111 Avenue Victor Hugo, Galerie Argentine, 75116 Paris, France
Contact: Thierry Lecourt

Military Fine Arts
5 Feversham Road, Salisbury, Wiltshire SP1 3PP
Contact: Guy Jennings-Bramly
Tel/fax: 01722 328523

Britain's leading British military art dealer. Access to all British publishers, artists and dealers including thousands of originals. 70 page catalogue £3.50 + A4 SAE; cheques payable to G Jennings-Bramly.

Multiprint
Unit 25, Llanelli Workshops, Trostre Road, Llanelli, Dyfed SA14 9UU, Wales
Tel/fax: 01554 775982

Your favourite military subjects, in full colour on a T shirt. Single figures a speciality, with a minimum order of one. Call and discuss your requirements.

Patrice Courcelle
38 Avenue Des Vallons, B-1410 Waterloo B-1410, Belgium
Contact: Patrice Courcelle
Tel/fax: (0033) 322 354 3607
Email: courcelle@linkline.be

Illustrator and painter. Main periods: American Revolution, French Revolution & Napoleonic wars. Publisher of the plate series 'Ceux Qui Bravaient l'Aigle' and of the 'Waterloo' prints.

Planches Pro-Patria
16 Rue Beaurepaire, 75010 Paris, France
Contact: Georges Courmontagne

The Pompadour Gallery
PO Box 11, Romford, Essex RM7 7HY
Contact: George Newark
Tel/fax: 01375 384020
Email: christopher.newark @virgin.net

Publishers and distributors of military books, postcards and reproduction cigarette cards.

Books: 'Kipling's Soldiers', 'Uniforms of the Foot Guards' and our latest book 'Uniforms of the Royal Marines 1664 to the Present Day'. Send SAE for full details of all our publications.

R I G O / Le Plumet
Louannec, 22700 Perros-Guirrec, France

R M K Fine Prints
13 Keere Street, Lewes, Sussex (E.) BN7 1TY
Contact: Roald Knutsen
Tel/fax: 01273 477959

Carefully researched fine limited edition prints: Japanese Samurai (54 diferent titles), European Medieval, Civil War and Military: colour and line. SAE for catalogue.

Scorpio Gallery
PO Box 475, Hailsham, Sussex (E.) BN27 2SH
Contact: Hilary Hook

Agents for resale of original paintings by leading illustrators, eg Richard Hook, Angus McBride, Gerry Embleton etc.

Thistle Miniatures
Findon Croft, Findon, Aberdeen, Grampian AB1 4RN, Scotland
Tel/fax: 01224 571831

Manufacturer of white metal/resin figures in 1/24th, 1/20th and 1/16th scales. They are mainly Scottish, in ranges: modern barrack dress, The Great War, personalities, and mess dress. Also produce prints depicting the kilts and trews of the British Army. Available by mail or telephone order.

Tony Jackson & Associates
Forge View, Lindsey Tye, Suffolk IP7 6PP
Contact: Tony Jackson
Tel: 01449 741609
Fax: 01449 741805

Publishers of 'Insignia Cards', a classic series of collector's cards featuring elite military and aviation units of WW2, together with their insignia, awards and decorations.

V S-Books
PO Box 20 05 40, D-44635 Herne, Germany
Contact: Carl Schulze
Tel: (0049) 2325 73818
Fax: (0049) 201 421396

Postcards with re-enactment photos in brilliant quality; also postcards for your museum, group or society etc. made to order.

RE-ENACTMENT GROUPS & SOCIETIES

112e Infanterie de Ligne (Nap)
10 Rue de la Paix, B-6200 Gosselies, Belgium

Contact: Eugene Grotard
Tel/fax: (0032) 71340668

116th Panzer Division
33 Peebles Close, Lindley, Huddersfield, Yorkshire (W.) HD3 3WD

Contact: Phillip Cadogan
Tel/fax: 01484 317387

The 116th Panzer Division is one of the major units of the WWII Living History Association, Europe's largest and most authentic WWII re-enactment society.

12th Light Dragoons (Nap)
Shepherds Cottage, Fernhill, Glemsford, Suffolk CO10 7PR

Contact: Martin Render
Tel/fax: 01787 280077

British cavalry regiment of the Napoleonic Association. A mounted battle re-enactment and 'living history' unit, emphasising high standards of authenticity and horsemanship.

1471
Quendon, Manor Road, Barnet, Hertfordshire EN5 2LE

Contact: Clive Bartlett

15th Hussars (Nap)
Rose Cottage, Caledonia, Winlanton, Tyne & Wear NE21 6AX

Contact: Neil Leonard

Group recreating the 15th (King's) Light Dragoons (Hussars) of the Napoleonic era. Mounted and dismounted displays.

1680s Re-enactments
61 Belmont Lane, Stanmore, Middlesex HA7 2PU

Contact: Philip Nickson

Covers the Monmouth Rebellion, the Glorious Revolution, and the First Jacobite Rebellion. Re-enactments comprise skirmishes, battles, drills, and 'living history'.

17th Century Living History Heritage Ctr
25 Connaught Street, Northampton, Northamptonshire NN1 3BP

Contact: Mrs J H Thompson

18e Infanterie de Ligne (Nap)
Dobereiner Strasse 12, 07745 Jena, Germany

Contact: Rolf Peter Graf

18th Missouri Infantry
98 Winstanley Road, Wellingborough, Northamptonshire NN8 1JF

Contact: John Hopper
Tel: 01933 442213
Fax: 01933 273318
Email: kshelt@globalnet.co.uk

The 18th Missouri Infantry re-enactment group portray soldiers of the Union Army of Tennessee 1864. Part of the Southern Skirmish Association. Family oriented group.

1ere Armée Francaise (WWII)
21 Rue du Bois Saint-Maur, 77120 Samoreau, France

Contact: Pierre Cornebise

1ere Tirailleurs-Grenadiers (Nap)
9 Rue du Chemin Vert, 91360 Epinay sur Orge, France

Contact: Christian Scoupe
Tel/fax: (0033) 169090735

The society re-enacts a section of this Young Guard regiment. Members participate in all European events of quality, on Napoleonic themes.

1st Arkansas Volunteer Infantry
136 Snakes Lane, Woodford Green, Essex IG8 7HZ

Contact: Michael Freeman
Tel/fax: 0181 505 0662

1st Foot (Royal Scots)
57 Broomhall Drive, Corstorphine, Edinburgh, Lothian EH12 7QJ, Scotland

Contact: David Bullions

1st Foot Guards
6 Arden Avenue, Brounstone, Leicestershire

1st Royal Dragoons
Upper Woodhouse Farm, Holmbridge, Yorkshire (W.) HD7 1QR

Contact: Mike Grove
Tel/fax: 01484 685615

The heavy cavalry unit of the Napoleonic Association's British division. Representing The Royals from 1797 to 1897, with emphasis on the Peninsular War period.

20e BCA (WWII)
15 Impasse Remi Cocheme, 51100 Reims, France

Contact: Jean Paul Lebailly
Tel/fax: (0033) 26361672

21e Infanterie de Ligne (Nap)
22 Swallow Street, Oldham, Lancashire OL8 4LD

Contact: Christopher Durkin
Tel/fax: 0161 652 1647

A regiment of the Napoleonic Association, recruiting nationwide, dedicated to the study and recreation of this French unit during the Napoleonic era.

22e Demi-Brigade de Ligne
Memmelsdorfer Str 102, 96 052 Bamberg, Germany

Contact: Hans-Karl Weiss
Tel/fax: 0049 95133458

The group portrays French line infantry 200 years ago and has a special interest in bicentennials. Approximately 50 members in Belgium, England, Germany and Italy.

23e Chasseurs (Nap)
15 Sentier Desire, 94350 Visire sur Marne, France

Contact: Daniel Hubert
Tel/fax: (0033) 149302225

23e Dragons (Nap)
53bis Rue Claude Terrasse, 75016 Paris, France

Contact: Joel Levieux
Tel/fax: (0033) 145244819

23rd Regiment of Foot
7 Arcadia Way, Trevethin, Pontypool, Gwent NP4 8DX, Wales

Contact: Allan Jones
Tel/fax: 01495 750479

Recreates Light Company of the Royal Welch Fusiliers, Napoleonic period.

24th Foot Grenadier Company
1 Cheviot Close, Risca, Gwent NP1 6RH, Wales

Contact: Colin Pearce

29th Division French Historical Association
9 Clos des Saules, 77860 Roissy en Brie, France

Contact: Jean Pierre Rigot

Promotion and preservation of the 29th US Infantry Division's history during WWII; ceremonies at memorials, re-enactments.

2e Hussards (Nap)
Route de Paris, 77171 Sourdun, France

Tel/fax: (0033) 64001955

2nd (Queen's) Regiment of Foot
18 Lilac Close, Bellfields Estate, Guildford, Surrey GU1 1PB

Contact: G. Brown
Tel/fax: 01483 574455

Formed in 1992 to recreate the early Peninsula period; families welcome.

2nd US Artillery
"Thickets", Odstock, Salisbury, Wiltshire SP5 4JE

Contact: Michael Boyd Camps
Tel/fax: 01722 329712

36th Regiment of Foote (1740-1760)
33 Battle Road, Tewkesbury Park, Tewkesbury, Gloucestershire GL20 5TZ

Recreating a British line regiment of the mid-Eighteenth century.

3rd Btn., 1st Foot Guards (Nap)
Waterloo Museum, Crow Hill, Broadstairs, Kent CT10 1HN

Contact: D P Saunders

Foot Guards re-enactment unit (1812-1816).

3rd New York Regiment
19 Treharne Road, Barry, Glamorgan (S.) CF6 7QY, Wales

Contact: Michael Bailey
Tel/fax: 01446 745936 (pm)

Recreation of the times of Captain Bruyn's Company of the Regiment. Emphasis on 'living history' displays including tactical demonstrations. The group is associated to The Society of the American Revolution and The Eagle Society.

40th Regiment of Foot
37 Lee Avenue, North Springvale, Melbourne, Victoria 3171, Australia

Contact: Terence Young

42nd Royal Highland Regiment
21 Lilian Gardens, Woodford Green, Essex IG8 7DN

Contact: Walker
Tel/fax: 0181 5053842

Formed some 15 years ago to recreate the Black Watch of the Napoleonic period.

45e Infanterie de Ligne (Nap)
29 Bayley Court, Winnersh, Wokingham, Berkshire RG11 5HT

Contact: David Prior
Tel/fax: 01734 772696

Napoleonic French Fusiliers group portraying c 1808-09 troops and their ladies; 'living history' camps and battle re-enactments; recruiting nationally.

45th (1st Notts) Regiment of Foot
62 Fairway, Keyworth, Nottingham, Nottinghamshire NG12 5DU

Contact: D A Kemp
Tel/fax: 0115 937 5232

Based on a Grenadier flank company, we put together living history displays and cameos using period tents and equipment.

46e Infanterie de Ligne (Nap)
Bower House, Park Lane, Sulgrave, Oxfordshire OX17 2RX

Contact: Mike Crawshaw
Tel: 01295 760747
Fax: 01295 760705

Represents the Grenadier Company, 1st Battalion, and attached Regimental Artillery; uniformed according to 1793 regulations.

47th Regiment of Foot (1770s)
189 Cowan Way, Widnes, Cheshire WA8 9BW

Contact: Andrew Cowley

5 Westfalisches Landwehr-Regt (NAP)
98 Winstanley Road, Wellingborough, Northamptonshire NN8 1JF

Contact: John Hopper
Tel: 01933 442213
Fax: 01933 273318
Email: kshelt@globalnet.co.uk

English contingent of a German based re-enactment unit of the highest quality. We operate in Europe, at least once per year. Recruiting for Leipzig 1998.

5 Westfalisches Landwehr-Regt (NAP)
Am Wiesgraben 2, D-69190 Walldorf, Germany

Contact: Dietrich Pott
Tel/fax: (0049) 62274859

'Living history' group depicting German militia of the Napoleonic period. High standards required.

55th Virginia Regiment
Flat 11, Grove Lodge, Crescent Grove, London SW4 7AE

Contact: Richard O'Sullivan
Tel/fax: 0171 622 4109

Exacting American Civil War 'living history' and drill display unit requires recruits. Must be over 15 years old, and have £500 for uniform and rifle.

68th Display Team (Nap)
213 Bishopton Road West, Fairfield, Stockton, Cleveland TS19 7LU

Contact: Tony Parker
Tel/fax: 01642 580868

Group re-enacting drill and tactics of the 68th (Durham) Light Infantry of the Napoleonic Wars, to high standards of turn-out and dedication. 'Living history' displays by male and female membership; research activities.

6e Infanterie Légère
H Hauschildt-Strasse 26, 04445 Liebertwolkwitz, Germany

Contact: Wolf-Dieter Schmidt

71st Highland Light Infantry
24 Highcroft Green, Parkwood, Maidstone, Kent ME15 9PN

Contact: M. Foreman
Tel/fax: 01622 763362

Recreates the regiment in the period 1810-1815.

7th Foot (Royal Fusiliers) (Nap)
55 Glencoe Road, Chatham, Kent ME4 5QD

Contact: Bob Willingham

916th Grenadier Regiment (WWII LHA)
86 Meredith Road, East Worthing, Sussex (W.) BN14 8EE

Contact: Tony Dudman
Tel: 01903 207514
Fax: 01903 725379

Unit portraying the ordinary WWII German infantryman of the North-West Europe Campaign, June 1944-May 1945. Politics play no part in our activities.

93rd Sutherland Highland Regiment of Foot Living History Unit
PO Box 100011, Fort Worth, Texas 76185, USA

Contact: Benton Jennings
Tel: (001) 817 927 7412
Fax: (001) 817 927 2454
Email: ninety3rd@aol.com

A leading Napoleonic re-enactment group in the USA, recruiting nationally. Some recreation of Crimean and Mutiny periods. Member, Napoleonic Association, VMS, North America British Brigade.

95th Rifles Regiment of Foot Living History Society
Grasshoppers, 42 Nunnery street, Castle Hedingham, Halstead, Essex C09 3DW

Contact: Les Handscombe
Tel/fax: 01787 461433

As members of the Napoleonic Association we recreate this elite regiment with period camp, drill and battle displays to a very high standard.

9e Demi-Brigade d'Infanterie Légère
11 Loring Road, London N20 0EJ

Contact: Martin Lancaster
Tel: 0181 361 6225
Fax: 0181 446 9943
Email: 101332.567 @compuserve.com

Marches, camps, displays, battle tactics and films; regimental membership of the bi-centennial 27e Division Militaire; affiliations with many quality Napoleonic groups. Enquiries welcome.

9th Foot (E Norfolk Regt)
Plum Tree Cottage, Winslow Road, Granborough, Buckinghamshire MK18 3NJ

Contact: Keith Phillips
Tel/fax: 01296 670324
Email: 100560.3544 @compuserve.com

Recreates a centre company during the Peninsula period. Re-enacts drill, camp life and battles. Part of Napoleonic Association. New recruits always welcome – contact us today.

A D 500
52 Harley Terrace, Gosforth, Tyne & Wear NE3 1UL

Contact: Chris Halewood
Tel/fax: 0191 284 7865

A D 95
95 Brindle Heath Road, Hednesford, Cannock, Staffordshire WS12 4DR

Contact: Alan Orton
Tel/fax: 01543 422084

A G "Live" (ACW)
Visier Magazine, Olgastrasse 86, 7000 Stuttgart, Germany

Contact: Dr David Schiller

Aasvogeli
Imladris, 224 Coatham Road, Redcar, Cleveland TS10 1RA

Contact: J Watson
Tel/fax: 01642 489227

Celtic society depicting lifestyle and combat style of the Celtic races from 500BC to 1600AD. Member of DELHC.

Acorns of Albion (Dark Ages)
39 Penistone Road, Park End, Middlesbrough, Cleveland TS3 ODJ

Adventurers Guild
53 Nelson Road, Basildon, Essex SS14 5QQ

Tel/fax: 01268 284806

Aescesdun Lethang (N F P S)
36 Ferndale Road, Gorsehill, Swindon, Wiltshire SN2 1EX

Contact: Mark Browne
Tel/fax: 01793 525626

The Age of Penda
1 Churchfield Avenue, Tipton, Midlands (W.) DY4 9NF

Contact: Andy Colley
Tel/fax: 0121 522 4405

Dark Age and Medieval combat re-enactment group. New troops and camp followers always welcomed.

Airlanding Battalion (WWII)
18 Rue Clovis Cappon, 76190 Yvetot 76190, France

Contact: Bruno Bourgine
Tel/fax: (0033) 35954759

American Civil War Society
PO Box 52, Brighouse, Yorkshire (W.) HD6 1JQ

Contact: Chris Wood
Tel/fax: 01625 265226

For 'living history' and re-enactment country-wide.

Amounderness Herred (N F P S)
26 Ashton Road, Blackpool,
Lancashire FY1 4QA

Contact: Jim Gillbanks
Tel/fax: 01243 294169

An Dal Cuinn Clan
Clan Resource Centre, B4D,
Balbutcher Lane, Dublin 11,
Co.Dublin, Ireland

Contact: Eric O'Cuinn

Ancient Arts Fellowship
21 Horsley Crescent, Melba,
ACT 2615, Australia

Contact: Jeffrey Davidson
Tel: (0061) 2 62587657
Fax: (0061) 2 62644488
Email: arioch@citrax.net.au

AAF is Canberra's medieval
re-enactment group. AAF has a
fighting arm (Dragonhead)
covering Anglo-Saxons, Vikings
and Normans, and Anglo-Saxon
'living history' display.

Anmod Dracan
Imladris, 224 Coatham Road,
Redcar, Cleveland TS10 1RA

Contact: J Watson
Tel/fax: 01642 489227

'Living history' society depicting
civilian and military lifestyle of
Viking Age England. Combat
demonstrations and many craft
demonstrations. Member of
DELHC.

Archers of the Black Prince
3 Bryn Castell, Caergwrle,
Wrexham, Clwyd LL2 9HB,
Wales

Contact: Paul Harston
Tel/fax: 01978 762313

Arms and Archery
The Coach House, London Road,
Ware, Hertfordshire SG12 9QU

Contact: Terry Goulden
Tel: 01920 460335
Fax: 01920 461044

Quality armour and weaponry,
chainmail: string metal plated or
real metal rings. Always in stock.

Arthurian Society
5 Manor Street, Hinckley,
Leicestershire LE10 0AS

Artilleurs de Marine (Nap)
19 Rue Tonerre-Bailly, 62470
Calonne-Ricouart, France

Contact: Frederic Gilliot

Ask (Dark Ages)
Lyngballevej 9, Herskind Hede,
DK-8464 Galten, Denmark

Contact: Oleg Zacharov
Tel/fax: (0045) 86954184

**Assoc Britannique de la
Garde Impériale**
8 Pippins Green Avenue,
Kirkhamgate, Wakefield,
Yorkshire (W.) WF2 0RX

Contact: Derek Mellard
Tel/fax: 01924 381820

**Assoc Napoleonienne du
Boulonnais**
97 Rue Louis Duflos, 62200
Boulogne sur Mer, France

Contact: Michel Lamesh
Tel/fax: (0033) 21804966

Association 1944
35 Avenue Paul Deroulede,
94300 Vincennes, France

Tel/fax: (0033) 1436508

**Association Big Red One
(WWII)**
44 Rue de Bretagne, 76600 Le
Havre, France

Tel/fax: (0033) 35430374

Association Corbineau (Nap)
2 Rue du 8 Mai 1945, 59215
Abscon, France

Contact: Regis Surmont
Tel/fax: (0033) 27362046

Association D-Day
52 Avenue du General Leclerc,
77320 La Ferte Gaucher, France

Contact: Gerard Signac
Tel/fax: (0033) 64033654

**Association of Crown Forces
1776**
Kennel House, 150 Capel Street,
Capel le Ferne, Folkestone, Kent
CT18 7HA

Contact: Alan Haselup

Coldstream Guards and 4th
Battalion Royal Artillery,
demonstrating drill and firing of
muskets and cannon, and
equipment used by soldiers in
American War of Independence.
Contact Membership Secretary at
above address.

**Austrian Napoleonic
Re-Enactment Grp**
Brugger Strasse 71, 6973
Hochst/Vorarlberg, Austria

Contact: Helmut Huber

Avalon
19 Windsor Road, Normanby,
Middlesborough, Cleveland TS6
0RB

Contact: Neil Tranter
Tel/fax: 01642 468511

The Baggage Trayne
24 Pye Croft, Bradley Stoke
Worth, Bristol, Avon BS12 0ED

Contact: Martin Braine

Balden Villingen Garde (Nap)
Rietstrasse 38, 7730 VS Villingen,
Germany

Contact: Wolfgang Kunle

Bangor Herred (N F P S)
12 Cefnfaes Street, Bethesda,
Bangor, Gwynedd LL57 3BW,
Wales

Contact: Russell Scott

Barn Door Bowmen
7 Cedar Court, Totteridge Road,
High Wycombe,
Buckinghamshire HP13 6DN

Contact: Graham Povey
Tel/fax: 01494 526620

**Bataillon des Marins de la
Garde (Nap)**
19 Rue de la Gilette, B-5646
Stave, Belgium

Contact: Jean Gerard
Tel: (0032) 71728375
Fax: (0032) 71887813

Group recreating Sailors of the
French Imperial Guard of the
Napoleonic period, including
officers, drum major, standard
bearer, petty officers, sailors, and
a Cantiniere; campaign dress and
parade uniforms represented.

Battalion Washington Artillery
c/o 95 Dalby Road, Melton
Mowbray, Leicestershire LE13
0BQ

Contact: Raymond Clowes
Tel/fax: 01664 61517

ACW – Confederate 3-gun
battery. Available for
re-enactments, displays, fetes;
c/w authentic encampment.
Suppliers of flags, uniforms etc.
Please send SAE for further
details.

Beaufo Household
65 Spencer Avenue, Yarnton,
Oxford, Oxfordshire OX5 1NQ

Contact: Penny Roberts
Tel: 01865 378140
Fax: 01865 272821
Email: pmr@bodley.ox.ac.uk

A small local group representing
the household and combatants of
an Oxfordshire family; mainly
15th Century but also some earlier
events.

Bills and Bows
23 Overhill Road, Burntwood,
Walsall, Midlands (W.)

Contact: Richard Dunk
Tel/fax: 01543 684726

A group specialising in the
portrayal of military and civilian
life in the 16th century.

**Black Knight Medieval
Entertainers**
c/o 47 Burton Street, Melton
Mowbray, Leicestershire LE13
1AF

Tel/fax: 01664 501010

The Black Legion
14 Keep Hill Drive, High
Wycombe, Buckinghamshire
HP11 1DU

Tel/fax: 01494 440277

**Blood Eagles (Blod Paa
Ryggen)**
34 Cambria Street, Bolton,
Lancashire

Contact: D A Parlett
Tel/fax: 01204 665148

Boston Svieter
63 Norfolk Street, Boston,
Lincolnshire

Contact: Steve Dobbs
Tel/fax: 01205 65055

The Bowmen of Woodland
11 Dovecroft, New Ollerton,
Newark, Nottinghamshire NG22
9RG

Contact: Stephen Button

**Bractwo Rycerskie Zamku
Gniewskiego**
Muzeum Archeologiczne, Ul.
Mriacka 25/26, 80-958 Gdan'sk,
Poland

Contact: Barbara Gostyn'ska

Brandenburger Artillerie
Kurz Strasse 2, 1000 Berlin 28,
Germany

Contact: Freia Cuppers

**Braunschweigisches
Feldcorps (Nap)**
Freiherr vom Stein 22, 55774
Baumholder 55774, Germany

Contact: Daniel S Peterson

Brigantia
44 Iron Mill Close, Fareham,
Hampshire PO15 6LB

Contact: Mrs J Smith
Tel/fax: 01329 42055

Bristol Militia (Med)
9A Gloucester Street, Eastville,
Bristol, Avon BS5 6QF

Contact: Brian Hicks
Tel/fax: 0117 9519120

Britannia (Arthurian)
13 Ardleigh, Basildon, Essex
SS16 5RF

Contact: Dan Shadrake
Tel/fax: 01268 544511

Re-enactment 'living history'
society covering the later Roman
& Arthurian era. Specialists in
recreating pre-gunpowder era
fight scenes for film and television.

Brotherhood of Noble Knights
2 Cynthia Close, Parkstone,
Poole, Dorset
Contact: Ross Pacifico
Tel/fax: 01202 715368

Buckingham Household
14 Keep Hill Drive, High
Wycombe, Buckinghamshire
HP11 1DU
Tel/fax: 01494 440277

Burgundian Mercenaries
20 Melrose Road, Pitsmoor,
Sheffield, Yorkshire (S.) S3 9DN
Contact: Adrian Elliston
Tel/fax: 0114 2738542

Cambridge Lethang (N F P S)
3 Benson Street, Cambridge,
Cambridgeshire
Contact: Paul Murphy

Camelot Medieval Society
33 Claret Gardens, South
Norwood, London SE25 6RP
Contact: Steve Hoey
Tel/fax: 0181 653 6606

**Celtic Tribes of the Rhine
Valley**
Durkheimer Strasse 1a, 67549
Worms, Rhein, Germany
Contact: Peter Hoch

**The Celts (Oxford Iron Age
Society)**
39 Old Road, Wheatley,
Shotover, Oxfordshire OX33
1NV
Contact: Clive Shorter
Tel/fax: 01865 875648

**Centre Equestre d'Attelage
(Nap)**
Domaine de la Roussie, 33140
Cadaujac, France
Contact: M Castelot

Cestria (Med)
62 Cheyney Road, Chester,
Cheshire CH1 4BT
Contact: Dan Osbaldeston
Tel/fax: 01244 3712694

Chapter of St Bartholomew
20 Melrose Road, Pitsmoor,
Sheffield, Yorkshire (S.) S3 9DN
Tel/fax: 0114 2738542

**Chasseurs à cheval de la
Garde (Nap)**
(10 Escadron), 48 Rue Chapon,
75003 Paris, France
Contact: Michel Fourrey
Tel/fax: (0033) 146662549

**Chasseurs à pied de la Garde
(Nap)**
8 Rue de l'ancien hopital, Le Sans
Souci, 21200
Gevrey-Chambertin, France
Contact: Vincent Bourgeot
Tel/fax: (0033) 80343707

Chasseurs Alpins (WWII)
Mandrion, 73100 Trevighin,
France
Contact: Laurent Demouzon

Chester City Guard
62 Cheyne Road, Chester,
Cheshire CH1 4BT
Contact: Dan Osbaldeston
Tel/fax: 01244 372694

**Les Chevaliers d'Ile de France
(Med)**
25 Rue Roland Garros, 78140
Velizy Villacoublay, France
Contact: Daniel Georges

**Les Chevaliers de
Franche-Comte (Med)**
4 Chemin des Chalettes, 39150
Morez, France
Contact: Jean-Claude Crotti

Chevaliers de Guise
46 St John's Court, Princess
Crescent, Finsbury Park, London
N4 2HL
Contact: Philip Burthem
Tel/fax: 0181 800 4013

Circa 1265 / Chivalry in Action
104 Clark Road,
Wolverhampton, Midlands (W.)
WV3 9PB
Contact: A Westmancoate
Tel/fax: 01902 26879

Clan Mackenna
61 Bexley Drive, Normanby,
Middlesbrough, Cleveland TS6
0ST
Contact: Neil Tranter
Tel/fax: 01642 456295

Highlander re-enactment: battles
/ 'living history' and museum
work undertaken. Authentic kit
and equipment used.

Clan Scottish Heritage Society
6 Guthrie Place, East Mains, East
Kilbride, Strathclyde G74 4EX,
Scotland
Contact: Jan Murdoch
Tel/fax: 013552 41048

**Clan Wallace/ Highlander
Images**
Clan Studios, 1 Shearer Street,
Kingston, Glasgow, Strathclyde
G5 8TA, Scotland
Contact: Seoras Wallace
Tel: 0141 429 6915
Fax: 0141 429 6968

The Wallace Clan Trust is a
modern clan system that can trace
its indigenous roots back to
300AD and is a Registered Charity.

Clarence Household
26 Halifax Old Road, Birkby,
Huddersfield, Yorkshire (W.)
HD1 6EE
Contact: Lindy Pickard
Tel/fax: 01484 512968

Cohors I Batavorum
69 The Warren, Hardingstone,
Northamptonshire NN4 0EP
Contact: Leonard G Morgan
Tel/fax: 01604 763136

Cohors V Gallorum
Arbeia Roman Fort and Museum,
Baring Street, South Shields,
Tyne & Wear NE33 2BB
Contact: Alexandra Croom
Tel/fax: 0191 4544093

Early 3rd century Roman
auxiliary and civilian
re-enactment and research society.

**Col Edward Montagu's
Regiment of Foot**
92 Station Road, Kippax, Leeds,
Yorkshire (W.) LS25 7LQ
Contact: Simon Lister

**Col George Fenwick's
Regiment of Foot**
7 Aintree Drive, Chippenham,
Wiltshire SN15 0FA
Contact: Keith Champion
Tel/fax: 01249 65054

**Col George Monck's
Regiment of Foot**
The Civil Wardrobe, Newtown
Road, Newbury, Berkshire
RG14 7ER
Contact: John Litchfield
Tel/fax: 01635 43806

Re-enactors of the early history of
the Coldstream Regiment of Foot
Guards, 1650-1815. English Civil
Wars. We seek recruits. We
provide training in period drill.

**Col John Bright's Regiment of
Foot**
219 Greenhill Avenue, Sheffield,
Yorkshire (S.) S8 7TJ
Contact: Kevin Fisher
Tel/fax: 01742 748922

**Col John Foxe's Regiment of
Foot**
Fountains Hall, Fountains Abbey
Estate, Ripon, Yorkshire (N.)
HG4 3DZ
Contact: Mark Newman

**Col Nicholas Devereux's
Regiment of Foot**
382 Woodend Road,
Wednesfield, Wolverhampton,
Midlands (W.) WV11 1YD
Contact: Chris Sheldon
Tel/fax: 01902 839261

A regiment of the Roundhead
Association within the English
Civil War Society based on the
New Model Army – for further
details contact the above.

**Col Ralph Weldon's Regiment
of Foot**
19 Charles Street, Colchester,
Essex CO1 2BL
Contact: Steve Petrie
Tel/fax: 01206 798846

**Col Robert Overton's
Regiment of Foot**
Broadview, Sowerby Bridge
Road, Sowerby Bridge, Yorkshire
(W.) HX6 1LQ
Contact: Steve Foster
Tel/fax: 01422 836420

Col Samuel Jones's Regiment
18 Saxel Close, Aston, Bampton,
Oxfordshire OX8 2EB
Contact: D Jones

Unit of Sealed Knot Society.

**Col Valentine Walton's
Regiment of Foot**
6 Weller Close, Amersham,
Buckinghamshire HP6 6LJ
Contact: Ian Tindle
Tel/fax: 01494 727451

**Col William Sydenham's
Regiment of Foot**
70B Kensington Gardens Square,
London W2 4DG
Contact: Steve Baker
Tel/fax: 0171 727 0035

**Colchester Historical
Enactment Society**
Newgrange, Cannon Road,
Colchester, Essex CO1 2EW
Contact: Margaret Kay
Tel/fax: 01206 791147

Colchester Roman Society
9 Glisson Square, Colchester,
Essex CO2 9AJ
Contact: Zane Green
Tel/fax: 01206 561276

Coldstream Regiment of Foot Guards
The Civil Wardrobe, Newtown Road, Newbury, Berkshire RG14 7ER

Contact: John Litchfield
Tel/fax: 01635 43806

Re-enactors of the early history of the Coldstream Regiment of Foot Guards, 1650-1815. Culloden and Waterloo Campaigns. We seek recruits. We provide training in period drill.

Colquhoun's Artillery Company
17 Rose Avenue, Horsforth, Leeds, Yorkshire (W.) LS18 4QE

Contact: Graham Hartley

Commission of Array (1257-1327)
8 Kingswick Drive, Sunninghill, Berkshire SL5 7BQ

Contact: Bruce MacLellan

Compagnie de la Rose
5 rue des Granges, Estavayer le Lac CH 1470, Switzerland

Contact: Katia Laurent

Compagnie Médiévale Mac'htiern
Ancienne Ecole, 35190 St Thual, France

Contact: Eric Magnini

Les Compagnons de Gabriel (Med)
Chateau de Murols, 12600 Murols, France

Companie of Knights Batchelor
PO Box 100, Albert Street, Brisbane, Queensland 4002, Australia

Contact: Terry Fitzsimmons

Companions of the Black Bear
60 Upper Park Street, Cheltenham, Gloucestershire GL52 6SA

Contact: David Cubbage
Tel/fax: 01242 510289

Companions of the Black Prince
c/o Tutbury Castle, Tutbury, Staffordshire DE13 9JF

Contact: Barrie Vallans
Tel/fax: 01283 812129

Company Ecorcheur
8 Burnlea Grove, West Heath, Birmingham, Midlands (W.) B31 3LT

Contact: Andrea Vance
Tel/fax: 0121 628 3127

We provide an authentic and entertaining portrayal of 15th century military life including foot combat, pole-arm drill and role plays, within an accurate period encampment.

Company of Chivalry
9a Gloucester Street, Eastville, Bristol, Avon BS5 6QF

Contact: Brian Hicks
Tel/fax: 0117 9519120

Company of Knights
Durendart House, Coles Lane, Capel, Surrey RH5 5HU

Tel/fax: 01306 711141

Company of Ordinance
12 Cape Avenue, Stafford, Staffordshire ST17 9FL

Contact: Keith Piggott
Tel/fax: 01785 227532

15th century company of archers, hand-gunners. Tudor Yeoman of the Guard. 'Living history' 13th to 16th centuries. Museum; historical interpretation, craft demonstrations; Tudor players.

Company of St Edmund
Wolffe Hall, 410 Beccles Road, Lowestoft, Suffolk NR33 8HL

Contact: David Pye
Tel/fax: 01502 500435

Company of St George
4 Rue Coulon, CH-2000 Neuchatel, Switzerland

Contact: John Howe
Tel/fax: (0041) 24711775

Late 15th century artillery company, also halberdiers, archers, general 'living history' activities. High standard of authenticity and discipline; multi-national membership; events in Europe. British contact: Victor Shreeve, Vern Path, Melplash, Bridport, Dorset, DT6 3UD.

Company of the Black Rose
3 New Lane, East Morden, Wareham, Dorset BH20 7DN

Contact: Eduard Duval
Tel/fax: 01929 45310

Company of the Falcon
73 East Holton, Holton Heath, Poole, Dorset BH16 6JN

Contact: Sue Cross
Tel/fax: 01202 625213

Company of the Golden Phoenix
73 Kinson Park Road, Kinson, Bournemouth, Dorset BH10 7HG

Contact: Andre de Flaville

Company of the Hawk
412 Foxhill Road, Foxhill, Sheffield, Yorkshire (S.) S6 1BP

Contact: Andrew Jarvis
Tel/fax: 0114 2344359

Company of the White Boar
10 Eastern Road, Sutton Coldfield, Midlands (W.) B73 5NU

Contact: Joanne Marriott
Tel/fax: 0121 355 3873

Authentic 15th century battle re-enactment group. Performs 'living history' display, scenarios and school talks. Bookings and new members welcome: please send SAE for further details.

Company of the White Wolf
30 Payne Road, Wooton, Bedfordshire MK43 9JL

Contact: Suzanne Tibbenham
Tel/fax: 01234 766808

Confederate & Union Re-enactment Section
28 Warkworth Avenue, Whitley Bay, Tyne & Wear NE26 3PS

Contact: E J Hindle

Confederate High Command
29 Grierson Close, Calne, Wiltshire SN11 8JJ

Confederation of Independent ACW Re-enactors
10 West Parade, Sea Mills, Bristol BS9 2JU

Contact: J. Lane

Conisbrough Castle Garrison (Med)
33 Sherburn Avenue, Billingham, Cleveland TS23 3PX

Contact: David O'Brien
Tel/fax: 01642 565977

Conroi de Burm Lethang
77 Phillimore Drive, Saltley, Birmingham, Midlands (W.) B8 1PT

Contact: Rob Roberts

The Constables
345 Spring Bank West, Hull, Humberside (N.) HU3 1LD

Tel/fax: 01482 448510

Corridors of Time Ltd
11 Mulberry Court, Pagham, Sussex (W.) PO21 4TP

Contact: Alan Jeffery
Tel: 01243 262291
Fax: 01243 266721

Create the Mood
Redcote, 228 Sydenham Road, Croyden CR0 2EB

Contact: Frances E Tucker
Tel/fax: 0181 684 1095

Music and costume from the past. We work with you to bring your historical project alive – Norman to 1930, one strolling minstrel to households of famous characters.

Crimdon Sword & Shield Society
58 Redworth Road, Hillingham, Cleveland TS23 3FI

Contact: S D Henderson

Based in late 13th century, we try to recreate a medieval tournament. Combat is hand to hand.

Croix du Nord
11 Dunkeld Street, Lancaster, Lancashire LA1 3DQ

Contact: Neil Harrison
Tel/fax: 01524 35869

Cwmmwd Ial (N F P S)
26 Windsor Avenue, Cargwrle, Clwyd, Wales

Contact: Ian Grant
Tel/fax: 01978 760795

Cylch O Ddur (Med)
2 Rock Villa, Criccieth, Gwynedd
LL52 0ED, Wales

Contact: Roger Clark
Tel/fax: 01766 522089

13th Century group re-enacting
the rising of Madog Ap
Llewelwyn, from the sacking of
Caernarfon to the Battle of
Aberconwy, 1294-95.

**Czechoslovak ACW
Association**
Macha Mosnova 2, 1500 Praha 5
Smichov, Czech Republic

Contact: Mr Zbynek

**D Company, 505 PIR, 82nd
Airborne Division**
18 Carterville Close, Marton,
Blackpool, Lancashire FY4 5BD

Contact: Alan Smith
Tel/fax: 01253 762819

A unit of the WW2 Living History
Association Limited. Dedicated to
faithful re-creations of combat,
parade and training at major
events in Britain and Europe.

D D R Collective
67 West End Road, Mortimer,
Reading, Berkshire RG7 3TQ

Contact: Andy Limming
Tel/fax: 01256 64415

Formed in 1992; East German
uniforms and vehicles, military
and civil. Members needed for
guard display or uniform
displays; male and female
welcome. Mobile tel. 01850 507202.

**Dahrg De Belne Lethang
(N F P S)**
43 Croft Road, Yardley,
Birmingham, Midlands (W.) BS26
1SQ

Contact: John Sheard
Tel/fax: 0121 784 6408

Danegeld
Ground Floor Flat, 9 Markham
Road, Bournemouth, Dorset
BH9 1HY

Tel/fax: 01202 54673

Danelaw Dark Age Society
109 Tweendikes Road, Sutton on
Hull, Humberside (N.) HU7 4XJ

Contact: Andrew Wiles
Tel/fax: 01482 826617

Danelaw Mercenaries
56 Hibbert Street, Luton,
Bedfordshire LU1 3UX

Contact: Jim Sizer
Tel/fax: 01582 458109

Dark Ages Society
20 Highwood Close, Shaw,
Newbury, Berkshire RG13 2EJ

Contact: Rosanna Day
Tel/fax: 01635 32447 (pm)

Recreation of the late 9th century
– the time of Alfred the Great and
the Viking invasions. Small,
independent group concentrating
on accurate recreation of the
period, with a balance between
warfare and everyday living.

De Thorpe Retinue
6 Park Terrace, Chippenham,
Wiltshire SN15 1NU

Contact: Craig Nicholas
Tel/fax: 01249 657434

Diehard Company (V M S)
21 Addison Way, North Bersted,
Bognor Regis, Sussex (W.) PO22
9HY

Contact: Tim Rose
Tel/fax: 01243 860036

Part of the Victorian Military
Society, we re-create the
Middlesex Regiment of 1883 in
full home service dress, armed
with Martini Henry rifles.
Nationwide arena shows, period
drill displays, musketry drill, and
authentic 'living history' tent lines.

Dobunni (Celtic Iron Age)
20 Mendip View, Wick, Bristol,
Avon BS15 5PY

Contact: Steve Harrill-Morris
Tel/fax: 0117 9374059

Dorset Levy
8 New Street, Puddletown,
Dorchester, Dorset DT2 8SF

Contact: Jason O'Keefe
Tel/fax: 01305 848148

Double Time
1 Golden Noble Hill, Colchester,
Essex CO1 2AG

Contact: Alison Rodham
Tel/fax: 01206 768517

Dragon Svieter
39b Seamoor Road, Westbourne,
Bournemouth, Dorset BH4 9AE

Contact: Peter Power
Tel/fax: 01202 751729

**Dragon's Eye Living History
Consortium**
Imladris, 224 Coatham Road,
Redcar, Cleveland TS10 1RA

Contact: John Watson
Tel/fax: 01642 489227

Research and parent organisation
to several active re-enactment
groups listed separately –
membership indicated by letters
DELHC in descriptions.

The Dragonlords
132 Spen Lane, West Park,
Leeds, Yorkshire (S.) LS6 3NA

Tel/fax: 0113 2784536

Dragoon/Anvil (WWII)
118 avenue Ziem, 06800 Cagnes
sur Mer, France

**Drayton Bassett Medieval
Society**
27 Tynings Lane, Aldridge,
Walsall, Midlands (W.) WS9 0AS

Contact: Kate Hayman

**Dryhtenfyrd Re-enactment
Society Inc**
45A St George's Crescent,
Faulconbridge, New South Wales
2776, Australia

Contact: Michael Mills

Dunholm Lethang
8 Carrs Way, Carville,
Co.Durham DH1 1BB

Contact: Rachel Lowerson
Tel/fax: 0191 386 6758

Durham 17th Century Society
c/o 24 Tollhouse Road,
Crossgate Moor, Durham City,
Co.Durham DH1 4HU

Tel/fax: 0191 384 2724

A local branch of the English Civil
War Society. Members welcome
from both Royalist and
Parliamentary factions. Monthly
meetings for discussion and
mutual assistance at various
locations.

Durotriges (Celtic)
New Barn Field Centre,
Bradford Peverel, Dorset DT2
9SD

Contact: David Freeman
Tel/fax: 01305 267463

**Earl of Dumbarton's
Regiment of Foot**
98 Suffolk Road, Barking, Essex

Contact: Steve Payne
Tel/fax: 0181 594 9958

**Earl of Essex's Regiment of
Foot**
53 Reeves Way, Wokingham,
Berkshire RG41 2PS

Contact: Steve Long
Tel/fax: 01189 792730
Email: slong@cix.compulink.co.uk

Essex can provide displays of
historical arms drill, 'living
history', small-scale skirmish
events and weapons-experienced
film extras for the English Civil
War period.

**Earl of Loudon's Regiment of
Foot**
43 Cairnsmoor Way, Irvine,
Ayrshire KA11 1HR, Scotland

Contact: David Cree
Tel/fax: 01294 214471

**Earl of Northampton's
Regiment of Foot**
51 Kestrel Road, Morton, Wirral,
Chester, Cheshire L46 6BN

Contact: John Reneham
Tel/fax: 01516 770368

Earldom of Wessex
9 Durleigh Close, Headley Park,
Bristol, Avon BS13 7NQ

Contact: J K Siddorn
Tel/fax: 0117 9646818

Recreation of the life and times of
the peoples of Western England at
the turn of the First Millennium.
They have resident specialists
who research to the highest
standards. Battle re-enactments
between Vikings, Saxons and
Normans complement the 'living
history' exhibit, creating living
images of those times.

East Anglian Dark Age Society
27 Greenacre Close, Brundall,
Norwich, Norfolk NR13 5QF

Contact: John Gibson
Tel/fax: 01603 715649

**English Civil War Society
(Admin body)**
70 Hailgate, Howden, East Riding
of Yorkshire DN14 7ST

Contact: Jonathan Taylor
Tel: 01430 430695
Fax: 01405 430695
Email: ecws@jpbooks.com

The Society, with 3,500 members,
authentically recreates both the
military and civilian aspects of
17th century battles and sieges as
well as 'living history' displays
and demonstrations of drill. The
King's Army, Roundhead
Association, and individual
regiments may be contacted
through this central address.

English Companions
38 Granworth Road, Worthing,
Sussex (W.) BN11 2JF

Contact: Janet Goldsborough
Tel/fax: 01903 207485

Ermine Street Guard
Oakland Farm, Dog Lane,
Witcombe, Gloucestershire GL3
4UG

Contact: Chris Haines
Tel/fax: 01452 862235

Leading British-based Roman
Army study, reconstruction, and
display society (Flavian period),
formed in 1972; full programme of
public displays, etc.

Escafeld Medieval Society
5 Mountain Row, Crane Moor
Nook, Sheffield, Yorkshire (S.)
S30 7AN
Contact: G Rhodes
Tel/fax: 0114 2884707

Medieval tournaments, 13th
century England.

**Excalibur Medieval Combat
Society**
10 Bowhays Walk, Eggbuckland,
Plymouth, Devon
Contact: Mike Pearce
Tel/fax: 01752 786839

**Excalibur Productions
(Jousting)**
1 Burkitts Court, Burkitts Lane,
Sudbury, Suffolk CO10 6HB
Contact: Ron Heath
Tel/fax: 01787 70856

Falcons Display Team
110 Trafalgar Road, Portslade,
Sussex (E.) BN4 1GS
Contact: Wayne de Strete
Tel/fax: 01273 411862

Fallschirmjager-Regiment 6
c/o 191 Limpsfield Road,
Sanderstead, South Croydon,
Surrey CR2 9LJ
Contact: R Schreiber
Tel/fax: 0181 657 7877

Fanfare du 9e Hussards (Nap)
L'Avenir de Ste Colombe, 4 Rue
de la Baude, 77650 Sainte
Colombe, France
Contact: Olivier Rousell
Tel/fax: (0033) 64001243

Far Isles Medieval Society
2 Birchwood Drive, Lightwood,
Surrey GU18 5RX
Tel/fax: 01276 479818

Medieval society, 500-1600AD;
archery, craft guilds, feasts etc.
Not a show group. Individual and
society membership available.

**Fedn of Dark Age Societies
(Admin body)**
(FODAS), 69 Intwood Road,
Norwich, Norfolk NR4 6AA
Contact: Paul Scruton
Tel/fax: 01603 54844

Feldbataillon Lauenburg (Nap)
Maxgrund 26, 21481 Lauenberg,
Germany
Contact: Gunter Kuster

**Fellowship of the Green
Arrow**
43 Woodlands Road, Stafford,
Staffordshire ST16 1QP
Contact: Michael Tallent
Tel/fax: 01785 212085

Fidei Defensor (1509-1558)
1/84 Church Road, Netherton,
Midlands (W.) DY2 0JJ
Tel/fax: 01384 238193

Fire and Steel
57 Wilandra Crescent, Windale,
New South Wales 2306, Australia
Contact: Darren Delaney

**First Lifeguard (1815)
Regiment**
6 Kinross Road, Ipswich, Suffolk
IP4 3PL
Contact: C Trench
Tel/fax: 01473 723359

British heavy guard cavalry unit
of the Napoleonic period.
Mounted and dismounted
displays, ceremonial a speciality.
Full family involvement
encouraged – something for all.

La Flamme Impériale (Nap)
BP 80, 83602 Frejus Cedex,
France
Contact: Roger Romero

**Fort Newhaven Military
Display Team Ltd.**
42 Janes Lane, Burgess Hill,
Sussex (W.) RH15 0QR
Contact: Richard Hunt
Tel/fax: 01444 233516

A 'living history' society
specialising in authentically
portraying the British, American
and German forces of the Second
World War. Units depicted
include: 1st Airborne Division,
Royal Artillery, Royal Sussex
Regiment, ATS, Military Police,
Home Guard and Commandos.
Events presented range from
battle re-enactments and drill
displays to 1940s dances; also
displays of restored military
vehicles, artillery and support
weapons. The members have
assembled much equipment and
relevant militaria over the years
and can construct various set
pieces from a barrack room
inspection scene with kit layouts
to a complete British field
headquarters. Enthusiastic new
members are always welcome.

France 1940
72C Rue de Coulommes, 77860
Quincy Voisins, France
Contact: Didier Coste

The Free Company
10 Gloucester Street, Brighton,
Sussex (E.)
Contact: Paul Hull
Tel/fax: 01273 683834

An association of 'living history'
groups offering late pre-Roman
Iron Age, Dark Age, Medieval,
Wars of the Roses, and

Renaissance (Landsknecht)
displays, combat. High standards
of authenticity. Film and TV extra
work undertaken.

Freie Gilden
Alte Schmelz, Aumenau 65606,
Germany
Contact: Manfred Struben

**Ft Cumberland & Portsmouth
Mil Soc**
c/o 49 Lichfield Road,
Portsmouth, Hampshire PO3
6DD
Contact: David Quinton
Tel/fax: 01705 668981

HQ & museum in Southsea
Castle. Re-enactment group: Fort
Cumberland Guard, Royal
Marines 1835-40. Musket, cannon
& drum displays. Admission fee
to castle.

Le Garde Chauvin (Nap)
14 Rue des Ouches, Echillais,
17620 Echillais, France
Contact: Daniel Dieu

Garde Impériale
Brandenburger Strasse 53, 07980
Finsterwalde, Germany
Contact: Hans-Michael Hillebrand

Garde Impériale
69 Culverden Park Road,
Tunbridge Wells, Kent TN4 9RB
Contact: Dave Paget

**Gen Sir Thomas Fairfax's
Battalia**
Broad View, Sowerby New
Road, Sowerby Bridge, Yorkshire
(W.) HX6 1LQ
Contact: Stephen Foster
Tel/fax: 01422 836420

English Civil War re-enactment
group representing military and
civilian activities of a company of
the New Model Army: drill
displays, 'living history', battle
re-enactments.

**George Stanley's Retinue
(Med)**
10 Morley Street, Sutton in
Ashfield, Nottinghamshire NG17
4ED
Contact: Tim Morley
Tel/fax: 01623 552491

Glasgow Herred (N F P S)
78 Auckland Place, Dalmuir,
Clydebank, Strathclyde G81 4JZ,
Scotland
Contact: Ian Whitehouse
Tel/fax: 0141 952 7553

Gloucester Household
8 Burnlea Grove, West Heath,
Birmingham, Midlands (W.) B31
3LT
Contact: Mark Vance
Tel/fax: 0121 628 3127

Portray a retinue of household
retainers to Richard Duke of
Gloucester (later King Richard III)
during the Wars of the Roses.

Goddams (15th C.)
1/84 Church Road, Netherton,
Midlands (W.) DY2 0JJ
Tel/fax: 01384 238193

Godolghan Companie
Star House, Star Corner, Breage,
Cornwall TR13 9PJ
Contact: Sally Herriett
Tel/fax: 01326 562908

Historical education, 'living
history' and public displays for
the periods 1450-1650, with
authentic tentage and artillery for
these periods.

**Golden Eagle Archery Display
Troop**
7 Huddleston Road, Tufnell Park,
London N7 0RE
Contact: Roger Summers
Tel/fax: 0171 609 8552

Golden Lions of England
Wincott, Kings Street, Sancton,
Yorkshire (N.) YO4 3QP
Contact: Khrys Yuen
Tel/fax: 01430 827185

Gosport Living History Society
88 Park Road, Gosport,
Hampshire PO12 2HH
Contact: Ros Teague
Tel/fax: 01705 580480

La Grande Armée
2 rue de Mai 1945, 59215
Abscon, France
Contact: Regis Surmont

Admin body for Napoleonic
re-enactment in France.

Grantanbrycq
23 Hillway, Linton, Cambridge,
Cambridgeshire CB1 6JE
Contact: Thomas Barker
Tel/fax: 01223 892920

11th century roleplaying group,
in-period roleplay by parties or
individuals, involving military,
economic, agricultural aspects.
Authentic-looking clothing /
weaponry.

Great War Society
18 Risedale Drive, Longridge, Lancashire PR3 3SB

Contact: Geoff Carefoot
Tel/fax: 01772 782551

First World War 'living history' group portraying British infantry units of the period. Available for public displays, parades, presentations and specialist advice.

Great War Society 1914-18 (Southern Sec)
64 Willowmead Close, Goldsworth Park, Woking, Surrey GU21 3DW

Contact: Nick Barnes
Tel/fax: 01483 747184

First World War 'living history' group that portrays British Infantry units of the period. Available for public displays, parades, presentations and specialist advice.

Grenadierkorps Villingen 1810 e.V.
Hochstrasse 44, 78048 Villingen, Germany

Contact: Wolfgang Kunle

Les Grenadiers du 27e (Nap)
11 Rue Courtepee, 21000 Dijon, France

Contact: Michel Jardel
Tel/fax: (0033) 80576670

Grey Goose Wing Trust (Archery)
59 Netley Street, Farnborough, Hampshire GU14 6AT

Contact: Craig Townend
Tel/fax: 01252 510534

Grimwood (Far Isles Medieval Society)
15 St Catherine's Cross, Bletchingley, Surrey RH1 4PX

Contact: Michelle Dennis
Tel/fax: 01883 742143

Grimwood is a small part of a larger medieval group researching combat, crafts and history in practical ways. We are not a 'show group'.

Le Grognard Lingon (14e de Ligne/Nap)
241 Les Pervenches, 52200 Langres, France

Contact: Guy Bouvier
Tel/fax: (0033) 25870979

Half Pay Officers Dining Society
The Beadles House, 24 Todmorden Road, Bacup, Lancashire

Contact: Major Stoate
Tel/fax: 01706 878278

Harlech Medieval Combat Society
2 Rock Villa, Criccieth, Gwynedd LL52 0ED, Wales

Contact: Roger Clark
Tel/fax: 01766 522089

The society aims to portray in hand-to-hand combat events highlighting Edward I's campaigns in North Wales around 1294/95.

Harrington Household
49 Washington Grove, Bentley, Doncaster, Yorkshire (S.) DN5 9RJ

Contact: Sally Ann Chandler
Tel/fax: 01302 876343

Medieval and Tudor re-enactment group representing an archery unit focusing on 'living history' in the 14th, 15th and 16th centuries.

Hazard's Artillery Company
19 Stanley Road, Poole, Dorset BH15 1QX

Contact: Dai Griffiths
Tel/fax: 01202 672647

Hessische Artillerie DG/FLG
Alt Griesheim 34, 65933 Frankfurt, Germany

Contact: Dieter Schule

Hessische Jaeger DG/FLG
Ringstrasse 3, 56739 Bermel, Germany

Contact: Klaus Westphalen

Highland Brigade (42nd & 79th Regts)
Postfach 1450, 65795 Hattersheim, Germany

Contact: Jurgen Weber

Historical Artillery (HA/UK)
9B Frome Road, Southwick, Trowbridge, Wiltshire BA14 9QB

Contact: R Barton
Tel/fax: 01225 768781

Muzzle-loading artillery research, collecting and demonstration, any period. Design and construction of working reproduction weapons, including cannons; renovation of original weapons. Emphasis on technical aspects.

Historical Re-enactment Co-operative Ltd
53 Lalors Road, Healsville, Melbourne, Victoria 3777, Australia

Contact: Joan Freestone

Historische Spielgruppe Jena e.V.
Brandstromstrasse 11, 07749 Jena, Germany

Contact: Bastian Hertwig

History Re-enactment Workshop
47 Chelmsford Road, South Woodford, London E18 2PW

Contact: Roger Wilson
Tel/fax: 0171 930 6727

Histrionix Living History Group
Bromley Cottage, Overthorpe, Banbury, Oxfordshire OX17 2AD

Contact: David Edge
Tel/fax: 01295 712677

Eighteenth century 'living history' group re-enacting social, domestic and military life, both as entertainment and to serve an educational function for the public, especially schoolchildren. Members donate their leisure time, working for expenses only. Current venues include Sulgrave Manor and English Heritage properties. Membership £5 per annum.

HMS Bellerophon (HM Navy 1793-1815)
29 Beechwood Avenue, Plymouth, Devon PL4 6PW

Contact: W Bertram

Members required: Sailors for RN warship (1793-1815) shore landing party. Directive: To inform the public about our second-to-none maritime heritage, but also the brutality of the regime.

Holger Danske
125 Narrow Lane, Halesowen, Birmingham, Midlands (W.) B62 9PE

Contact: Charlie Peterson
Tel/fax: 0121 421 3569

The Horde
(Lowestoft Dark Age Society), 412 London Road South, Lowestoft, Suffolk NR33 0BH

Contact: Jim Ward
Tel/fax: 01502 500925

Hounds of the Morrigan
66 Mullins Close, Oakridge, Basingstoke, Hampshire RG21 2QY

Contact: Max Bennet
Tel/fax: 01256 63337

House of Bayard (1150-1250)
43 Penhill, Marsh Farm, Luton, Bedfordshire LU3 3LL

Contact: Lisal Haines

House of Lions (Med)
33 Claret Gardens, South Norwood, London SE25 6RP

Contact: Steve Hoey
Tel/fax: 0181 653 6606

Howard's Household
96 Walkley Road, Sheffield, Yorkshire (S.) S6 2XP

Contact: John Ansari
Tel/fax: 0114 2338062

Huddersfield University Medieval Society
26 Halifax Old Road, Birkby, Huddersfield, Yorkshire (W.) HD1 6EE

Contact: David Rushworth
Tel/fax: 01484 512968

The Huscarls
87 Lewis Street, Maryville, New South Wales 2293, Australia

Contact: Ian Dixon

Huskarla
19 Tufnell Park Road, Holloway, London N7 0PG

Contact: Cormac O'Neill
Tel/fax: 01892 510637

Les Hussards de Lasalle (Nap)
7 Impasse des Balmes, F-78540 Villepreux, France

Contact: Jean-Pierre Mir
Tel/fax: (0033) 130562416

Re-enactment 7th and 5th Hussars Regiment, 2nd Horse Artillery with one Gribauval 4 pound gun.

Hwicce Lethang
36 Ferndale Road, Swindon, Wiltshire SN2 1EX

Contact: Mark Browne
Tel/fax: 01793 525626

Hy-Breasail
19 Chance Street, Tewkesbury, Gloucestershire GL20 5RF

Contact: Mike Edwards
Tel/fax: 01684 292940

I G Lutzower Freicorps 1813
Wilhelm-Michel-Strasse 13, 04249 Leipzig, Germany

Contact: Frank Zetzsche

International Jousting Association
Post Office Cottage, Cowesby Village, Thirsk, Yorkshire (N.) YO7 2JJ

Contact: Alan Beattie
Tel/fax: 01845 537431

IJA provides a free advisory service to organisations and associations. Provides training in jousting skills up to Level V (Instructors) Certificates of Competence including Pony Clubs.

International Jousting Federation
British Jousting Centre, Tapeley Park, Instow, Bideford, Devon EX39 4NT

Contact: J Diamond
Tel/fax: 01271 861200

Iron Brigade
161 Waddington Road,
Nuneaton, Warwickshire

Contact: Martin Harbour
Tel/fax: 01203 345886

Islamic Guard (Med)
55 Bowie Close, London SW4
8HB

Contact: Richard Kemp
Tel/fax: 0181 671 5301

Izum Hussar Regiment (Nap)
Flat 7, Leninst 132, Tulchin,
Vinitza Region 288300, Ukraine

Contact: Vladimir Beltser

**Jacobite Regiment
(17th/18th C)**
67 Avondale, Sligo, Co. Sligo,
Ireland

Contact: Noel Connolly

**John de Vere's (Earl of
Oxford's) Troop**
33 Sebastian Close, Greenstead,
Colchester, Essex CO4 3SH

Contact: Alex Booth
Tel/fax: 01206 869918

Jomsberg Elag Herred
46 St.John's Court, Princess
Crescent, Finsbury Park, London
N4 2HL

Contact: Philip Burthem
Tel/fax: 0181 800 4013

Jomsvikings
11a Waddicor Avenue,
Ashton-under-Lyne, Lancashire
OL6 9HE

Contact: Mark Harrison
Tel/fax: 0161 344 1324

Specialising in the Viking and
Saxon peoples, we provide
authentic re-enactments for events
throughout the UK. Jomsvikings
provide main arena battles and
Viking village life displays. Video
and information pack available.
Write for details.

Jotunheim Herred
157 Rochester Avenue,
Rochester, Kent

Contact: Andy Wilkinson
Tel/fax: 01634 840143

Junior Jousting Club
Combwell Priory, London Road,
Flimwell, Kent TN5 7QD

Contact: Jeremy Richardson
Tel/fax: 01580 87754

**K & K IR 4 "Hoch und
Deutschmeister"**
49 Belsize Park, London NW3
4EE

Contact: Ian Castle

Members of the Napoleonic
Association; Britain's only
Austrian re-enactment regiment.

Keltoi (Iron Age Celtic)
Clwt Melyn, Pen Lon,
Newborough, Anglesey,
Gwynedd LL61 6RS, Wales

Contact: Paul Curtis
Tel/fax: 01248 79500

Kemysii (Celts)
Romney, 14 Aston Close,
Kempsey, Hereford & Worcester
WR5 3JR

Contact: Jane & Derek Stevens
Tel/fax: 01905 821148

Kiev Druzhina (Med)
94 Fremantle Crescent, Sutton
Estate, Middlesbrough, Cleveland
TS4 3HP

Contact: Graham Birkbeck

**The King's Army (Admin
body)**
26 Woodnook Garth, Leeds,
Yorkshire (W.) LS16 6PH

Contact: Adrian Richardson

Administrative HQ for the
Royalist units of the English Civil
War Society.

King's German Legion
Zum Rott 19, 49078 Osnabruck,
Germany

Contact: Karl-Heinz Lange

**The King's Grenadiers circa
1759**
11 Poole Crescent, Bilston,
Midlands, (W.) WV14 8SR

Contact: David Whitehouse
Tel/fax: 01902 492183
Email: mitre@mail.on-line.co.uk

Period: Seven Years War 1756-63.
The recreated grenadier company
of the 12th Regiment of Foot
(Napier's).

**King's Pageant Medieval
Re-enactment Society**
87 Eggbuckland Road, Hartley,
Plymouth, Devon PL3 5JR

Contact: Chris Thomas
Tel/fax: 01752 789629

Period 1250-1300. We aim for
accuracy and safety while still
having fun, mostly at local events.
New members welcome for
combat, crafts, music etc.

The Kingmakers
7 Chapmans Crescent, Chesham,
Buckinghamshire HP5 2QU

Contact: Duke Henry Plantagenet
Tel/fax: 01494 784271

Small, informal and friendly, the
Kingmakers do not attend shows,
but have good parties. Email:
duke@calltoarms.com

Knights in Battle
96 Walkley Road, Sheffield,
Yorkshire (S.) S6 2XP

Contact: John Ansari
Tel/fax: 0114 2338062

**Knights in Battle Medieval
Society**
133 Hadfield Street, Sheffield S6
3RS

Contact: Chris Felton
Tel/fax: 0114 232 4085

Early 13th (reign of King John)
and late 15th-century (Wars of the
Roses) tournaments, battles,
murder mysteries and living
history events. EH 'Living
History' approved.

Knights of Outremer
39 Cutthorpe Road, Cutthorpe,
Chesterfield, Derbyshire S42
7AD

Contact: Joanna Woodhall
Tel/fax: 01246 221154

**Knights of St John Medieval
Society**
58 Welton House, Netherfields,
Middlesbrough, Cleveland TS3
0TQ

Contact: James Phoenix

Medieval period covered from
1050 to 1250 (chain mail period).

Knights of the Crusades
14 Berkley Waye, Heston,
Hounslow, Middlesex TW5 9HL

Contact: Michael Ellis
Tel/fax: 0181 570 9073

Knights of the Dragon
Imladris, 224 Coatham Road,
Redcar, Cleveland TS10 1RA

Contact: J Watson
Tel/fax: 01642 489227

Small group of knights depicting
foot combat circa 1250 AD.
Member of DELHC.

Knights of the Realm
11a Park Lane, Tutbury, Burton
on Trent, Staffordshire DE13 9JQ

Contact: Richard Bradbury
Tel/fax: 01283 812641

Knights of the Rose
(Northampton Medieval Society),
6 Simons Walk, Spring Boroughs,
Northamptonshire NN1 2SR

Contact: Joe Joyce
Tel/fax: 01604 31442

Knights of Theoc
115 Queen's Road, Prior's Park,
Tewkesbury, Gloucestershire
GL20 5EN

Contact: Mike Cash
Tel/fax: 01684 294778

Medieval entertainment:
hand-to-hand combat, dancing,
wrestling, fire-eating. Sixty miles
radius of Tewkesbury only; for
information phone after 6pm.

Knights Templar
39 Penistone Road, Park End,
Middlesbrough, Cleveland TS3
0DJ

Kowethas Bran Gwyn
203 Bicester Road, Aylesbury,
Buckinghamshire HP19 3BD

Contact: Tom Green
Tel/fax: 01296 87518

**Kurmarkisches
Landwehr-Regiment (Nap)**
Wisbyer Strasse 27a, 10439
Berlin, Germany

Contact: Hans-Jurgen Klingner

Lancastrian Royal Household
10 Morley Street, Sutton in
Ashfield, Nottinghamshire NG17
4ED

Contact: Tim Morley
Tel/fax: 01623 552491

Legio II Augusta
288 Copnor Road, Portsmouth,
Hampshire PO3 5DD

Contact: David Richardson
Tel/fax: 01705 790617

Roman 'living history' society,
including ladies' group. School
visits and events by arrangement.
New members welcome.

Legio VI Victrix Pia Fidelis
8 Leechmere Way, Ryhope,
Sunderland, Tyne & Wear SR2
0DH

Contact: Eddie Barrass
Tel/fax: 0191 523 6377

Legio VIII Augusta
31 Llwyn Menlli, Ruthin, Clwyd
LL15 1RG, Wales

Contact: Ken Evans
Tel/fax: 01824 704250

Legio VIIII Hispana
3 Murphy Street, Blaxland, New
South Wales 2774, Australia

Contact: Wayne Robinson

Legio X Gemina
Pharus 309, 1503 Zandam,
Netherlands

**Legio XIIII Gemina Martia
Victrix**
Freiherr vom Stein 22, 55774
Baumholder, Germany

Contact: Daniel S Peterson

**Leibgarde-Grenadier-Regiment
1813**
Heinrichsruher Weg 3, 03238
Finsterwalde, Germany

Contact: Uwe Bergmann

Lewes Reavers
20 Bradford Street, Eastbourne,
Sussex (E.) BN21 1HC

Tel/fax: 01323 33225

Lincoln Castle Longbowmen
54 Grantham Road, Waddington, Lincolnshire LN5 9LS

Contact: Paul Mason
Tel: 01522 720507
Fax: 01522 720269
Email: Angela@polonium. demon.uk

15th century English 'Living History' featuring archery, crossbows, arquebuses, kitchen, trades & researched scenario & timetable based in an authentic tented camp site of the Wars of the Roses.

Lincolnshire Military Preservation Soc
Memories, 20 Market Place, Alford, Lincolnshire LN13 9EB

Contact: Trevor Budworth
Tel/fax: 01507 462541

Established 1981, and dedicated to WWII social, dances, military vehicles, battle re-enactment. British, American, German units with own field HQ at former RAF station. New recruits, ex-service, and associated groups welcome. Charity events undertaken.

Lion Rampant
2 Hillside, Totteridge, High Wycombe, Buckinghamshire HP13 7LG

Contact: Izzy Legg
Tel/fax: 01494 473129

Livery and Maintenance (Admin body)
Flat B, 10 Alexandra Road, Newport, Gwent NP9 2GY, Wales

Contact: Chris & Vicki Howell
Tel/fax: 01633 221857

An umbrella group comprising 15th century re-enactment groups with a high standard of authenticity.

Living History Association Inc
PO Box 1389, Wilmington, Vermont 05363-1389, USA

Contact: James Dassatti
Tel/fax: (001) 802 464 5569

World's most diverse re-enactment organisation, specialising in re-enactor personal injury and liability insurances for units and events.

Longbow Presentations
29 Batley Court, Oldland, South Gloucestershire BS15 5Y2

Contact: Veronica Soar
Tel/fax: 0117 9323276

Mobile exhibition interpreted by experts providing fascinating and informative insight into the craft of the traditional English bowyer and fletcher: including many interesting antique artifacts.

Lord Ferrers Household
412 Foxhill Road, Foxhill, Sheffield, Yorkshire (S.) S6 1BP

Contact: Andrew Jarvis
Tel/fax: 0114 2344359

The Lord General's Staff
17 Gatehead Croft, Delph, Saddleworth, Lancashire OL3 5QB

Contact: Rod Lawson

An innovative and friendly group, with long and varied experience of organising and participating in civil/military 'living history' of the mid 17th Century.

Lord Gloucester's Retinue
Quendon, Manor Road, Barnet, Hertfordshire EN5 2LE

Contact: Clive Bartlett

Lovell Household
35 Longfield Road, Tring, Hertfordshire HP23 4DG

Contact: Mike Woodhouse
Tel/fax: 01442 824941

Lutzow'sches Freicorps (Nap)
Sander Strasse 3a, 21029 Hamburg, Germany

Contact: Ulf Kretschmann

Lygtun Raiders
76 Trident Drive, Houghton Regis, Dunstable, Bedfordshire

Tel/fax: 01582 864010

Lys et Lion
40 rue Voltaire, 59800 Lille, France

Contact: Dom Delgrange

'Living history' 1450-1465 – the Autumn of the Middle Ages.

Maj Gen Lawrence Crawford's Battalia
219 Greenhill Avenue, Sheffield, Yorkshire (S.) S8 7TJ

Contact: Kevin Fisher
Tel/fax: 01742 748922

Maj Gen Sedenham Pointz's Battalia
92 Station Road, Kippax, Leeds, Yorkshire (W.) LS25 7LQ

Contact: Simon Lister
Tel/fax: 0113 287 0191

Major Surgery's Brigade
The Bungalow, 150 Thornton Road, Thornton Heath, Surrey CR7 6BB

Contact: Malcolm McDonald
Tel/fax: 0181 683 1128

Major Surgery's Brigade is a unit within the Napoleonic Association that re-enacts field surgery of the 1800s as part of the British Brigade.

Manareafan Lethang (N F P S)
1 Leecroft Close, Yiewsley, Middlesex UB7 8AQ

Contact: Pete James
Tel/fax: 01895 445076

Marcher Knights
490 Bryn Road, Ashton in Makerfield, Wigan, Lancashire WN4 8AN

Contact: Anna Calder
Tel/fax: 01942 715724

Max Diamond Enterprises Ltd
The British Jousting Centre, Chilham Castle, Chilham, nr Canterbury, Kent CT4 8DB

Tel/fax: 01227 730704

Mayhem Dark Age & Medieval Society
668 Chappel Street, Broken Hill, New South Wales 2880, Australia

Contact: Paul Green

Medieval & Renaissance Society
291 Roberts Road, Greenacre, New South Wales 2190, Australia

Contact: Jim Adams

Medieval Heritage Society
42 Shelburne Road, Calne, Wiltshire SN11 8ES

Contact: Brian Stokes
Tel/fax: 01249 815369

Medieval Siege Society
70 Markyate Road, Dagenham, Essex RM8 2LD

Contact: Phil Fraser
Tel/fax: 0181 592 3621

Mainly 15th century re-enactment society with approximately 250 members. Accurately recreating England's military past. Family membership very welcome.

Medieval Society
90 Crouch House Road, Edenbridge, Kent

Contact: Michael Loades
Tel/fax: 01732 865550

Medieval Times
524 Aylestone Road, Leicester, Leicestershire LE2 8JB

Contact: John Marnell
Tel/fax: 0116 2831993

Mercenarie Gild
24 Cricketers Close, Ockley, Dorking, Surrey RH5 5BA

Contact: Michael Hickling
Tel/fax: 01306 627796

14th century 'living history' group of artisans including, amongst others, cook, cordwainer, fletcher, falconer, plying their trades in period surroundings and costume using authentic techniques.

Mercenaries of Lower England
7 Cedar Court, Totteridge Road, High Wycombe, Buckinghamshire HP13 6DN

Contact: Graham Povey
Tel/fax: 01494 526620

Mercia Herred
Rosemary Cottage, Camp Road, Canwell, Sutton Coldfield, Birmingham B75 5RA

Contact: Paul Craddock
Tel/fax: 0121 323 4309

Dark Ages re-enactment, village 'living history', and battles, Iron Age to late Norman periods but mainly 8th to 10th centuries. Vikings, Saxons and Celts welcome.

Military Music Re-enactors' Society
17 Booth Street, Handsworth, Birmingham, Midlands (W.) B21 0NG

Contact: T Horne

Period style brass band and fife and drum corps of the Napoleonic and American Civil War periods; both marching and concert displays, and research.

Milites Litoris Saxoni
82 London Road, Faversham, Kent ME13 8TA

Contact: John Harris

Late Roman period reconstruction and re-enactment group.

Militia
13 Elm Grove, Great Clacton, Clacton on Sea, Essex CO15 4DL

Contact: Iain Cockburn
Tel/fax: 01255 424567

Mindgames
105a Queen Street, Maidenhead, Berkshire SL6 1LR

Tel/fax: 01628 770676

Mortimer Household
Lower House, Burrington, Ludlow, Shropshire SY8 2HT

Contact: D Woodward

Musique de la Garde (Nap)
Avenue Prince Charles 18, B-1410 Waterloo, Belgium

Contact: Pierre Grapin
Tel/fax: (0032) 23548280

N V G Dun Loughaire Garrison
442 Latrobe Terrace, Newtown, Victoria 3220, Australia

Contact: Andrea Redden

An early medieval re-enactment association, the New Varangian Guard (N V G) recreates the Scandinavian bodyguard of the Byzantine Emperors, their allies and opponents.

N V G Miklagard Garrison
4/85 Croydon Street, Lakemba,
New South Wales 2195, Australia

Contact: Martin Baker

N V G Mountains Garrison
91 Russell Avenue, Valley
Heights, New South Wales 2777,
Australia

Contact: Graham Walker

N V G Riverina Garrison
113 McKell Avenue, Wagga
Wagga, New South Wales 1347,
Australia

Contact: Ian Kelleher

N V G Rusland Garrison
120 Fletcher Parade, Bardon,
Queensland 4065, Australia

Contact: Patrick Urquhart

N V G Sarkland Garrison
PO Box 264, Sebastopol, Victoria
3356, Australia

Contact: Julie Heron

N V G Vlachernai
PO Box 4284, Melbourne
University, Melbourne, Victoria
3052, Australia

Contact: Darren Robinson

**Napoleonic Association
(Admin body)**
3 Steadman Avenue, Cosby,
Leicestershire LE9 5UZ

Contact: Adrian Proudfoot

Co-ordinating body for
re-enactment and research
throughout Britain and Europe;
individual affiliated units of
horse, foot, and guns listed
separately.

**Napoleonische Gesellschaft
(Admin body)**
Zum Rott 19, D-49078
Osnabruck, Germany

Contact: Karl-Heinz Lange
Tel: (0049) 541444350
Fax: (0049) 54148070

Admin body for Napoleonic
re-enactment in Germany; some
20 affiliated regiments include
Prussian Line and Landwehr,
KGL, Brunswickers, French, and
Confederation of the Rhine units.

**National Association of
Re-enactment Societies**
PO Box 1218, Swindon,
Wiltshire SN2 6ZG

Contact: May Griffiths
Tel/fax: 01793 524465

The society for reputable
re-enactors, acting as a forum for
historical re-enactment societies in
Britain and providing guidelines
on safety and authenticity. We
represent members' interests and

liaise with Government
departments, encouraging contact
between societies and other
bodies.

National Jousting Association
Combwell Priory, London Road,
nr.Flimwell, Kent TN5 7QD

Contact: Jeremy Richardson
Tel/fax: 01580 87754

Neville Household
35 Longfield Road, Tring,
Hertfordshire HP23 4DG

Contact: Mike Woodhouse
Tel/fax: 01442 824941

Society of 20 members specialising
in 'living history' of period
1450-1487, including weapons,
armour, tentage, cooking, crafts,
dancing and fighting displays.

**New England Colonial Living
History Grp**
Wychwood, MSF 2007,
Armidale, New South Wales
2350, Australia

Contact: Keith H Burgess
Tel/fax: (0061) 67 755292

Period: 1756-1776 Colonial
America. Muzzle-loading,
tomahawk and knife throwing,
camping, trekking, colonial living
skills, survival skills. Fully
equipped 18th century trading
post fort.

**New Zealand Medieval
Jousting Soc Inc**
8/23 Amesbry Street, Palmerston
North 5301, New Zealand

Contact: John Billings

**Newcastle Early Anglian
Society**
Student Union Building, Kings
Walk, Newcastle upon Tyne,
Tyne & Wear NE1 8QB

Contact: Peter Cook

Newcastle's Foote
28 Chapelfield Way, Thorpe
Hesley, Rotherham, Yorkshire
(S.) S61 2TL

Contact: Simon Wright
Tel/fax: 0114 2454982

A Royalist foot regiment of the
Sealed Knot, recreating
Newcastle's Whitecoats, famous
for their last stand at Marston
Moor, 1644. The regiment recruits
nationwide.

Norfolke Trayned Bandes
38 Richmond Road, Lincoln,
Lincolnshire LN1 1LQ

Contact: Richard Coxon
Tel/fax: 01522 528995

**The Normandy Arnhem
Society**
22 Cousin Lane, Illingworth,
Halifax, Yorkshire (W.) HX2 8AF

Contact: David P Mitchell
Tel/fax: 01422 256891

The Normandy Arnhem Society is
a living, breathing museum of
remembrance dedicated to
keeping history alive by
recreating, as accurately as
possible, the life and times of
German and British soldiers
during the latter half of WWII.
Comprises 9./SS-Pz.Gr.Rgt.20 and
9 Field Company RE. Specialises
in 'living history' displays for
museums.

Normanitas
Crosby's Caravan Court, Broken
Head Road, Byron Bay, New
South Wales 2481, Australia

Contact: James Dempsey
Tel/fax: (0061) 66 858844

Normanitas provide Dark Age
and early Middle Ages
re-enactment displays for public
and private show, educational
talks and workshops can also be
arranged. Displays featuring
combat and 'living history'.

Norsemen of Kent
33 High Street, Herne Bay, Kent
CT6 5LJ

Contact: Dylan Hampshire

North Historical (ACW)
41 Avenue Gabriel Peri, 91550
Paray-Vielle-Poste, France

Contact: Jean-Claude Renaudin

**Northampton Medieval
Society**
6 Simons Walk, Spring Boroughs,
Northamptonshire NN1 2SR

Contact: Joe Joyce
Tel/fax: 01604 31442

Northern Allied Axis Society
39 Rosendale Grove, Spring Bank
West, Hull, Humberside (N.)

Contact: B Nuttall
Tel/fax: 01482 566144

NAAS are battle re-enactors and
also offer a mobile military
museum for events.

Northland Mercenaries
33 Stanley Gardens, Herne Bay,
Kent CT6 5SQ

Contact: G Schofield
Tel/fax: 01227 360891

Northumbria Herred
2 Oswald Terrace, Easington
Colliery, Peterlee, Co. Durham
SR8 3LB

Contact: Allan Beneke
Tel/fax: 0191 527 0937

**Norwich & Norfolk Medieval
Assoc**
33 Chestnut Avenue, Spixworth,
Norwich, Norfolk NR10 3QD

Contact: H Appleton

Norwich Svieter (N F P S)
4 Rachel Close, West Earlham,
Norwich, Norfolk NR5 8JF

Contact: Lisanne Norman
Tel/fax: 01603 55032

Nottingham Herred
34 Sedgley Avenue, Sneinton,
Nottinghamshire NG2 4HZ

Contact: Peter Carss
Tel/fax: 0115 979 9554

O'Rourke Clan (Med)
The Blue House, Courtglancy,
Co.Leitrem, Ireland

Odin's Chosen
12 Navistock Crescent,
Woodford Green, Essex

Contact: Frank Jones
Tel/fax: 0181 506 1581

Order of the Black Prince
107 Mongeham Road, Great
Mongeham, Deal, Kent CT14 9LJ

Contact: Alex Summers
Tel/fax: 01304 381699

A small 14th century 'living
history' group portraying garrison
life both military and civilian with
emphasis on correctness of dress,
equipment, etc.

Ormsheim Herred (N F P S)
Darkwood, West Slaithwaite,
Huddersfield, Yorkshire (W.)
HD7 5XA

Contact: Neil Shepherd
Tel/fax: 01484 843968

Outremer
7 Camelia Grove, Gymea, New
South Wales 2227, Australia

Contact: Sean Turkington

Paladins of Chivalry
Top Flat, 4 Park Road,
Wallington, Surrey SM6 8AH

Contact: Alison Salmon
Tel: 0181 773 1544
Fax: 0181 776 2146

Fourteenth century re-enactment
of peasants' life, Edward III's
court and tournament. Includes
dancing, longbow archery, sword
and other weapons. High costume
standard. Educational and
entertaining.

Le Passepoil (WWI)
5 Ruis des Juifs, 51000 Chalons,
France

Contact: P Romary

Pax Normanni (1066-1200)
Flat 3, 153 Hucknall Road,
Carrington, Nottinghamshire
NG5 1FA

Contact: Jonathan Preston
Tel/fax: 0115 9857561

**Peterborough Lethang
(N F P S)**
34 Scotney Street, New England,
Peterborough, Cambridgeshire
PE1 3NS

Contact: Colin Owrid
Tel/fax: 01733 341224

**Peterborough Medieval
Jousting Society**
Greenways, New Road,
Woodwalton, Huntingdon,
Cambridgeshire PE17 5YT

Contact: Maggie Stansfield
Tel/fax: 01487 773449

**Pike & Musket Society of
New South Wales**
57 John Street, Petersham, New
South Wales 2049, Australia

Contact: David Green
Tel/fax: (0061) 2 560 8527

Re-enactment group. Infantry
Company of Pikemen and
Musketeers modelled on the
London Trained Bands, English
Civil War period.

Plantagenet Medieval Society
122 Hartland Avenue, Coventry,
Midlands (W.) CV2 3ES

Contact: D Rees
Tel/fax: 01203 453673

**Pleasley Hill Raiders
(American)**
551 Chesterfield Road North,
Pleasley, Mansfield,
Nottinghamshire NG19 7RD

Contact: R Houghton
Tel/fax: 01623 811913

Le Poilu de la Marne (WWI)
10 Rue Bourg de Vesles, 51480
Fleury la Riviere, France

Contact: Didier Blanchard
Tel/fax: (0033) 26546478

**Portsdown Artillery
Volunteers**
Cliff House, Portsdown Hill,
Drayton, Portsmouth,
Hampshire PO6 1BS

Contact: Ian Maine
Tel/fax: 01705 327710

Based at Fort Nelson as an
Artillery Volunteer unit of 1890.
Activities include heavy gun drill,
Martini-Henry carbine firing and
horse drawn artillery.

Prepare to Die (Med)
25 Egleston Road, Morden,
Surrey SM4 6PN

*Contact: Diana & Colin Lempriere-
Knight*
Tel/fax: 0181 395 3729

**Preuss.
Lieb-Infanterie-Regiment**
Kaempchen Strasse 16, 45468
Mulheim/Ruhr, Germany

Contact: Gunter Berker

Preussen von Mockern 1813
Hans-Eisler-Strasse 30, 04318
Leipzig, Germany

Contact: Bernd Baumbach

**Preussische Felddruckerei
1813**
Davidstrasse 3, Leipzig 04109,
Germany

Contact: Peter Mechler

Provincial Lethang
565 Tong Street, Bradford,
Yorkshire (W.) BD4 6NL

Contact: Steve Ward
Tel/fax: 01274 680908

Prytani (Iron Age Celtic)
8 Long Street, Gerlan, Bethesda,
Gwynedd LL57 3SY, Wales

Tel/fax: 01248 602772

'Living history' and re-enactment
of the period 50BC-50AD.
Recreating lifestyles of period in
Britain prior to and immediately
after the Claudian invasion.
Available for all events including
film and media. Regional contacts:
South Wales, 01291 422146, Mike;
South-West, 01272 394059, Steve.

Queen's Lifeguard of Foote
66 Bramble Close, Shrublands,
Shirley, Croydon, Surrey

Contact: C Steele

Ragged Staff Medieval Society
35 Longfield Road, Tring,
Hertfordshire HP23 4DG

Contact: Mike Woodhouse
Tel/fax: 01442 824941

Rampant Bears
181 Fremfield Road, Brighton,
Sussex (E.) BN2 2YE

Contact: Fergus Harris
Tel/fax: 01273 600878

**Ratatoska Heim & Warager
(Dark Ages)**
Uhlandstrasse 3, 30851
Langenhagen, Niedersachsen,
Germany

Contact: Michael Stadtler

**Reavers Dark Age Combat
Society**
14 Wren Close, Heathfield,
Sussex (E.) TN21 8HG

Contact: K M Bradfield
Tel/fax: 01435 863993

Dark Age combat, up to 11th
century; archery, weapon & craft
exhibition, birds of prey, medieval
tents. Terms negotiable. New
members welcome.

**Red Gauntlet Re-enactment
Society**
25 Rowan Close, Scarborough,
Yorkshire (N.) YO12 6NH

Contact: Gary Hughs

A small, multi-period society
specialising in 10th, 15th, 16th and
17th Century non-combat displays
and events.

Regia Anglorum (Admin body)
9 Durleigh Close, Headley Park,
Bristol, Avon BS13 7NQ

Contact: J K Siddorn
Tel/fax: 0117 9646818

An international society of over
five hundred people, dedicated to
the authentic re-creation of
Vikings, Saxons and Normans
(approximately 950 to 1066). We
have a very extensive working
'living history' & craft exhibit.
Early music and many crafts are
available. Our wooden 'Viking'
ship replica is over forty feet long,
with its own experienced,
costumed & equipped crew.
Battle re-enactments involving
over 200 people are available, as
are KS2 school visits. Branches
everywhere!

**Regiment de la Chaudière
(WWII Can)**
24 Rue de l'Audience, 27100 Le
Val de Reuil 27100, France

Contact: Philippe Frances

Rhyfel Gwynedd
12 Cefnfaes Street, Bethesda,
Bangor, Gwynedd LL57 3BW,
Wales

Contact: Russell Scott

Right Royal Revels
22 Sutton Wick Lane, Drayton,
Abingdon, Oxfordshire OX14
4HJ

Contact: G & M Jones
Tel/fax: 01235 31610

Ring of Steel
2 Rock Villa, Criccieth, Gwynedd
LL52 0ED, Wales

Contact: Roger Clark
Tel/fax: 01766 522089

13th Century group re-enacting
period 1220-1350, with all the
sights and sounds of this period.

Robin Hood Society
37 Moorsham Drive, Wollaton,
Nottinghamshire

Contact: Mary Chamberlain
Tel/fax: 0115 285204

Rochdale Herred (N F P S)
147-149 Market Street,
Broadley, Whitworth, Rochdale,
Manchester, Gt. OL12 8RU

Contact: Peter Campbell
Tel/fax: 01706 343205

**Roman Military Research
Society**
69 The Warren, Hardingstone,
Northamptonshire NN4 0EP

Contact: L G Morgan
Tel/fax: 01604 763136

Rosa Mundi (Med)
34 Raby Road, Redcar, Cleveland
TS10 2HE

Contact: Lesley Thurston
Tel/fax: 01642 471749

**Rossiter's Artillery Company
& Trayne**
74 Heatherley Drive, Basford,
Nottinghamshire NG6 0FN

Contact: Dennis Radnell
Tel/fax: 01602 785502

**Roundhead Association
(Admin body)**
55 New Writtle Street,
Chelmsford, Essex CM2 0SB

Contact: David Blackmore
Tel/fax: 01245 268407

Co-ordinating body for the
Parliamentarian units of the
English Civil War Society. Fight
for the good old cause – join one
of the friendliest re-enactment
societies around.

Royal Horse Artillery (Nap)
16 Adelaide Road, Blacon,
Chester, Cheshire CH1 5SY

Contact: G Lee

Royal Household
(White Company West), Gravel
Hill, Stretford Bridge,
Leominster, Hereford &
Worcester HR6 9DQ

Contact: R S Brown
Tel/fax: 01568 88352

Royal Marine Artillery
1 Andrew Cescent,
Waterlooville, Portsmouth,
Hampshire PO7 6BE

Contact: Tony Oatley

Ruadain Reivers Gaelic Warband
85 Healey Road, Ossett, Yorkshire (W.) WF5 8LT

Contact: Sarah Mawson

Celtic enactment: battles, lifestyle and crafts 100BC-1000AD. All equipment is researched and authentic. We educate and entertain while celebrating our heritage.

Russisch-Deutsche Legion (I R Nr 31)
Prager Strasse 345, 04289 Leipzig, Germany

Contact: Uwe Meyer

Rye Medieval Society
Half House, Military Road, Rye, Sussex (E.) TN31 7NY

Tel/fax: 01797 223404

S C U M
67 Elm Grove, Brighton, Sussex (E.)

Tel/fax: 01273 623441

S C U M West
20 Mendip View, Wick, Bristol, Avon BS15 5PY

Contact: Steve Harrill-Morris
Tel/fax: 0117 9374059

Sachsiche Feldpost
Blumenstrasse 12, 04445 Liebertwolkwitz, Germany

Contact: Peter Hainke

Sachsische Leichte Infanterie
Zolaweg 27, 04289 Leipzig, Germany

Contact: Jorg Hensel

Sachsisches I R Prinz Clementz
Rathausstrasse 40, 04416 Markkleeberg, Germany

Contact: Otto Schaubs

The Salon
480 Chiswick High Road, London W4 5TT

Contact: Dawn Wood
Tel/fax: 0181 742 0730

The Salon is a civilian 'living history' group, re-enacting life in the 18th and 19th centuries, with demonstrations. Sponsors include English Heritage.

Saltburn Archery Club
5 West Avenue, Saltburn by Sea, Cleveland

Contact: Peter Furbank

Saxon Village Crafts
Delbush, Whatlington Road, Battle, Sussex (E.) TN33 0JN

Contact: P J Butler
Tel/fax: 014246 2351

Costumed crafts demonstrations for events and schools 500 AD to 1600 AD. Cordwaining, lipwork, narroware, heraldry, bone and antler working and many other displays available.

Scarborough Archers
6 Beverley Close, Cayton, Scarborough, Yorkshire (N.) YO11 3SN

Contact: P J Lilley
Tel/fax: 01723 582784

Schinderhannes-Bande
Lucas-Cranach-Strasse 2, 65527 Niedernhausen, Germany

Contact: Jutta Seliger

Schlesische Landwehr 1813 e.V.
Mühlstrasse 50 (Im Museum), 04435 Schkeuditz, Germany

Contact: Joerg Rojahn
Tel/fax: (0049) 34204 14267
Email: jrojahn_2 @compuserve.com

Schwertbruder (c1476)
30 Marlborough Close, Clacton-on-Sea, Essex CO15 2AL

Contact: Chris Fallows
Tel/fax: 01255 429968

Scrope's Household
34 Raby Road, Redcar, Cleveland TS10 2HE

Contact: Lesley Thurston
Tel/fax: 01642 471749

Sea Wolves
24 Grove Road, Chadwell Heath, Essex RM6 4AG

Contact: John Quadling
Tel/fax: 0181 597 0570

Sealed Knot Society (Admin body)
49 Stagsden, Orton Goldhay, Peterborough, Cambridgeshire PE2 5RW

Contact: J Crawford
Tel/fax: 01733 239508

With around 6,500 members, the Sealed Knot is Europe's largest and most experienced battle re-enactment society, founded in 1968 with the object of promoting interest in the English Civil War period. The society recruits nationally; contact us for details of regionally organised regiments.

Service Commemoratives Ltd
PO Box 173, Dromana, Victoria 3936, Australia

Tel/fax: (0061) 359 810201

Military service commemorative medals and a range of over 100 different area and specialist clasps/bars available for: combatant service; sea service; foreign service; army service; aviation service; volunteers; national defence; Antarctica service; and POW liberation 1945.

Service de Santé
21 Yeomans Way, Plympton, Plymouth, Devon PL7 3JW

Contact: H Wills
Tel/fax: 01752 338898

Small unit recreating surgeons and orderlies of the Grande Armee c 1812, and demonstrating procedures by using modern theatrical techniques.

Seventeenth Century Heritage Centre
3 Penley Avenue, Prestatyn, Clwyd, Wales

Contact: John Carter

Shire of Lough Devnaree
44 Bancroft Avenue, Tallaght, Co.Dublin 24, Ireland

Contact: Lesley Grant

The Siege Group
Flat 11 Marlborough Court, Marlborough Hill, Harrow, Middlesex HA1 1UF

Contact: Dennis Wraight
Tel/fax: 0181 861 0830

English Civil War period, both military and civilian aspects of 17th century life, comprising battle re-enactment, 'living history' and role-playing.

Silures Iron Age Celtic Society
4 Sunnybank Court, Bracla, Bridgend, Glamorgan, Mid-, Wales

Tel/fax: 01656 766490

Silver Branch (Scoran Arghans)
Barley Splatt, Panters Bridge, Mount, Bodmin, Cornwall

Tel/fax: 01208 821336

Sir Edward Hungerford's Regt of Horse
111 Grove Park, London SE5 8LE

Contact: Hop Stephens
Tel/fax: 0171 733 0065

Sir Gilbert Hoghton's Company of Foot
10 Winfield Avenue, Withington, Manchester, Gt.

Contact: Andy Pettifer
Tel/fax: 0161 434 6730

Sir John Astley's Retinue
11A Park Lane, Tutbury, Burton on Trent, Staffordshire DE13 9JQ

Contact: Richard Bradbury
Tel/fax: 01283 812641

Sir William Waller's Regiment of Foot
28 Wydean Rise, Belmont, Hereford & Worcester HR2 7XZ

Contact: David Kemester
Tel/fax: 01432 263687

Society for Creative Anachronism
98 Mullins Close, Basingstoke, Hampshire RG21 2QZ

Contact: Mike Lacy
Tel/fax: 01256 23141

Society for Creative Anachronism (USA)
PO Box 360743, Milpitas, California 95036, USA

Society of British Fight Directors
26 Goldhurst Terrace, London NW6 3HT

Contact: Penelope Lemont

Society of the American Revolution
44 Chapel Street, Bishop's Itchington, Leamington Spa, Warwickshire CV33 0RB

Contact: John Harris
Tel/fax: 01926 613990

'Living history' society comprising units representing both sides in the American War of Independence; recruits nationally; events throughout UK and some abroad.

Somerset Household
Basement Flat, 6 Buckingham Road, Brighton, Sussex (E.)

Contact: Jacqueline Beevis
Tel/fax: 01273 206123

Sons of Penda Medieval Society
8 Himley Avenue, Dudley, Midlands (W.) DY1 2QP

Contact: Alex S Kay
Tel/fax: 01384 256438

Southern Medieval Society
8 New Street, Puddletown, Dorchester, Dorset DT2 8SF

Contact: Jason O'Keefe
Tel/fax: 01305 848148

We are a dedicated and enthusiastic band of Medieval re-enactors who recreate tournaments as well as taking part in larger battles.

The Southern Skirmish Association
PO Box 485, Swindon, Wiltshire SN2 6BF

Contact: May Griffiths
Tel/fax: 01793 524465

Founded 1968. American Civil War re-enactment society, specialising in battle re-enactments, 'living history' displays, research and lectures. The Association can also supply museum exhibits. A member of the National Association of Re-enactment Societies.

Spectacle et Chevalerie (Med)
La Ferte-Clairbois, 53270 Chammes, Maine 53270, France

Contact: Gilles Raab
Tel: (0033) 43014215
Fax: (0033) 43014531

Promotion and demonstration of the Knightly ideal and associated virtues, by means of historical reconstruction and artistic celebration.

Spirit of England Medieval Theatre Company
2 Haddon Street, Sherwood, Nottingham, Nottinghamshire NG5 2HN

Contact: Joanne Marriott
Tel/fax: 0115 903 6209

Nottingham's foremost Robin Hood enactors, we also perform King Arthur and Ivanhoe shows as well as portraying other famous historical characters and periods.

St Cuthbert's Land (Regia Anglorum)
32 Balliol Square, Merryoaks, Durham, Co.Durham DH1 3QH

Contact: Martin Williams
Tel/fax: 0191 384 3273

Durham-based, Saxon group of a national re-enactment society. Covers the period 950-1086. Military and civilian activities authentically recreated. Women warriors are welcome. Events held nationally. Regular meetings.

St Petersburg Mil Hist Assoc (Nap)
Fontanla 118, 198005 St Petersburg, Russia

Contact: Oleg Sokolov

Stafford Household
6 Frew Close, Stafford, Staffordshire ST16 3FB

Contact: James Wilson
Tel/fax: 01785 223633

We are dedicated to the re-enactment of military and social history of the latter half of the 15th century – emphasising all aspects of the period.

Steel Bonnets
61 Bexley Drive, Normanby, Middlesbrough, Cleveland TS6 0ST

Contact: Neil Tranter
Tel/fax: 01642 456295

Border Reiver re-enactment for schools, museums, films, battles etc. Authentic kit for 'living history' projects. Full training for new members in weaponry / crafts.

Sussex Levy (1450-1500)
171 High Street, Hurstpierpoint, Sussex (W.)

Contact: Richard Fitch
Tel/fax: 01273 834822

Suthur Viking-Lag
36 Aglaia Road, Worthing, Sussex (W.) BN11 5SW

Contact: Will Matthews
Tel/fax: 01903 249588

Swords of Albion
59 Vere Road, Brighton, Sussex (E.) BN1 4NQ

Contact: Spike Medland
Tel/fax: 01273 888053

Medieval combat/re-enactment society, covering all periods 1100-1500. Also re-enactment horses supplied for combat or parades.

Swords of Pendragon
1 Hazel Way, Shrublands Estate, Gorleston, Gt.Yarmouth, Norfolk NR31 8LP

Contact: Dave Holman
Tel/fax: 01493 665506

Tête de colonne, Garde Grenadiers (Nap)
BP 1510, Hotel de Ville, 21033 Dijon Cedex, France

Tel/fax: (0033) 80676226

Thor's Bund
4 Ashford Road, Withington, Manchester, Gt. M20 9EH

Contact: Jim Torrance
Tel/fax: 0161 434 5620

Tirailleurs Corses (Nap)
Villa Erbajola, Chemin du Finosello, 20090 Ajaccio, Corsica

Contact: Rene Chauvin

Tower Hamlets Trayned Bandes (SK)
Windrush House, Burford Road, Minster Lovell, Oxfordshire 0X8 5R2

Contact: Paul Eaglestone
Tel/fax: 01993 702562

Recreating London Trained Bands and Scots Covenant identities of the English Civil Wars. Working for improved standards in Living History, siege, garrison, drill, battle re-enactment.

Towton Battlefield Society
79 Barleyhill Road, Garforth, Leeds, Yorkshire (W.) LS25 1AU

Contact: Andrew Boardman
Tel/fax: 0113 286 5711

Traditionscorps 1813 Finsterwalde
Brandenburger Strasse 53, 03238 Finsterwalde, Germany

Contact: Hans Michael Hillebrandt

The Troop of Shew (ECW)
120 Penrhyn Road, Delapre, Northamptonshire NN4 9ED

Contact: Stan Watts
Tel/fax: 01604 706501

The Troop
Lower Cavendish, 28 Heatherdale Road, Camberley, Surrey GU15 2LT

Contact: Alan Larsen

The Troop is a specialist cavalry reconstruction group, covering all periods. Consultancy and supply of horses, riders, weapons and saddlery for historical shows, skill at arms displays, video and publishing projects.

Trybe Bilkskirnir
35 Cripsey Avenue, Shelley, Ongar, Essex CM5 0AI

The Trybe
1 Kimberley Street, Wymondham, Norfolk NR18 ONU

Contact: Val Cartmell
Tel/fax: 01953 601326

A small, independent group covering 1st to 5th centuries AD, in combat, archery, 'living history', and authentic trading.

Tuatha Na Gael Clan
59 Ballyneety Road, Ballyfermot, Dublin 10, Co.Dublin, Ireland

Contact: Peter O'Connor

Tudor Household
Basement Flat, 46 Buckingham Road, Brighton, Sussex (E.) BN1 3RQ

Contact: Jacqueline Beevis

Tudor Household (Dragon Company)
20 Mendip View, Wick, Bristol, Avon BS15 5PY

Contact: Steve Harrill-Morris
Tel/fax: 0117 937 4059

Tudor Household (Greyhound Company)
66 Mullins Close, Oakridge, Basingstoke, Hampshire RG21 2QY

Contact: Max Bennett
Tel/fax: 01256 63337

Tutbury Castle Medieval Society
Tutbury Castle, Tutbury, Staffordshire DE13 9JF

Tel/fax: 01283 812129

Tyrslith Herred (N F P S)
51 Kedleston Road, Evington, Leicestershire LE5 6HY

Contact: Guy Raynor
Tel/fax: 0116 273 5624

Vikings of middle England; combat and 'living history' displays, archery, lectures plus PR and advertising service. UK and continental (brochure on request). Contact Linda Edwards (administrator) 0116 2735624.

U S Army Reconstitution Group (WWII)
9 Rue Desmazieres, 59110 La Madeleine, France

Tel/fax: (0033) 20518213

US Navy 'Seabees'; GIs of the 29th Infantry Division; 1st Bn. Gordon Highlanders of 51st HD; or French Fusiliers-Marins, all with correct drill.

U S M C (WWII)
89 Les Fontinettes, 59620 Leval, France

Contact: Bertrande De Jonghe
Tel/fax: (0033) 27674325

Ulfstahm (N F P S)
27 Primrose Road, Norwich, Norfolk

Contact: Danni Ware
Tel/fax: 01603 661808

Performers of Dark Age re-enactment, battles and village crafts. Meetings held at Coach and Horses, Norwich, Tuesday nights. Please contact us for membership or booking details.

Union of European Military-Historical Clubs
Schonburgstrasse 50/17, A-1040 Wien, Austria

Contact: Friedrich Nachazel
Tel/fax: (0043) 15034894

The Union embraces all respective groups in Europe for common fostering of military traditions in Europe, currently in 16 countries.

Val Winship's Jousting Tournament
The Lazy 'W', Yapton Lane, Walberton, Sussex (W.) BN18 0AS

Contact: Val Winship
Tel/fax: 01243 551405

Valhalla Herred
20 Snowbell Square, Ecton
Brook, Northampton NN3 5HH

Contact: Mike Haywood
Tel/fax: 01604 412672

We are a Viking re-enactment
group within the larger national
society The Vikings (NFPS). We
perform 'living history' of the 10th
century (crafts and combat).

Valhalla Vikings
5 Booth Place, Eaton Bray,
Dunstable, Bedfordshire LU6
2DR

Tel/fax: 01525 221230

Valiant Knights
36 Shirley Road, Moordown,
Bournemouth, Dorset BH9 1SL

Contact: John Shave

Vanaheim Herred
15 Melrose Close, Whitefield,
Manchester, Gt. M25 6WZ

Contact: Cath Davies
Tel/fax: 0161 796 6264

Venta Silurum 456
18 Sandy Lane, Caldicot, Gwent
NP6 4NA, Wales

Contact: Angie & Mike Day
Tel/fax: 01291 422146

Vexillatio Legionis Geminae
Henneth Rhun, 69 The Warren,
Hardingstone, Northamptonshire
NN4 0EP

Contact: L G Morgan
Tel/fax: 01604 763136

Victorian Military Society
20 Priory Road, Newbury,
Berkshire RG14 7QN

Contact: Dan Allen
Tel/fax: 01635 48628

The leading society in this field
promotes research into all aspects
of military history 1837-1914,
wargames, re-enactment, etc; and
publishes a quartery journal.

Victory Association (WWII)
BP 36, 01480 Jassans-Riottier,
France

Contact: Pascal Raymond
Tel/fax: (0033) 474609714

The Victory Association
specialises in films, re-enactments,
etc.in France and abroad. We are
also looking for new members in
Europe, the USA and the
Commmonwealth.

**Victory In Europe
Re-enactment Assoc**
64 Barrington Avenue, Hull,
Humberside (N.) HU5 4BD

Contact: Andy Marsh
Tel/fax: 01482 448134

Public re-enactments, private
battles; nationwide membership;
new members, venues, always
welcome – ring or write for details
of our British, German, American,
Partisan units of World War II. .

Viking Re-enactment Society
PO Box 259, St Agnes, South
Australia 5097, Australia

Contact: Ray Self

**The Vikings (N F P S)
(Admin body)**
2 Stanford Road, Shefford,
Bedfordshire SG17 5DS

Contact: Sandra Orchard
Tel/fax: 01462 812208

As the original re-enactment
society, The Vikings – Norse Film
and Pageant Society – captures the
atmosphere of the Dark Ages with
a unique blend of authenticity and
humour. Our Vikings, Celts,
Saxons and Normans encapsulate
the whole flavour of Medieval life,
from the simple craftsman to the
professional warrior. We have
performed at home and abroad,
and with equal ease for film crews
and schoolchildren. Contact your
local Hird now.

**The Vikings (N F P S)
(Authenticity)**
12 Cefnfaes Street, Bethesda,
Gwynedd LL57 3BW, Wales

Contact: Russell Scott
Tel/fax: 01248 600605

Viroconium Militia
104 Clark Road,
Wolverhampton, Midlands (W.)
WV3 9PB

Contact: Nick Marshall

Volkerschlacht Leipzig 1813
Im Alten Rathaus, Markt 1,
04109 Leipzig, Germany

Contact: Stefan Poser

**Volontaires Nationaux
DG/FLG (Nap)**
Koenigsberger Strasse 7, 97222
Rimpar, Wurzburg, Germany

Contact: Dieter Heller

Volund's Svieter
The Mission Hall, Holbrook
Road, Stutton, Ipswich, Suffolk
IP9 2RY

War Machine
BCM Box 220, London
WC1N 3XX

Contact: Stuart Andrews
Tel/fax: 0181 809 6119

14th to 16th century mercenary
artillery unit, incorporating
archers, swordsmen & catapults.
Pyrotechnics arranged. Combat
instructors loaned. We are
available for hire.

Warlords
14 Wren Close, Heathfield,
Sussex (E.) TN21 8HG

Contact: K M Bradfield
Tel/fax: 01435 863993

Medieval re-enactment and
display of weapons, costumes,
crafts; medieval tents. Displays of
archery, combat, birds of prey can
be arranged. Terms negotiable.

Warriors of the Sussex Table
21 The Close, Great Dunmow,
Essex CM6 1EW

Contact: J Fitzsamuel-Nicholls
Tel/fax: 01371 875838

**Wars of the Roses Fedn
(Admin body)**
25 Rippon Crescent, Malin
Bridge, Sheffield, Yorkshire (S.)
S6 4RG

Contact: George Heeley
Tel/fax: 0114 233 2155

Re-enactment of 15th century
military and domestic history; we
recruit nationally.

Wessex Dark Age Society
Freshfields, Chieveley, Berkshire
RG20 8TF

Contact: Julia & Jon Day
Tel/fax: 01635 248711

Dark Age re-enactment society
based in the 9th century. Events
are mixed private and public
combat with banquets through the
year.

Wessex Silver Daggers
The Volunteer Inn, Seavington St
Michael, Ilminster, Somerset
TA19 0QE

Contact: Rebecca Lloyd-Jones
Tel/fax: 01460 240126

Westworld
48 Athlestone Road, Harrow,
Middlesex HA3 5NZ

Contact: John Pacey
Tel/fax: 0181 861 0339

**The White Company
(Admin body)**
4 Westlea Close, Wormley,
Cheshunt, Hertfordshire EN10
6JG

Contact: Paul Thompson
Tel/fax: 01992 443636

The largest medieval group in the
country, recruiting nationally into
regional Households.

Wogan Household
25 Rowan Close, Scarborough,
Yorkshire (N.) YO12 6NJ

Contact: Gary Hughs
Tel/fax: 01723 367746

**Wolfbane Historical Society
(Med)**
14 Keep Hill Drive, High
Wycombe, Buckinghamshire
HP11 1DU

Tel/fax: 01494 440277

Wolfguard (Med)
7 Manor Crescent, Newport, Isle
of Wight PO30 2BL

Contact: Julie Clothier
Tel/fax: 01983 527677

Wolfshead
Romney, 14 Aston Close,
Kempsey, Hereford & Worcester
WR5 3JR

Contact: Jane & Derek Stevens
Tel/fax: 01905 821148

Wolfshead Bowmen
Rosemary Cottage, 15 Tas
Combe Way, Willingdon,
Eastbourne, Sussex (E.) BN20 9JA

Contact: Heath Pye
Tel/fax: 01323 503666

Authentic medieval
entertainment; archery displays,
combat; cookery; encampment,
mounted knights and archers,
various crafts; equipment and
extras available for film and
television.

**WWII Living History
Association**
25 Olde Farm Drive, Darby
Green, Camberley, Surrey
GU17 0DU

Contact: D J Bennett
Tel/fax: 01252 875412

Oldest and largest WWII
re-enactment group in the UK
(previously known as the WWII
Battle Re-enactment Association).
Public shows and private battles
mounted throughout the year by
British, German and American
units. Subscription £15.

Wychwood Warriors
42 Argyll Street, Oxford,
Oxfordshire OX4 1SS

Contact: M Lyssen

RE-ENACTMENT SERVICES & SUPPLIES

Y Dalir Lethang (N F P S)
35 Harold Street, Prestwich,
Manchester, Gt M25 7HY

Contact: Paul Windett
Tel/fax: 0161 773 7023

Y Draig Herred (N F P S)
8 Long Row, Caverwall, Stoke,
Staffordshire ST119EJ

Contact: Fon Matthews
Tel/fax: 01782 399285

York City Levy
54 Broadwell Road, Easterside,
Middlesbrough, Cleveland
TS4 3NP

Contact: Paul Morris

Re-enact late 15th century military
life. Our performance includes
combat, drill, archery, mumming,
talks and roleplay in an authentic
encampment. We are a friendly
group.

Yorks Horses
140 Duncan Road, Aylestone,
Leicestershire LE2 8ED

Contact: Neil Elverson
Tel/fax: 0116 2839791

15th century arms and archery,
'living history' re-enactment,
mounted archers. Period clothing
made to order.

Zouche Household
10 Morley Street, Sutton in
Ashfield, Nottinghamshire
NG17 4ED

Contact: Tim Morley
Tel/fax: 01623 552491

A R Fabb Bros Ltd
29/31 Risborough Road,
Maidenhead, Berkshire 5L6 7YT

Contact: J R Fabb
Tel: 01628 23533
Fax: 01628 22705

Manufacturers since 1887 of gold
wire embroidery for uniforms – all
periods and countries. Also
standards, banners, tabards, drum
banners, sashes and
accoutrements.

A R L H O
PO Box 12325 Brisbane,
Elizabeth Street, Brisbane,
Queensland 4002, Australia

Contact: Anthony Cryan

A S Bottomley
The Coach House, Huddersfield
Road, Holmfirth, Yorkshire (W.)
HD7 2TT

Contact: Andrew Bottomley
Tel: 01484 685234
Fax: 01484 681551

Established 30 years with clients
overseas and in the UK. A fully
illustrated mail order catalogue
containing a large range of
antique weapons and military
items despatched world wide.
Every item is guaranteed original.
Full money back if not satisfied.
Deactivated weapons available.
Valuations for insurance and
probate. Interested in buying
weapons or taking items in part
exchange. Business hours Mon-Fri
9am – 5pm. Mail order only. All
major credit cards welcome.
Catalogue UK £5, Euorpe £7, rest
of world £10.

Aarrgg Armoury
1 Churchfield Avenue, Tipton,
Midlands (W.) DY4 9NF

Contact: A S Colley
Tel/fax: 0121 5224405

Dark Age and Medieval weapons,
armour and iron work made to
order.

Abbeyhorn of Lakeland
Holme Mills Industrial Estate,
Holme, Carnforth, Lancashire
LA6 1RD

Contact: H McKellar
Tel/fax: 01524 782387

We specialise in making goods
from ox horn, stag antler and bone;
our range includes kitchenware,
shoehorns, walking sticks, soldiers'
mugs, and much more.

Ages of Elegance (Costumes)
480 Chiswick High Road, London
W4 5TT

Contact: Dawn Wood
Tel/fax: 0181 742 0730

Maker of replica costumes, all
periods; clients include English
Heritage, Bromley Museum,
Franco Zeffirelli. Shop premises
also stock virtually everything the
re-enactor needs. Historical
weddings organised; lecturers
available.

Agnes Grymme (Seamstress)
35 Longfield Road, Tring,
Hertfordshire HP23 4DG

Tel/fax: 01442 824941

Albion Armouries
59 Vere Road, Brighton, Sussex
(E.) BN1 4NQ

Contact: Spike Medland
Tel/fax: 01273 699050

Albion Small Arms
Unit 4, AML Industrial Estate,
Rugeley Road, Hednesford,
Staffordshire WS12 5QW

Contact: Ron Curley
Tel/fax: 01543 426113

Amyris (Costume)
16 Treesdale Road, Harrogate,
Yorkshire (N.) HE2 0LX

Contact: Jon Beavis-Harrison
Tel: 01423 560583
Fax: 01423 522603

Ancestral Instruments
Tudor Lodge, Pymoor Lane,
Pymoor, Ely, Cambridgeshire
CB6 2EE

Contact: David Marshall
Tel/fax: 01353 698084
Email: http://www.gmm.co.uk/ai

Individually crafted musical
instruments for re-enactment –
reedpipes, hornpipes, historical
bagpipes, medieval fiddles,
rebecs, crwth, lyres, etc. Advisory
service; music cassettes; video.

**Andrew Butler Militaria &
Insignia**
10 Avebury Avenue, Ramsgate,
Kent CT11 8BB

Contact: Andrew Butler
Tel: 01843 592529
Fax: 01843 582216

U.S. insignia and uniforms. WW2
to Vietnam. Specialising in rarer
items, 8th AAF, Special Forces,
etc. Re-enactment items also
stocked.

Andrew Kirkham
60 Leedham Road, Rotherham,
Yorkshire (S.) S65 3EB

Contact: Andrew Kirkham
Tel/fax: 01709 540390

Medieval craftsman – supplier to
re-enactment groups. Quality
longbows, crossbows,
hand-forged arrowheads, fire
equipment, ironwork and knives
at competitive prices. Latest
catalogue and further informaton
on request.

Annart Swords & Accessories
490 Bryn Road, Ashton in
Makerfield, Wigan, Lancashire
WN4 8AN

Tel/fax: 01942 274655

Anne Laverick
Vale Head Farm, 52 Pontefract
Road, Knottingley, Yorkshire
(W.) WF11 8RN

Contact: Anne Laverick
Tel/fax: 01977 677390

Historical costumier, and supplier
of specialised fabrics for all
periods. Cottons £1.75 yd, linen
£3.75 yd, wool from £4.50 yd.
Postal service available.

Archery Centre
Highgate Hill, Hawkhurst, Kent
TN18 4LG

Contact: Tom Foy
Tel/fax: 01580 752808

Armour
307 Walsall Wood Road,
Aldridge, Walsall, Midlands (W.)
WS9 8HQ

Contact: Derek Harper
Tel/fax: 01922 57689

Armour & Accessories
The Mill House, 2 Stowe Mill,
Knighton, Powys LD7 1NB,
Wales

Contact: Christopher Franklin
Tel/fax: 01547 520484

Armour Class
Unit 8, Block 17, Abbeymill
Business Centre, Paisley,
Strathclyde PA1 2AP, Scotland

Tel/fax: 0141 889 0688

Arms of Legend
39 Penistone Road, Park End,
Middlesbrough, Cleveland
TS3 0DJ

Art & Archery
Swains Mill, Crane Mead, Ware,
Hertfordshire SG12 9PY

Contact: Terry Goulden
Tel/fax: 01920 460335

**Artemis Archery/Longbow
Presentations**
29 Batley Court, Oldland,
Gloucestershire (S.) BS15 5Y2

Contact: Veronica Soar
Tel/fax: 0117 9323276

Archery related lectures, talks &
displays given. Research.
Instruction: coaching. Traditional
longbow a speciality. Pre-war
artifacts identified. Conservation
advice available. Items purchased.
Book list (SAE).

Attleborough Accessories
White House, Morley St Peter,
Wymondham, Norfolk NR18 9TZ

Contact: C E Pearce
Tel: 01953 454932
Fax: 01953 456744

Militaria and memorabilia,
including: Sheffield-made
Fairbarin Sykes 3rd pattern and
commemorative knives; K-Bar
USMC knives; knife-making
supplies, etc. Large SAE for full
lists.

Bruce Bagley
33 Nevill Road, Rottingdean,
Sussex (E.) BN2 7HH

Tel/fax: 01273 306533

Cast iron cauldrons in a range of
sizes and prices. Personal callers
welcome by appointment only.

**Bainbridge Traditional
Bootmakers**
The Square, Timsbury, Bath BA3
1HY

Contact: David McCabe
Tel/fax: 01761 471430

Makers of high quality
reproduction and traditional
footwear for museums and 'living
historians' worldwide. Clients
include: Museum of London,
National Army Museum, Tower
Armouries.

Battle Orders Scotland Ltd
76 Coburg Street, Leith,
Edinburgh, Lothian EH6 6HJ,
Scotland

Contact: Angus Neilson
Tel: 0131 538 8383
Fax: 0131 555 2071

Suppliers of replica/re-enactment
arms and armour; medieval,
Scottish Highlander and 20th
century weaponry a speciality for
stage/re-enactment; instructional
videos.

Bilbo the Trader
52 Barnards Yard, Norwich,
Norfolk NR3 3DS

Contact: B Dunion
Tel/fax: 01603 766959
Email: b.dunion@netcom.co.uk

Specialist in Dark Age jewellery;
show attendance sales only – will
also attend Medieval shows. No
stock list as stock changes
constantly.

Birkfield (Heraldic flags)
Birkfield, Rumbling Bridge,
Kinross-shire KY13 7PT, Scotland

Contact: Dr Patrick Barden
Tel/fax: 01577 840598

Heraldic banners, standards,
gonfannons, pipe-banners etc. No
bogus heraldry. Hand-painted on
polyester, bunting, satin etc.
Commissions only – no stock kept.

**Bjarni's Boots (Handsewn
Footwear & Leathergoods)**
The Craft Court, Royal
Armouries Museum, Leeds,
Yorkshire (W.) LS10 1LT

Contact: Mark Beabey
Tel/fax: 0113 245 8824

We offer hand-cut and hand-sewn
work made up on a bespoke basis
and constructed to exacting
standards of historical accuracy.
An extensive portfolio of work
covering 2,000 years of footwear
styles. Leather goods include:
armour, vessels, scabbards, cases,
holsters, saddlery and harness.
Clients include: Historic Royal
Palaces, Museum of London,
Royal Armouries, National Army
Museum, MOD (Household
Cavalry).

Black Hat Woodcraft
59 Bridge Street, Downham
Market, Norfolk PE38 9DW

Contact: Matthew Champion
Tel/fax: 01366 388740

Bodgerarmour
129 Kent Road, Mapperley,
Nottinghamshire NG3 6BS

Contact: Dave Hodgson
Tel/fax: 0115 9525711

The Button Lady
16 Hollyfield Road South, Sutton
Coldield, Midlands (W.) B76 1X

Contact: Pauline Walker
Tel/fax: 0121 329 3234

Buttons and clasps made from
natural materials – pewter, china,
shell, wood, horn etc. Spinning
wheels, books, fibres and supplies.

Caduceus
624 Lea Bridge Road, Leyton,
London E10 6AP

Contact: Morgana
Tel/fax: 0181 539 3569

We make an extensive range of
pagan and historical jewellery in
gold and silver; hand-crafted
swords and knives. Catalogue
£1.45; shop open 9.30am-5.30pm,
closed Thursdays.

Call to Arms
7 Chapmans Crescent, Chesham,
Buckinghamshire HP5 2QU

Contact: Duke Henry Plantagenet
Tel/fax: 01494 784271

The Worldwide Directory of
Historical Re-enactment Societies
and Traders. Published twice a
year with continuous updating
service – fax and email
supported – entry in our listings
is free. For sample copy send
£2.50 (UK), £3.00 (airmail
Europe), or £3.50 (airmail rest of
world) – sterling only. Listings
are uniquely annotated with
society size / activity data. Also
contains high quality articles of
news, research and development.
Free and friendly advice on-line
to subscribers, a 'to-your-door'
update service planned. W W
Web pages planned. Email for
details. To find out anything
about Historical Re-enactment
and Living History, first you buy
'Call to Arms'. Get your copy
now. Email: duke
@calltoarms.com

Carol Leathercraft-Fletcher
Craft Cottage, Bookham Lodge
Stud, Cobham Road, Stoke
d'Abernon, Surrey KT11 3QG

Contact: Carol Edwards
Tel/fax: 01932 865181

Mediaeval and modern archery
equipment: wood arrows in
eco-friendly Scandinavian pine,
plus leatherwork to order.
Longbow tuition – details on
request.

Cath Davies Leathercraft
15 Melrose Close, Whitefield,
Manchester, Gt. M25 6WZ

Tel/fax: 0161 796 6264

Celtic Lodges
Dan-y-Coad, Gwynfe, Llangadog,
Dyfed, Wales

**The Chapman's Pad
(Costume, etc)**
Bwlch, Beguildy, Powys LD7
1UG, Wales

Contact: Barry Carter
Tel/fax: 01547 510289

**Chiltern Open Air Museum
(Costume)**
Gorelands Lane, Chalfont St
Giles, Buckinghamshire HP8 4AD

Tel/fax: 01494 871117

Chris Blythman
The Flat, Brook House Farm,
Middleton, Ludlow, Shropshire
SY8 2DZ

Tel/fax: 01584 878591

Quality hand forged military and
domestic ironwork, for museums,
re-enactment, TV and film.

Cinque Pots
26 The Drive, Worthing, Sussex
(W.) BN11 5LL

Contact: Marcus Dole
Tel/fax: 01903 246108

**The Civil Wardrobe
(Costume Sales)**
Newtown Road, Newbury,
Berkshire RG14 7ER

Contact: Debbie Goldsmith
Tel/fax: 01635 43806

Manufacturers of practical period
costume, from museum display
items to good quality
re-enactment costume. Please
send 3 loose 2nd class stamps for
our catalogue.

Classic Images
PO Box 1863, Charlbury,
Chipping Norton, Oxfordshire
OX7 3PD

Contact: Jef Savage
Tel/fax: 01608 676635

American Civil War specialists for
'Gettysburg' – the epic motion
picture; re-enactment videos,

books, prints, music, newspapers,
maps, clothing patterns, toy
soldiers and product search
service.

Clink Armoury
1 Clink Street, London SE1 9DG

Tel/fax: 0171 403 6515

Coeur de Léon Costumes
10 Eastern Road, Sutton
Coldfield, Midlands (W.)
B73 5NU

Contact: Joanne Marriott
Tel/fax: 0121 355 3873

Makers of period costumes and
accessories – specialists in
Medieval period – for re-enactors,
collectors, museums and film
companies. Send 50p & SAE for
details.

Cotswold Forge
2 Exmouth Street, Leckhampton,
Cheltenham, Gloucestershire
GL53 7NS

Contact: Terry Andrews
Tel/fax: 01242 242754

The Crashing Boar (Cloth,etc)
18 Bosworth Close, West
Bletchley, Milton Keynes,
Buckinghamshire MK3 7UB

Contact: Brian Biddle
Tel/fax: 01908 642400

Create the Mood
Redcote, 228 Sydenham Road,
Croydon, Surrey CR0 2EB

Contact: Frances E Tucker
Tel/fax: 0181 684 1095

Historic haberdashery. Frances'
stall, crammed with unusual
things old and new, creates a
bustling atmosphere at costume
and heritage events, 1066–1930.
Invitations accepted.

**Crown Forge Historical
Reproductions**
Moss Lane Farm, Moss Lane,
Warmingham, Crewe, Cheshire
CW1 4RW

Contact: Thomas Phillips
Tel/fax: 01270 583900

D H Simons (Cooper)
Craft Workshops, 56/57
Catherine Street, Swansea,
Glamorgan (W.) SA1 4JS, Wales

Tel/fax: 01792 653871

Danza Antiqua
3 Mill Close, Chadlington,
Oxfordshire OX7 3PA

Contact: Nancy J Walker
Tel/fax: 01608 676635

Improve your period impression
with professional dance and
etiquette instruction by period
dance specialist. Dance and

etiquette workshops for English
and American Civil Wars,
Napoleonic and Renaissance
periods.

Daphne Hilsdon Tebbutt
118 High Street, Winslow,
Buckinghamshire MK18 3DQ

Contact: Daphne Hilsdon Tebbutt
Tel/fax: 01296 713643 eves.

Period dress Roman to 1920s
made. Also demonstrations and
displays of dress, cookery and
social history.

Darr Publications
Thorshof, 106 Oakridge Road,
High Wycombe,
Buckinghamshire HP11 2PL

Contact: Thorskegga Thorn
Tel: 01494 451814
Fax: 01494 784271

Large number of historical
booklets, both theoretical studies
and 'how-to-do-it' manuals.
Heavily researched but
inexpensive, practical and easy to
read. SAE for details. Email:
thorskegga@calltoarms.com

Darr Spinning & Braidmaking
106 Oakridge Road, High
Wycombe, Buckinghamshire
HP11 2PL

Tel/fax: 01494 451814

David Brown Arms & Armour
c/o Gravel Hill, Stretford Bridge,
Leominster, Hereford &
Worcester HR6 9DQ

Tel/fax: 01568 720352

**Dawn of Time Crafts
(Hist eqpt)**
9 Arnold Drive, Colchester,
Essex CO4 3YZ

Contact: Ivor Lawton

Derek Harper (Armour)
307 Walsall Wood Road,
Aldridge, Walsall, Midlands (W.)
WS9 8HQ

Tel/fax: 01922 57689

Downland Sword Supplies
16 Shelley Close, Lewes, Sussex
(E.) BN7 1RG

Contact: Don Blenkinsop
Tel/fax: 01273 472712

Dressed to Kill
The Garden Bothy, Holdenby
House, Holdenby,
Northamptonshire NN6 8DJ

Contact: Mark Taylor
Tel/fax: 01604 770150

Makers of high quality
reproduction arms and armour –
'Real armour for real re-enactors'.
Visit our showrooms (ring first).

**The Drop Spindle
(Costume, etc)**
35 Cross Street, Upton,
Pontefract, Yorkshire (W.)
WF9 1EU

Tel/fax: 01977 647647

Earthchild (Costume, etc)
39 Penistone Road, Park End,
Middlesbrough, Cleveland
TS3 0DJ

**Edward Henry Fox
(Wheelwright)**
Homestead Farm, Lonning Eye,
Reading, Berkshire

Tel/fax: 01734 342532

English Armourie
Department 10, 1 Walsall Street,
Willenhall, Midlands (W.) WV13
2EX

Contact: Alan Jones
Tel/fax: 01902 870579

ECW specialist; steel armour,
matchlock and doglock firearms.
De-activated guns.

**English Heritage SEU
(Admin body)**
Room 101, Keysign House, 429
Oxford Street, London W1R 2HD

Contact: Howard Giles
Tel: 0171 973 3457
Fax: 0171 973 3430

Creation and direction of outdoor
events, specialising in authentic
historical re-enactments and living
histories. English Heritage sites
only. Also contact Thomas
Cardwell or Karen Cooper.

Ensign Embroidery
Kilcreggan, Dunbartonshire,
Scotland

Tel/fax: 01436 842581/842716

**Excalibur Enterprises
(Costume)**
9 Dodbrooke Road, West
Norwood, London SE27 0PF

Contact: Adrian Lucas

Excalibur Enterprises specializes
in high-quality Dark Ages and
Medieval costumes, made in the
finest quality materials, with due
attention given to the customer's
needs. To obtain further details of
our full range send £1.00 plus SAE
for our catalogue and spending
voucher.

First Class Mail
9 Miller Street, Warrington,
Cheshire WA4 1BD

Contact: Tony Whittaker
Tel/fax: 01925 659756

Riveted and butted mail. Made to
your requirements, various types
of links available. Please write for
further enquiries.

Flying Spirit Nose Art Company
81 Charles Street, Epping, Essex
CM16 7AX

Contact: Simon Boultwood
Tel/fax: 01992 574554

Specialises in reproduction WWII
U.S.A.A.F. 'A2' and U.S.N. 'G1'
flying jackets, with nose-art and
insignia. Jacket painting available
and insignia sold separately. Mail
order only. Trade and film
company orders accepted.
Discounts offered to re-enactment
groups and ex-services members.
Catalogue available £3.50 or SAE
for free list.

Funn Stockings Traditional Hosiery
PO Box 102, Steyning, Sussex
(W.) BN44 3DS

Contact: Graham Huntley
Tel/fax: 01903 892841

Traditional silk, cotton, wool
stockings; also cotton Fustian, and
over-the-knee opaque cotton
stay-ups, as used in over 1000
plays, films, musicals etc.

Gallery (Leather, Wood etc)
21 The Close, Great Dunmow,
Essex CM6 1EW

Contact: David Fitzsamuel-Nicholls
Tel/fax: 01371 875838

Gaunt d'Or (Weaponsmith)
58 Springfield Road,
Wolverhampton, Midlands (W.)
WV10 0LJ

Contact: Brian Gunter
Tel/fax: 01902 683875
Email: 106135.2771
@compuserve.com

For authentic reproduction
swords, daggers and other
weapons from the Dark Ages
through to Renaissance. Suitable
for re-enactment, display or wall
hangings.

The Good Old Cause (Hist eqpt)
73 Longden, Coleham,
Shrewsbury, Shropshire

Contact: Adam Johnston
Tel/fax: 01743 271227

Grey Goose Wing Trust (Archery)
59 Netley Street, Farnborough,
Hampshire GU14 6AT

Contact: Craig Townend
Tel/fax: 01252 510534

Guild of Master Craftsmen
166 High Street, Lewes, Sussex
(E.) BN7 1XU

Contact: Information Officer
Tel: 01273 478449
Fax: 01273 478606

The Guild is a trade association
for skilled crafts-people and
companies. It covers over 400
different trades; the helpline can
provide selective lists of members.

Guildhouse & Appleby Designs (Costume)
16 Treesdale Road, Harrogate,
Yorkshire (N.) HE2 0LX

Contact: Jon Beavis-Harrison
Tel/fax: 01423 560583

Handweavers Studio & Gallery
29 Haroldstone Road, London
E17 7AN

Tel/fax: 0181 521 2281

Spinning/weaving supplies, books,
tuition; fibres, fleece, spinning
wheels, spindles, carders, yarns,
looms, shuttles, heddles.

Hector Cole Ironwork
The Mead, Great Somerford,
Chippenham, Wiltshire SN15 5JB

Contact: Hector Cole
Tel: 01666 825794
Fax: 01249 720485

Specialist in medieval ironwork
techniques with particular
reference to arrow and blade
smithing. All London Museum
type arrowheads forged to order.
SAE for details.

Heritage Arms
Units 58/59 Clocktower Centre,
Hollingwood, Chesterfield,
Derbyshire S43 2PE

Contact: J & P Chester
Tel/fax: 01246 475782

Herts Fabrics
11 Brickfield, Hatfield,
Hertfordshire AL10 8TN

Contact: M Archer
Tel/fax: 01707 265815

Specialist in exclusive 'Linen'
fabrics. Suppliers to re-enactment
societies. Very good colour range
and designs kept in stock. Mail
order service available.

Highlander Images (Film & TV)
1 Shearer Street, Kingston,
Glasgow, Strathclyde G5 8TA,
Scotland

Contact: Helen Craig
Tel: 0141 429 6915
Fax: 0141 429 6968

Can provide extensive period
wardrobe / weaponry / props /
accessories, fight / stunt
arrangement crews. Special extras
for film / TV industry,
specialising in the image of
historic Scotland.

Hightower Crafts (Weapons, etc)
Clwt Melyn, Pen Lon,
Newborough, Anglesey,
Gwynedd LL61 6RS, Wales

Tel/fax: 01248 79500

Hilltop Spinning & Weaving Centre
Unit 1, Hope Farm, Sellindge,
Kent TN25 6HH

Contact: Sue Chitty
Tel/fax: 01303 814442

Historic Costume Design
Sussex Farm Museum, Horam,
Heathfield, Sussex (E.) TN21 0JB

Contact: Julie Ede
Tel/fax: 01424 775590 eves.

Historical Management Assocs Ltd
117 Farleigh Road, Backwell,
Bristol, Avon BS19 3PG

Contact: Stuart Peachey
Tel/fax: 01275 463041

Provides consultancy, banqueting,
lectures, agricultural produce, and
displays of the period 1580-1660.
Also publishes books on the
English Civil War and
contemporary civilian topics.

Historical Uniforms Research
Unit 3, Enterprise Centre,
Emmet Road, Ballymote,
Co.Sligo, Ireland

Contact: John Durant
Tel: (00353) 71 83930
Fax: (00353) 71 82404

Manufactures and supplies
world's best reproduction
Waffen-SS reversible camouflage
items to museums, collectors,
re-enactors and film industry.
Smocks, helmet covers,
non-standard field tailored items.
All camouflages and construction
exact to originals. Examples have
been mistakenly judged genuine
by collectors and curators. Can
also supply genuine WSS items
ex-SS 'Skanderbeg' Division; also
1941-1960 Russian camouflages
and other kit used by Albanian
forces. Specialist sourcing, to
order, of Russian WWII weapons,
personal kit, communications
equipment, documents, manuals
etc. Will print to order any of over
200 camouflage patterns;
minimum 1000 metres. Dealer
enquiries welcome. Can also
arrange military interest tours to
Albania. Accept VISA, ACCESS,
MASTERCARD, EUROCARD.

Hoards of Albion (Hist eqpt)
Dark Wood, West Slaithwaite,
Huddersfield, Yorkshire (W.)
HD7 5XA

Contact: Lee & Anthony Gilbert

Ingram's Trading Post (Armourer, etc)
46 Stokesley Road, Seaton
Carew, Hartlepool, Cleveland
TS25 1EW

Intellectual Animals
Endeavour House, 6 Station
Road, Stoke d'Abernon, Surrey
KT11 3BN

Contact: Gerry Cott
Tel/fax: 01932 865412

International Viking Association
46 St John's Court, Princess
Crescent, Finsbury Park, London
N4 2HL

Contact: Colin Richards
Tel/fax: 0181 800 4013

Cultural resource of the Viking
Era. Promotes period research,
educational talks, experimental
archaeology, research consultancy
for films, props, 'living history',
trained extras, re-enactments.
Membership £12.50 pa.

J & A Leonard (Sheepskin shop)
Unit 23, Batten Road, Downton
Business Park, Downton,
Wiltshire

Tel/fax: 01725 22963

J A Morrison & Co (Armour, weapons etc)
The Old Vicarage, Main Street,
Tugby, Leicestershire LE7 9WD

Tel/fax: 01537 56337

Jack Greene Longbows
Oldwood Pits, Tanhouse Lane,
Yate, Bristol BS17 5PZ

Contact: Jack Greene
Tel/fax: 01454 227164

Utility English longbows in
degame/hickory or yew. Varied
for different periods, e.g. plain or
horn nocks etc. Linen strings.
Visitors welcome by appointment.

Jeremy Tenniswood
PO Box 73, Aldershot,
Hampshire GU11 1UJ

Contact: Jeremy Tenniswood
Tel: 01252 319791
Fax: 01252 342339
Email: 100307,1735@compuserve

Established 1966, dealing in
collectable firearms civil and
military, de-activated and for
shooters; also swords, bayonets,
medals, badges, insignia, buttons,
headdress, ethnographica; and
books. Regular lists of Firearms
and Accessories; Medals; Edged
Weapons; Headdress, Headdress
Badges and Insignia;
comprehensive lists Specialist and
Technical Books. Office open
9am-5pm, closed all day Sunday.
Medal mounting service.

Keith Lyon (Leather skins)
146 Balfour Road, Northampton,
Northamptonshire NN2 6JP
Tel/fax: 01604 717349

Kevin Garlick (Footwear)
21 South Street, Ventnor, Isle of
Wight PO38 ING
Tel/fax: 01983 854753

Kitty Hats
32 Birch Way, Chesham,
Buckinghamshire HP5 3JL
Contact: Catherine McLaren
Tel/fax: 01494 773147

Felt hats of most periods, Robin
Hood style, tall Tudor, Civil War
and cocked hats. Some straw hats
available. Also modern and hand
felted hats.

The Leatherworker
c/o The Civil Wardrobe,
Newtown Road, Newbury,
Berkshire RG14 7ER
Contact: Karl Robinson
Tel/fax: 01635 43806

Manufacture of reproduction
military leatherwork for
re-enactors and museums.

Legion Forge
Common Lane Industrial Estate,
Common Lane, Kenilworth,
Warwickshire CV8 2VL
Contact: Andrew O'Leary
Tel/fax: 01926 864667

Made to order re-enactment
weapons, medieval to modern
jewellery and metal tools for
'living history', plus general
blacksmiths work.

Linnet the Seamstress
19 Cowper Close, Mundesley,
Norfolk NR11 8JS
Tel/fax: 01263 721574

15th century costume, eg. shirts,
braies, hose, jerkins, kirtles,
gowns. Personal attention,
reasonable prices. List on request.
Other periods considered.
Accuracy & hand finishing
paramount. 'Living history' &
talks on request. Member of The
Company of the White Lion.

Little Things
4 Christopher Close,
Heckington, Sleaford,
Lincolnshire NG34 9SA
Contact: F L Adams
Tel/fax: 01529 461934

Childrens' 17th century clothing
plus cloaks and accessories such
as collars and coifs.
Haberdashery, ribbon, lace, braid,
etc.

**Living History Resources
(Costume, etc)**
43 Croft Road, Yardley,
Birmingham, Midlands (W.)
BS26 ISQ
Contact: R & J Sheard
Tel/fax: 0121 784 6408

Longship Trading Co Ltd
342 Albion Street, Wall Heath,
Kingswinford, Midlands (W.)
DY6 0JR
Contact: Ivor Wilcox
Tel/fax: 01384 292237

We supply most articles for living
and warfare during the Dark Ages
period. Free brochure available on
request.

M & F Firearms
Hauptstrasse 2, 55483
Schlierschied, Germany
Contact: M Murfin
Tel/fax: (0049) 67651279

M Champion Reproductions
The Thatched Cottage, Manor
Fram, Kerdiston, Reepham,
Norfolk NR10 4RY
Contact: M Champion
Tel/fax: 01603 870308

Specialising in reproducing
original artefacts in a range of
materials (including gold and
silver) for museums, re-enactment
societies and schools.

M J Hinchcliffe (Bladesmith)
73 Minterne Waye, Hayes,
Middlesex UB4 0PE
Contact: Martin Hinchcliffe
Tel/fax: 0181 561 5996

Maggie's Period Costumes
22 Garden Crescent, Pelsall,
Walsall, Midlands (W.) WS3 4NQ
Tel/fax: 01922 691228

The Mailman
Plague Pit Cottage, 71 Church
Street, Chesham,
Buckinghamshire HP5 1HY
Contact: Ken Polton
Tel/fax: 01494 776320

Marcus Music
Tredegar House, Newport,
Gwent NP1 9YW, Wales
Contact: W M Butler
Tel: 01633 815612
Fax: 01633 816979

Makers, repairers and suppliers of
rope-tensioned drums, hooped or
stitched skinned. Period, size,
colour to customer requirements.
Early stringed instruments also
available.

**Mayhem Photographics
(History in Camera)**
Basement Flat, 10 Alexandra
Road, St Leonards on Sea, Sussex
(E.) TN37 6LE
Contact: Dick Clark

Medi-Arm (Weaponsmith)
25 Rippon Crescent, Malin
Bridge, Sheffield, Yorkshire (S.)
S6 4RG
Contact: George Heeley
Tel/fax: 0114 2332155

Medieval Clothing Company
49 Washington Grove, Bentley,
Doncaster, Yorkshire (S.)
DN5 9RJ
Contact: Sally Ann Chandler
Tel/fax: 01302 876343

Authentic reproduction clothing
and uniforms for re-enactment,
schools and museums from the
Conquest to early 20th century
hand-made to order.

Medieval Tournament School
Austins Farm, Twytchells Lane,
Jordans, Buckinghamshire HP9
2RA
Tel/fax: 01494 871493

Megin-Gjord
12 Cefnfaes Street, Bethesda,
Gwynedd LL57 3BW, Wales
Tel/fax: 01248 600605

Supplier of authentic Dark Age
patterns for shoes, scabbards etc.
Also manufacturer of scabbards,
sheaths, belts, all kinds of fittings
in pewter, bronze, silver and
gold-gilt. Fitment of horns, tusks,
bear and wolf teeth.

The Merchant's House
1 Marsh Lane, Carlton Colville,
Lowestoft, Suffolk NR33 8BW
Contact: Ian Pycroft
Tel/fax: 01502 511093

Costumed interpretation and
living history displays. Pageants,
special events and corporate
hospitality. Research, consultation
and lecture services to schools.
Reproductions for museums and
education.

Mercia Sveiter
Rosemary Cottage, Camp Road,
Canwell, Sutton Coldfield,
Birmingham B75 5RA
Contact: Paul Craddock
Tel/fax: 0121 323 4309

Drinking horns, chainmail links,
pressed shield bosses, swords,
spears, sheepskins, period
jewellery, bronze castings, leather
goods. Celtic to Medieval
specialist.

Merlin Enterprises
24 Prices Lane, York, Yorkshire
YO2 IAL
Tel/fax: 01904 611537

Import-export & supply of
re-enactment weapons, period
arms and armour – fencing
equipment from A to Z. PX
upgrades, secondhand kit and
theatrical props.

Mick the Mail Maker
Windemere, Lake Road, Kinson,
Bournemouth, Dorset
Contact: Mick Goff

Military Collectables
93 Lynwood Crescent,
Pontefract, Yorkshire (W.)
WF8 3QX
Contact: P A Hampton
Tel/fax: 01977 792084

World War II British, US and
Allied uniforms, webbing
equipment, personal kit etc. Mail
order, telephone enquiries;
personal callers by appointment.

Military Metalwork
1 Almond Grove, Brentford,
Middlesex TW8 8NP
Contact: Andrew Clark
Tel/fax: 0181 568 3210

Specialists in reproduction
uniform and costume
accoutrements and regalia,
including leatherwork, circa
1600-1900. Napoleonic and
Victorian periods a speciality.
Electroformed shako plates and
gilding, cast metal badges, buttons
and fittings.

Minerva Trading Company
51 Market Crescent, Wingate,
Co.Durham TS28 5AJ
Contact: Ian & Linda Frances
Tel/fax: 01429 837787

**Mistress MacQueen
(Seamstress)**
51 King's Drive, Eastbourne,
Sussex (E.) BN21 2NY
Tel/fax: 01323 32983

**Mists of Time
(Historical consultant)**
67B Pembroke Road, Clifton,
Bristol, Avon BS8 3DW
Contact: David Lazenby
Tel/fax: 0117 9731860

Morgan (Braid)
147-149 Market Street,
Broadley, Whitworth, Lancashire
OL12 8RY
Tel/fax: 01706 343205

Mountainstone Forge & Armoury
8 Fairfield Road, Morecambe, Lancashire LA3 1ER

Contact: Peter Constantine
Tel/fax: 01524 401292

Viking, Norman, Medieval, fantasy. Swords, daggers, scrams, axes, helmets, shields, polearms, hearth furniture and fittings. Museums supplied, hotels etc.

Musket, Fife & Drum
27 Church Road Business Centre, Sittingbourne, Kent ME10 3RS

Contact: Phil Bleazey
Tel: 01795 470149
Fax: 01795 474109

Manufacturers of black powder weapons – 17th Century muskets, and earlier pieces (made to order).

The Needle's Excellency
Wynnstay Cottage, Lamin Gap Lane, The Fosse, Colgrave, Nottinghamshire NG12 3HG

Contact: Lesley Coddington
Tel/fax: 01949 81743

P Butler (Medieval craftsman)
Delbush, Whatlington Road, Battle, Sussex (E.) TN33 0JN

Tel/fax: 01424 772351

Accurate reproductions for museums, collectors and the discerning re-enactor. Boiled leather bottles, tankards, helmets and armour. Jewellery, combs, spectacles, dice, shoes, pattens and other items.

Paescod
38 Cowleigh Road, Malvern, Hereford & Worcester WR14 1QD

Contact: Phil Howard
Tel/fax: 01684 292940
*Email: paescod
@malverns.demon.co.uk*

A troupe of raggle-taggle musicians (12th to 17th centuries). Parades, marches, victory banquets and open-air performances. Website: www.malverns.demon.co.uk/paescod.

Past Lincs (Carpentry)
75 Lincoln Road, Ruskington, Sleaford, Lincolnshire NG34 9AR

Contact: E Chambers

Past Tents
Hill View Bungalow, Main Street, Clarborough, Retford, Nottinghamshire DN22 9NG

Contact: John Waterhouse
Tel/fax: 01777 869821

High quality reproduction tents to suit any period; we specialise in late Medieval, English Civil War, Napoleonic and American Civil

War designs. Extensively researched; made of waterproof and dry rot resistant grade A, 12 or 14oz cotton duck, with one year warranty. Repair service also available. For catalogue send A5 size SAE.

Past Unlimited (inc. Tan Your Hide)
78C Carleton Road, Tufnell Park, London N7 0ES

Contact: Cormac O'Neill
Tel/fax: 0171 607 6061

Paul Windett Design (Hist eqpt)
35 Harold Street, Prestwich, Manchester, Gt. M25 7HY

Tel/fax: 0161 773 7023

Pendragon (Craft goods)
The Glastonbury Experience, 2 High Street, Glastonbury, Somerset BA6 9DU

Tel/fax: 01458 832533

Period Crossbows
7 Alexandra Close, Milton Regis, Sittingbourne, Kent ME10 2JP

Contact: Robin Knight
Tel/fax: 01795 427461

Crossbows made to order for re-enactors and collectors. Dark Age to 19th century, including stone bows. Full range of historically accurate bolts available.

Peter Butler (Weapons etc)
Delbush, Whatlington Road, Battle, Sussex (E.) TN33 0JN

Tel/fax: 014246 2351

Petty Chapman
26 Halifax Old Road, Birkby, Huddersfield, Yorkshire (W.) HD1 6EE

Contact: David Rushworth
Tel/fax: 01484 512968

Phil Fraser Medieval Supplies
70 Markyate Road, Daghenham, Essex RM8 2LD

Contact: Phil Fraser
Tel/fax: 0181 592 3621

Maker of Medieval and Dark Age re-enactors' weapons, armour, leather accoutrements and archery equipment. 'Living history', role-play and fantasy items made to order.

Plantagenet Shoes
82 Cozens Hardy Road, Sprowston, Norwich, Norfolk NR7 8QG

Contact: Morgan Hubbard
Tel/fax: 01603 414045

Quartermasterie
Flat 1 Shelley Court, 4 Lovelace Road, Surbiton, Surrey KT6 6NP

R M W Historical Research Consultants
1 Shearer Street, Kingston, Glasgow, Strathclyde G5 8TA, Scotland

Contact: John Christison
Tel/fax: 0141 429 6915

Raven Armoury
Handleys Farm, Dunmow Road, Thaxted, Essex CM6 2NX

Contact: Simon Fearnham
Tel/fax: 01371 870486

Handcrafted swords, chainmail and armour for collectors and re-enactors. Sword range includes historical designs on display at The Royal Armouries Shop in The Tower of London. Fully guaranteed.

Redcoat (18th/19thC Mil Reproductions)
37 Lee Avenue, North Springvale, Melbourne, Victoria 3171, Australia

Contact: Terence Young

Richard Dunk (Armourer)
23 Overhill Road, Burntwood, Walsall, Midlands (W.) WS7 8SU

Tel/fax: 01543 684726

Manufacturer and restorer of arms, armour and period ironwork.

Richard Head (Longbows)
405 The Spa, Melksham, Wiltshire SN12 6QL

Tel/fax: 01225 790452

Manufacture and sale of longbows, arrows, accessories and materials. Longbow seminars and bowmaking demonstrations.

Roy King Historical Reproductions
Sussex Farm Museum, Manor Farm, Horam, nr. Heathfield, Sussex (E.) TN21 0JB

Tel/fax: 01435 33733

Reproduction armour, helmets, edged weapons and some associated goods. Theatrical and film props, cannons, siege engines etc. Pyrotechnic service for simulated battles available. Mail order, limited catalogue available – most items made to order. Visitors by appointment only. Surrounding fields and Farm Museum available for location hire.

Royal Swords & Daggers
490 Bryn Road, Ashton in Makerfield, Wigan, Lancashire WN4 8AN

Tel/fax: 01942 274655

Running Wolf Productions (Video etc)
Crogo Maws, Corsock, Castle Douglas, Kirkcudbrightshire DG7 3DR, Scotland

Contact: M Loades
*Email: mike
@runningwolf.softnet.co.uk*

Video production company – suppliers of 'Archery – Its History and Forms' and 'The Blow by Blow Guide to Swordfighting'. Mike Loades also available for lectures, workshops and fight arranging assignments.

Saemarr (Armour etc)
73 Howdale Road, Downham Market, Norfolk PE38 9AH

Contact: Peter Seymour
Tel/fax: 01366 384316

Dark Ages, Medieval clothing, weapons, armour, leatherware, domestic and military artwork and utensils. Specialists in heathen Germanic jewellery, regalia; free runelore tuition and divination service.

Sally Green Historical Costume
1 Lyng Lane, North Lopham, Diss, Norfolk IP22 2HR

Contact: Sally Green
Tel/fax: 01953 681676
Email: sally.green@calltoarms.com

12th to 19th century accurate, quality costume at reasonable prices. 17th and 15th century clothing for re-enactors a speciality. Other periods and requirements to order.

Sarah Juniper
109 Woodmancote, Dursley, Gloucestershire GL11 4AH

Tel/fax: 01453 545675

Sarah Thursfield Historical Costumes
Ashgrove, Overton Road, St Martins, Oswestry, Shropshire SY11 3DG

Tel/fax: 01691 778019

Patterns, demonstrations and technical advice; museum quality clothing and sewn accessories especially pre-1650. Capacity limited, only one pair of hands.

Sea-Side Productions (Costume etc)
4 Church Lane, Eccles on Sea, Norwich, Norfolk NR12 0SY

Contact: Tim Wilson
Tel/fax: 01850 414493

Armourer and weapon-smith. Armour made to measure; swords, knives, axes, spears etc. made to order, Dark Age to Late Medieval. Also Dark Age jewellery.

Service Commemoratives Ltd
PO Box 173, Dromana, Victoria
3936, Australia
Tel/fax: (0061) 359 810201

Military service commemorative
medals and a range of over 100
different area and specialist
clasps/bars available for:
combatant service; sea service;
foreign service; army service;
aviation service; volunteers;
national defence; Antarctica
service; and POW liberation 1945.

**Sligo Living History
(Costume, etc)**
67 Avondale, Sligo, Ireland
Contact: Noel Connolly

Streets of Time
19th Hole, 107 Fairway,
Keyworth, Nottinghamshire
Contact: Dave Brown

**Sue Benjafield Historical
Costumes**
4 Kilton Crescent, Worksop,
Nottinghamshire S81 0AS
Contact: Sue or Dave Benjafield
Tel/fax: 01909 480482

17th century costume specialists.
Some other periods from 1400
onwards. Special commissions no
problem. SAE for price list. Phone
for more details.

**Syntown Militaria
International**
Westway, Rectory Lane,
Winchelsea, Sussex (E.) TN36
4EY
Contact: Christopher Coxon
Tel: 01797 223388
Fax: 01797 224834

Stockists of high quality
reproduction and original
uniforms, equipment and soldiers'
personal small kit from WWII,
exact repro German camo
garments a speciality, Axis and
Allied lists quarterly £2.00 each.
Access, Visa and Diners Club by
post or phone.

Tarpaulin & Tent Company
101-103 Brixton Hill, London
SW2 1AA
Tel: 0181 674 0121
Fax: 0181 674 0124

Terry Abrams
3 Ongar Road, Margaret Roding,
Dunmow, Essex CM6 1QP
Tel/fax: 0124 531753

Dealer in de-activated firearms
including vintage military
handguns. Parts available for
modern and vintage weapons
including magazines and grips.
Callers by appointment only.

Tim Noyes (Armourer)
46 Eddington Lane, Herne Bay,
Kent CT6 5TS
Tel/fax: 01227 373663

The Time Lords (Med)
BCM Box 220, London
WC1N 3XX
Contact: Stuart Andrews
Tel/fax: 0181 809 6119

Specialists in presenting and
producing your re-enactment
event. Pyrotechnics and stunts
arranged; PA systems and sound
production; video and theatrical
events catered for; 15 years
experience.

Time Travellers (Hist eqpt)
19 The Arches Craft Arcade,
Granary Wharf, Leeds, Yorkshire
(W.) LS1 4BR
Tel/fax: 0113 2342375

**Tony Sayer (Armour,
weapons etc.)**
71 Lichfield Mount, King's Road,
Bradford, Yorkshire (W.)
BD2 1NX
Tel/fax: 01274 727925

Traditional Oak Carpentry
26 Cowslip Close, Ipswich,
Suffolk IP2 0NZ
Contact: Rick Lewis
Tel/fax: 01473 210679

Conservation of listed buildings,
bespoke carpentry for the heritage
industry and serious 'living
history' enthusiast. Chests, gun
carriages, benches, pavises etc. All
work fully researched and
accurately reproduced by hand.

Trenchart
PO Box 3887, Bromsgrove,
Hereford & Worcester B61 0NL
Tel/fax: 0121 453 6329

Suppliers to museums, gift and
surplus shops. Inert ammunition,
all calibres; bullet keyrings,

dogtags, souvenirs, replica
grenades, de-activated weapons,
MG belts and clips, large cartridge
cases, etc. Trade only supplied.

Two Js (Armourers)
c/o 32 Ashfield Drive, Anstey,
Leicestershire LE7 7TA
Contact: J Austin
Tel/fax: 0116 2363514

**Unicorn (Leatherwork,
Jewellery)**
15 Stevenson Place, Littleover,
Derbyshire DE23 7EX

Victor James (Tent maker)
427 Anglesey Road, Burton on
Trent, Staffordshire DE14 3NE
Contact: Victor James
Tel/fax: 01283 510285

Victor James (Tent Makers)
manufacture of 12oz duck cotton
tents to re-enactor's designs for all
periods, send SAE for catalogue.

Viking Crafts
20 Snowbell Square, Ecton
Brook, Northampton NN3 5HH
Contact: Mike Haywood
Tel/fax: 01604 412672

Makers of authentic style games,
jewellery and other artifacts of the
Viking age. Supplier of weapons,
armour, helmets and shields etc.

Vitae Lampadae
25 Rowan Close, Scarborough,
Yorkshire (N.) YO12 6NH
Contact: G Hughs
Tel: 01723 367746
Fax: 01723 342111

Historical education services to
schools, libraries and museums,
with full coverage of the national
curriculum. Costumed
interpreters, artifacts, work sheets.
SAE for further details.

**Warcraft (Armour,
weapons etc)**
35 Barnhill Gardens, Marlow,
Buckinghamshire SL7 3HB
Contact: Tony Masefield
Tel/fax: 016284 2372

Warrell, Adrian (Armourer)
32 St Andrews Road South, St
Anne's on Sea, Lancashire
FY8 1PS
Tel/fax: 01253 780401

Westwood Pewter Studio
116 Upper Westwood, Bradford
on Avon, Wiltshire BA15 2DN
Contact: Terry Robinson
Tel/fax: 012216 7754

Weyland Iron
176 Victoria Street, Hartshill,
Stoke on Trent, Staffordshire
ST4 6HD
Contact: A E Mason
Tel/fax: 01782 616381

Wyedean Weaving Company
Bridgehouse Mill, Haworth,
Yorkshire (W.) BD22 8PA
Tel: 01535 643077
Fax: 01535 646671

Manufacturers of braid, cords,
webbing, and other uniform
accoutrements. Special orders
made up for historical
reconstructions – British Army
regimental lace etc.

Wyte Stone Metalcrafts
2 Laggotts Close, Hinton
Waldrist, Oxfordshire SN7 8RY
Contact: Paul Ellison

Bronze Age to ECW: jewellery,
costume accessories, esoterica and
other properties to your
requirements. Film, stage and
fantasy. Illustrated catalogue
available.

TOY SOLDIER MANUFACTURERS & SUPPLIES

All the Queen's Men
The Old Cottage, Gilmorton,
Lutterworth, Leicestershire
LE17 5PN

Contact: D Cross
Tel: 01455 552653
Fax: 01455 557787

Designers & producers of military
miniatures (toy soldiers). 490 sets
available. Full colour catalogue
with descriptive lists. Plus unique
exquisite interlocking base
vignettes. Visits to our
showroom/museum by
appointment.

Arkova
29 Taw View, Fremington,
Barnstaple, Devon EX31 2NJ

Contact: Allan R Over

Armoury of St James
17 Piccadilly Arcade, Piccadilly,
London SW1Y 6NH

Contact: Richard Kirch
Tel: 0171 493 5082
Fax: 0171 499 4422

Specialists in Orders of Chivalry
of the world and hand painted
military and historical model
figures.

Army Supply Company
6 Old Bank, Ripponden, Halifax,
Yorkshire (W.) HX6 4DG

Contact: Michael Carter

Au Plat d'Etain
16 Rue Guisarde, 75006 Paris,
France

Avant Garde Models
22 Barcaldine Avenue, Chryston,
Glasgow, Strathclyde G69 9NT,
Scotland

Contact: Derek McCarron

Manufacturers of quality toy
soldiers; mail order only. Ranges
include ACW, WWII, Scottish and
Montenegrin figures.
Commissions undertaken. Send
SAE or IRC for list.

B F M Collectables
Nuthatches, Crown Gardens,
Fleet, Hampshire GU13 9PD

Contact: B Ford
Tel/fax: 01252 614362

Producers of superb historical
figure sets realistically sculpted
and painted in traditional style.
SAE for leaflet; trade enquiries
welcome.

Bastion Models
36 St. Mary's Road, Liss,
Hampshire GU33 7AH

Contact: Andrew Rose
Tel/fax: 01730 893478

Toy soldier designer of Bastion,
Wessex, and other ranges. Author
of 'Collector's All-Colour Guide to
Toy Soldiers'. C.B.G.Mignot
dealer.

Britains Petite Ltd
Chelsea Street, New Basford,
Nottingham, Nottinghamshire
NG7 7HR

Contact: Richard Hopkins
Tel: 0115 9420777
Fax: 0115 9420250

Continuing the classic tradition of
the best-known toy soldiers in the
world.

Campaign Miniatures
5 Barrowgate Road, Chiswick,
London W4 4QX

Contact: Peter Johnstone

Charles Hall Productions
Paisley Terrace, Edinburgh,
Lothian EH8 7JW, Scotland

Contact: Charles Hall

Cèard-Staoine
Liebigstr 8, D-91052 Erlangen,
Germany

Contact: Friedrich Frenzel
Tel/fax: (0049) 9131 34973

Private producer of flats: 30mm
Celtic and British history, 30mm
English literature, 10mm doll's
house tin figures, 20mm dioramas
in matchboxes.

Derek Cross (A Q M) Ltd
The Old Cottage, Gilmorton,
Lutterworth, Leicestershire LE17
5PN

Contact: Derek Cross
Tel: 01455 552653
Fax: 01455 557787

We are designers & producers of
military miniatures (toy soldiers);
over 373 sets available. Full colour
catalogue with descriptive lists.
Visits to our showroom/museum
by appointment.

**Dorset (Metal Model)
Soldiers Ltd**
PO Box 2112, Yeovil, Somerset
BA22 7YD

Contact: Giles Brown
Tel: 01935 851550
Fax: 01935 850234

Manufacturers of metal toy-style
soldiers, civilians, spare parts and
related items. Send SAE or 2 IRCs
for further details.

Drill Square Toy Soldiers
8 Orchard Close, Towcester,
Northamptonshire NN12 7BP

Contact: Dave Sparrow

Drumbeat Miniatures
4 Approach Road, Ramsey, Isle of
Man IM8 1EB

Contact: Peter Rogerson
Tel/fax: 01624 816667

A large and rapidly expanding
range of toy soldiers covering
many periods and unusual
subjects. Sets can be varied on
request and a sculpting service is
offered.

Ducal Models
5 Weavills Road, Eastleigh,
Hampshire SO50 8HQ

Contact: Thelma Duke
Tel: 01703 692119
Fax: 01703 602456

Makers and distributors of an
international range of hand
crafted and painted 54mm metal
ceremonial figures, mounted and
on foot. World wide mail order;
colour illustrated catalogue £3.95
(overseas postage extra) includes
badges and postcards. Visitors
welcome to see our extensive
display weekdays 9am-4.30pm;
also some Saturdays, by
appointment. Phone for directions.

Ensign Historical Miniatures
32 Scaitcliffe View, Todmorden,
Lancashire OL14 8EL

Contact: Paul Wood
Tel/fax: 01706 818203

Manufacturers of white metal
54mm traditional toy soldiers.
Ranges include: English Civil
War, Colonial, Indian Army and
Vikings. All figures available
painted or as castings.

F & S Scale Models
227 Droylsden Road,
Audenshaw, Manchester, Gt.
M34 5RT

Tel/fax: 0161 370 3235

Forbes & Thomson
Burgate Antiques Centre, 10c
Burgate, Canterbury, Kent

Contact: Rowena Forbes

Specialist dealers in old toy
soldiers by Britains, Johillco etc.
and related lead and tinplate toys.
Model soldiers & kits by Rose
Models and Chota Sahib. Painted
military miniature figures. Mail
order and office address, P.O. Box
375, South Croydon, CR2 6ZG.
Open Mon-Sat 10.00am-5.00pm.

G P M Models
Twiga Lodge, Victoria Avenue,
Kirkby-le-Soken, Essex CO13 0DJ

Tel/fax: 01255 673013

George Opperman
Flat 12, 110/112 Bath Road,
Cheltenham, Gloucestershire
GL35 7JX

Contact: George Opperman

Private collector with large
collection to sell. 50,000 lead
soldiers, military books and
magazines, post and cigarette
cards. Send SAE and 3 x 1st class
stamps for lists. Mail order only.

Glebe Miniatures
Retreat House, Dorchester Road, Broadwey, Weymouth, Dorset DT3 5LN

Contact: Peter Turner
Tel/fax: 01305 815300

Glebe Miniatures, in addition to their original range, now offer marching sets to complement action sets, and action sets to complement marching sets – all in the original Britains style. The first sets represent the Balkan Wars and the Russian/Japanese conflicts. There are new mounted officers and action artillery sets. Other period sets will follow.

Good Soldiers
246 Broadwater Crescent, Stevenage, Hertfordshire SG2 8HL

Contact: Alan Goodwin
Tel/fax: 01438 354362 (pm)

Producers of 'toy style' soldiers and figures, also cartoon characters and personalities. These are available painted and unpainted. SAE for lists. Trade enquiries welcomed.

Great Britain & Empire Toy Soldiers
The Cedars, 97 High Street, Coningsby, Lincolnshire LN4 4RF

Contact: Andrew Humphries
Tel/fax: 01526 342012

750 types of hand-painted traditional toy soldiers. Castings available; old toy soldiers repaired. Send for catalogue. Figures purchased direct by mail order.

H M of Great Britain (Hadencroft Ltd.)
Unit 8, 22 Leyburn Road, Sheffield, Yorkshire (S.) S8 OXA

Contact: Peter Kingsland
Tel/fax: 0114 209969

Manufacturers of finely sculpted models with a 'toy soldier' finish. Ranges: Field Day – action poses, full dress uniform; On Foreign Service – action poses, 1879-1900; Jubilee – parade figures, 1890-1910. Some European subjects planned.

Henley Model Miniatures
24 Reading Road, Henley on Thames, Oxfordshire RG9 1AB

Contact: David Hazell
Tel/fax: 01491 572684
Email: enquiries@toysoldier.co.uk

We stock the widest range of painted toy soldiers available in the UK, together with castings, paints and modelling materials. Mail order a pleasure.

J S Dietz
Monreith, 233 Alexandra Parade, Kirn, Dunoon, Argyll PA23 8HD, Scotland

Manufacturer of new toy soldiers. Culloden, Gallipoli, Palestine, Western Front, Peninsular, French & Indian Wars, New Orleans, 1812, soldiers & artillery. Free lists. Trade enquiries welcome.

Jean-Pierre Feigly
BP 66, 93162 Noisy le Grand, France

King & Country
20 Rockingham Way, Portchester, Fareham, Hampshire PO16 8QS

Contact: Mike Neville
Tel/fax: 01329 233141

Suppliers of fine quality, all-metal, handpainted 54mm military and civilian figures, plus handcarved 1:32 scale desk top display. Aircraft and vehicles. Mail order our speciality.

Laurie Granfield
20 Whittan Close, Rhoose, Glamorgan (S.) CF6 9FW, Wales

Lone Warrior
PO Box 16171, Glasgow G13 1YJ, Scotland

Contact: Les White
Tel/fax: 0141 959 9141

Manufacturer of quality metal 54mm toy soldiers, specialist in unusual military periods. Commissions undertaken. Supplier of new, reissue and orginal plastic figures. List available.

M D M Les Grandes Collections
9 Rue Villedo, 75001 Paris, France

Tel/fax: (0033) 142617601

Manufacturer of tin soldiers, miniatures for collectors; Napoleonics; French Regimental standards of US War of Independence; mail order service.

M J Models
15 Station Road, Carcroft, Nr Doncaster, Yorkshire (S.) DN6 8DB

Contact: S Murray

M K L Models
PO Box 32, Wokingham, Berkshire RG11 4XZ

Contact: Lynn Kenwood
Tel/fax: 01734 733690

Macs Models
133-135 Canongate, The Royal Mile, Edinburgh, Lothian EH8 8BP, Scotland

Tel/fax: 0131 557 5551

Mainly Military
61 Fore Street, Ilfracombe, Devon EX34 9DJ

Tel/fax: 01271 865093

Marksmen Models
(Dept MDS), 7 Goldsmith Avenue, London W3 6HR

Contact: Michael Ellis
Tel: 0181 992 0132
Fax: 0181 992 5980

High quality, low cost plastic figures from 25mm to 120mm, most periods from ancient to Korean War. Mostly cast from original Marx and Ideal moulds. Also sculpting and visualising for many leading manufacturers.

Militia Models
Rosedean, Gorsty Knoll, Coleford, Gloucestershire GL16 7LR

Contact: Esme Walker

Cottage industry still producing – after sudden death of founder Ken Walker – small number of new 54mm figures in limited edition action sets, 1870-1902 period.

Naegel
Stand B23, Grays Antique Mkt, 1-7 Davies Mews, Davies Street, London W1

Contact: Stephen Naegel
Tel: 0171 491 3066
Fax: 0171 493 9344

Largest display of lead toy soldiers in Europe. Books; spare parts; list available; mail order welcome. Open 10am-4pm, Mon-Fri.

New Cavendish Books
3 Denbigh Road, London W11 2SJ

Tel: 0171 229 6765
Fax: 0171 792 0027

Specialist publishers of quality illustrated books on toys and other collectables.

P & B England
119 Farnham Road, Guildford, Surrey GU2 5QE

The Parade Ground
6 Seventh Avenue, Bridlington, Humberside (N.) YO15 2LQ

Tel/fax: 01262 673724

Patricks Toys and Models
107-111 Lillie Road, Fulham, London

Tel/fax: 0171 385 9864

Pax Britannica
Tharn Cottage, 67 Malcolm Road, Peterculter, Aberdeen AB14 OXB, Scotland

Contact: T Brown

Manufacturer of hand painted toy soldiers in traditional style, specialising in Scottish units.

Piper Craft
4 Hillside Cottages, Glenboig, Lanarkshire ML5 2QY, Scotland

Contact: Thomas Moles
Tel: 01236 873801
Fax: 01236 873044

Manufacturer of white metal military and non-military figures designed to a general scale of 75mm. Suppliers to museums, places of historic interest, shops and collectors. Established 1985. Send SAE or 2 x IRC's for a complete illustrated list.

Pixi
25 Rue Amelot, 75011 Paris, France

Pride of Europe
Shamrock Villa, Southernhay, Clifton Wood, Bristol, Avon BS8 4TL

Contact: R J Dew

Pride of Europe produce an economically priced range of new 'old style' toy soldiers made of metal alloy and hand painted to a high standard.

Prince August UK Ltd
Dept DM, Small Dole, Henfield, Sussex (E.) BN5 9XH

Quality Model Soldiers
Hippins Farm, Blackshawshead, Hebden Bridge, Yorkshire (W.) HX7 7JG

Contact: G M Haley
Tel/fax: 01422 842484

Quality Model Soldiers is pleased to announce the re-launch of toy soldiers from the Franco-Prussian War. These finely cast colourfully painted and realistically animated models were first issued in the late 1970s and early 1980s and soon gained a devoted following. The new range with entirely new figures will be produced on a thematic basis representing regiments that figured prominently in actions from the campaign. Foot, cavalry, artillery and equipment will be made, production being limited to 150 sets, obtainable from the manufacturer only. There will be the option of obtaining limited edition single figures. In addition, a special display set will be available of 50 sets only for each action, featuring scenery or buildings. For further details on the first sets "Action at Schirlendorf" please contact us at the above address.

R A E Models
Unit 2, Service Road, (off Corrie Road), Addlestone, Surrey KT15 2LP

Rank & File
16 Oxburton, Stoke Gifford,
Bristol BS12 6RP

Contact: P Tarrant
Tel/fax: 01454 777278

Produce toy soldiers for the
connoisseur from the late 1800s.
Also high detail model waterline
ships of the Merchant and Royal
Navy. Bands of the county
regiments, rifle brigades and
fusiliers are available to order.
Send SAE for price list.

Red Box Toy Soldier Co
Manor Field House,
Aston-cum-Aughton, Sheffield,
Yorkshire (S.) S31 0XJ

Contact: Dennis Johnson

Replica Models
40 Durbar Avenue, Foleshill,
Coventry, Midlands (W.) CV6
5LU

Tel/fax: 01203 684338

S A C Ltd
Studio Anne Carlton, Flinton
Street, Hull, Humberside (N.)
HU3 4NB

Contact: M A Schofield
Tel: 01482 327019
Fax: 01482 210490

Manufacturers of the finest hand
made and hand decorated chess
sets in the world, many of them
featuring military campaigns and
battles.

Sarum Soldiers Ltd
2A Upper Tooting Park, London
SW17 7SW

Contact: Patrick Willis
Tel: 0181 767 1525
Fax: 0181 677 5503

Manufacturer and retailer of
Sarum Studio Figurines and
Sarum Traditional Soldiers in
54mm scale. Studio Figurines
include the 'History of the
Regiments' series designed by
Andrew C.Stadden, and the
'Chota Sahib' range by Sid
Horton. Toy soldiers include
'Armies of the Great Powers
1870-1914', and modern British
army in ceremonial uniforms.

Seabea Miniatures
26 Leicester Way, Leegomery,
Telford, Shropshire TF1 4UT

Contact: Kevin Seabrook
Tel/fax: 01952 260395

Manufacturers of 54mm white
metal toy soldiers and figurines.
Available hand painted, as
castings or kits. Suppliers to trade
and public. SAE for lists.

Simply Soldiers
A15 The Cowdray Centre,
Colchester, Essex CO1 1BH

Toy Army Workshop
The Hollies, Roe Downs Road,
Medstead, Alton, Hampshire
GU34 5LG

Contact: Graham Pettitt

Specialist in WWI vehicles,
equipment & soldiers in 54mm
scale in toy style. Send SAE for list.

Toytub
100a Raeburn Place, Edinburgh,
Lothian EH4, Scotland

Toyway
Unit 20 Jubilee Trade Centre,
Jubilee Road, Letchworth,
Hertfordshire SG6 1SG

Contact: Richard Morriss
Tel: 01462 672509
Fax: 01462 672132

Toyway manufacture and
distribute Timpo 54mm figures
and accessories.

Tradition of London Ltd
33 Curzon Street, Mayfair,
London W1V 7AE

Contact: Steve Hare
Tel: 0171 493 7452
Fax: 0171 355 1224

Largest range of painted and
unpainted figures in 54mm,
90mm, 110mm plus over 300
different 'Toy Soldier' style sets,
including 'Sharpe' range. Full
mail order, credit cards. Shop
open Mon–Fri 9.00–5.30, Sat
9.30–3.00.

Trafalgar Models
122 Lazy Hill Road, Aldridge,
Walsall, Midlands (W.) WS9 8RR

Trophy Miniatures Wales Ltd
Unit 4, Vale Enterprise Centre,
Sully, Penarth, Glamorgan (S.)
CF64 5SY, Wales

Tel: 01446 721011
Fax: 01446 732483

Toy soldier manufacturer;
catalogue, mail order. Factory
open for visitors – phone for
directions and appointment.

Under Two Flags
4 Saint Christopher's Place,
London W1M 5HB

Contact: Jock Coutts
Tel/fax: 0171 935 6934

Stockist of toy soldiers, model
kits, military books, painted
figures & dioramas. Open
Mon-Sat, 10.00am-5.00pm.

Ursine Supply
24 Tollhouse Road, Crossgate
Moor, Co.Durham DH1 4HV

Contact: David & Mary Reay
Tel/fax: 01911 3842724

Firm specialising in mail order
and supplying to overseas
customers. Write with
requirements.

V C Miniatures
16 Dunraven Street, Aberkenfig,
Bridgend, Glamorgan, Mid-
CF32 9AS, Wales

Contact: Lyn Thorne
Tel/fax: 01656 725006

Established 1987, manufacturer of
hand painted 54mm toy soldiers,
sculpted by Lyn Thorne and
specialising in the Victorian
period. In addition to sets of
traditional toy soldiers, military
and civilian vignettes and
individual figures are also
produced. Recently acquired are
Burlington Models range of
racehorses, jockeys and
showjumping figures.

W D Model Fairs
c/o Garden Flat, 75 Bouverie
Street West, Folkestone, Kent
CT20 2RL

Contact: Richard Windrow
Tel/fax: 01303 240006 (eves.)

Organisers of International Toy
Soldier Fair held annually on a
Saturday in March at a venue on

The Lees, Folkestone. All trade
stands (from £30), toy soldier
manufacturers and dealers; open
to public. Enquire for up-to-date
details.

Welsh Dragon Miniatures
32 Layton Crescent, Brampton,
Huntingdon, Cambridgeshire
PE18 8UT

Contact: George Humphrey
Tel/fax: 01480 458422

Manufacture and hand paint
model quality toy soldiers.

Whittlesey Miniatures
75 Mayfield Road, Eastrea,
Whittlesey, Peterborough,
Cambridgeshire PE7 2AY

Contact: Keith Over
Tel/fax: 01733 205131

Manufacturers and suppliers of
high quality, painted 'toy'
soldiers. Currently specialising in
Medieval and Ancient subjects in
54mm. Master and mould making
services are available.

Wolfe Toy Soldiers
445 Wisden Road, Stevenage,
Hertfordshire SG1 5JS

Contact: Ken George
Tel/fax: 01438 210230

British Army and colonial figures
c.1900. Sets of six painted, single
figures painted and castings. All
compatible with early Britains
figures, 54mm.

Yeomanry Miniatures
34 Vesey Close, Cove,
Farnborough, Hampshire
GU14 8UT

Contact: Brian Harrison
Tel: 01252 523943
Fax: 01420 563221

Yeomanry Miniatures produce
fully sculptured 'toy style' figures
for the connoisseur representing
the Yeomanry Cavalry and the
regular cavalry of the British
Army.

TRAVEL

Assegai Safaris
Oakridge, 1 Merryhill Close,
London E4 7PT
Tel/fax: 01670 366729

**Birmingham War Research
Society**
43 Norfolk Place, Kings Norton,
Birmingham, Midlands (W.) 30
3LB
Contact: Alex Bulloch
Tel: 0121 459 9008
Fax: 0121 459 8128

We operate trips by coach to
battlefields and war cemeteries in
Northern Europe.

**Galina International
Battlefield Tours**
1 Tokenspire Business Park,
Woodmansey, Beverley,
Humberside (N.) HU17 0TB
Contact: Barry Matthews
Tel: 01482 804409
Fax: 01482 809717

Official tour operators Normandy
Veterans' Association. Group,
school, guided coach tours,
self-drive tours. 1997 programme
includes Western Front, Dunkirk,
Normandy, Arnhem, Italy,
Gallipoli. Fully Escrow Bonded
Trustee Account company.

Grapeshot Tours
Bristow House, Castle Street,
Mere, Wiltshire BA12 6JF
Contact: Ann & Michael Hannon
Tel/fax: 01747 860149

In the footsteps of the Emperor –
tours which bring the whole
exciting story of the Napoleonic
Age to life, with lectures, visits,
superbly comfortable hotels and
transport. Write or ring for colour
brochure.

Historic Zululand Tours
49 Belsize Park, London NW3
4EE
Contact: Ian Castle

Join us on an extraordinary
adventure through Africa's
history as we explore the battle
sites of the 1879 Zulu War – the
experience of a lifetime.

**Holts' Tours
(Battlefields & History)**
15 Market Street, Sandwich,
Kent CT13 9DA
Contact: John Hughes-Wilson
Tel: 01304 612248
Fax: 01304 614930
Email: www.battletours.co.uk

Europe's leading military
historical tour operator, offering
annual world-wide programme
spanning history from the
Romans to the Falklands War.
Holts' provides tours for both the
Royal Armouries Leeds and the
IWM. Every tour accompanied by
specialist guide-lecturer. Send for
free brochure.

Midas Battlefield Tours Ltd
The Old Dairy, The Green,
Godstone, Surrey RH9 8DY
Contact: Alan Rooney
Tel: 01883 744955
Fax: 01883 744967
*Email: arooney
@midas.itsnet.co.uk*

Escorted tours to the battlefields
of the world, including Ancient,
Medieval, Napoleonic, ACW,
South Africa, India, Crimea, WWI,
WWII. Specialised group tours
organised. ATOL 3716.

**Middlebrook-Hodgson
Battlefield Tours**
48 Linden Way, Boston,
Lincolnshire PE21 9DS
Tel: 01526 342249
Fax: 01526 345249

Organisers of battlefield tours to
1914-18 Western Front, Ypres to
Verdun; Normandy; Gallipoli.
Small groups, individual personal
attention. Non-smoking.

Orientours Ltd
Kent House, 87 Regent Street,
London W1R 8LS
Tel/fax: 0171 734 7971

Silken East
36c Sisters Avenue, London
SW11 5SQ
Contact: Nicholas Greenwood
Tel/fax: 0171 223 8987

Burma: war tours to all areas,
including Arakan, Chindwin,
Myitkyina, Mandalay, Meiktila,
Maymyo, Lashio, Pegu,
Moulmein, Htaukkyant and
Thanbyuzayat cemeteries.

**Society of Friends of the
National Army Museum**
c/o National Army Museum,
Royal Hospital Road, London
SW3 4HT
Contact: Derek A Mumford
Tel/fax: 0171 730 0717

The Society of Friends of the
National Army Museum assist the
Museum in the acquisition of
significant militaria; and members
enjoy lectures, private views,
battlefield and Army
establishment excursions, and
newsletters. Annual subscription
£8.00; contact Secretary/Treasurer
above.

**War Research Society
(Battlefield Tours)**
27 Courtway Avenue, Maypole,
Birmingham, Midlands (W.) B14
4PP
Contact: Ian C Alexander
Tel: 0121 430 5348
Fax: 0121 436 7401

Battlefield tours, World Wars I &
II: France, Belgium, Holland and
Germany. Graves visited by
request; photographs taken if
unable to travel.

UNIFORMS, INSIGNIA, ARMOUR & GENERAL MILITARIA DEALERS

A & V A Hoffmann
Kolonnenstrasse 46, 10829
Berlin, Germany

A B I
118 Chapelon, Glascote Heath,
Tamworth, Staffordshire
B77 2EW

Contact: Douglas Preece
Tel/fax: 01827 286423

Mainly British World War II
equipment & uniforms bought
and sold; search service; callers by
appointment only. Airborne and
SOE equipment always wanted.

A S Bottomley
The Coach House, Huddersfield
Road, Holmfirth, Yorkshire (W.)
HD7 2TT

Contact: Andrew Bottomley
Tel: 01484 685234
Fax: 01484 681551

Established 30 years with clients
overseas and in the UK. A fully
illustrated mail order catalogue
containing a large range of
antique weapons and military
items despatched world wide.
Every item is guaranteed original.
Full money back if not satisfied.
Deactivated weapons available.
Valuations for insurance and
probate. Interested in buying
weapons or taking items in part
exchange. Business hours Mon-Fri
9am – 5pm. Mail order only. All
major credit cards welcome.
Catalogue UK £5, Euorpe £7, rest
of world £10.

Adler Militaria
Stb 4/1 All Saints Place, Oerbka
Lager, Fallingbostel, Germany

Contact: Kevin Greenhalgh

Specialist in WWII German,
British, American items and
battlefield relics. Send wants lists.

Adrian Forman
PO Box 25, Minehead, Somerset
TA24 8YX

Tel/fax: 01643 862511

Napoleonic books, prints,
postcards, documents, model
soldiers, buttons, artifacts, relics
(1792-1815 era). SAE for sample
list. Also small aero collection,
insignia, relics, artwork, books.

Alan Beadle Antique Arms
320 Upper Street, Islington,
London N1

AMAC (UK) Ltd
Unit 4, Crossley Park, Crossley
Road, Manchester, Manchester,
Gt M19 2SH

Contact: David Sykes
Tel: 0161 442 7224
Fax: 0161 442 4140

Sells all clothing, textiles and
tentage which was previously
sold by auction or tender by the
Ministry of Defence. House of
business 8.30am-5pm Mon-Fri.

Anchor Supplies Ltd
Peasehill Road, Ripley,
Derbyshire DE5 3JG

Contact: Barbara Merrett
Tel: 01773 570139
Fax: 01773 570537

Anchor Supplies, one of Europe's
largest genuine government
surplus dealers. Specialising in
clothing, tools, electronics,
domestic ware, furniture, watches,
military vehicles, you name it!
Goods are available mail order, or
visit our Derbyshire or
Nottingham depots. Please ring
for directions.

Anderson's
B12, Grays Antique Market,
1-7 Davies Mews, London W1

Tel/fax: 0171 491 3066

Full Dress British uniforms and
headdress; postcards and other
small ceremonial items. Open
Mon-Fri 10am-4pm.

**Andrew Butler Militaria &
Insignia**
10 Avebury Avenue, Ramsgate,
Kent CT11 8BB

Contact: Andrew Butler
Tel: 01843 592529
Fax: 01843 582216

U.S. insignia and uniforms. WW2
to Vietnam. Specialising in rarer
items, 8th AAF, Special Forces,
etc. Re-enactment items also
stocked.

Angels-One Five
16 Clover Avenue, Bedford
MK41 0TZ

Contact: Chris Chandler
Tel/fax: 01234 211026

Specialising in aviation and
military items, also collector of
USAAF 379 Bomb Group
memorabilia.

Anthony D Goodlad
26 Fairfield Road, Brockwell,
Chesterfield, Derbyshire S40 4TP

Tel/fax: 01246 204004

General militaria, with emphasis
on German World War I and II
items. Exhibitor at many UK arms
and militaria fairs. Personal callers
by appointment only.

Antique Armoury
Moorside Farm, Cauldon Lowe,
Stoke on Trent, Staffordshire
ST10 3ET

Contact: D R Cooper
Tel: 01538 702738
Fax: 01538 702662

Bought/sold: antique
arms/armour, British military
medals, headgear, badges,
uniforms etc. Specialising in
pre-1920, and militaria to Staffs
Regiments.

**Antique Militaria & Sporting
Exhibitions**
PO Box 194, Warwick,
Warwickshire CV34 5ZG

Contact: Chris James
Tel: 01926 400554
Fax: 01926 497340

Exhibition organisers for dealers
in antique arms, militaria, and
medals. Send SAE for details of
1998 programme.

Antiques
Main Street, Durham-on-Trent,
Nottinghamshire NG22 0TY

Contact: R G Barnett
Tel/fax: 01777 228312

Antique flintlock and percussion
pistols and long arms, swords,
daggers and armour our
speciality. Also buy and sell
general antiques, furniture, etc.

Arms Fairs Ltd
PO Box 2654, Lewes, Sussex (E.)
BN7 1BF

Tel/fax: 01273 475959

Organisers of London Arms Fairs
held in April and September each
year. Both fairs open for two days
– 140 tables.

SPECIALISTS IN HAND EMBROIDERED WIRE BLAZER BADGES, PATCHES, EMBLEMS, MILITARY INSIGNIAS, RANK & REGIMENTAL BADGES, CAP BADGES & VISORS Heraldry Coats-of-Arms, Scottish Clan Crests, Uniform Regalia, Accessories & Accoutrements, Police & Armed Forces Uniform Equipment, Banners, Pennants, Medals, Ribbons, Chevrons, Laces, Braids, Cords, Shoulderboards, Wings, Logos-WWI, WWII, American Civil Wars, Vietnam & Korean Wars, German Nazi Insignias. Also machine embroidered Patches available in all styles and sizes.

WHOLESALE EXPORT • TRADE ENQUIRIES WELCOME

THE BADGESMEN

P.O. BOX 1227, Sialkot 51310, Pakistan

Tel: (0432) 86921 Fax: 92-432-588417, 582978

Army Surplus
60 Tib Street, Manchester, Manchester, Gt M4 1LG

Contact: Vince Spencer
Tel: 0161 834 6118
Fax: 0161 834 7483
Email: lode@stone.u-net.com

Supply of Swedish/British ex-government surplus, textiles, vehicles and goods plus fold-a-cup and messkit from Sweden plus footwear, boots and shoes.

Army Surplus Store
(Props) BYC Summerscales, 1063 Thornton Road, Bradford, Yorkshire (W.) BD8 0PA

Contact: Charles Summerscales
Tel/fax: 01274 816945

A full range of military and outdoor clothing, footwear, hats, caps, knitwear, equipment, camping, flags, insignia, medals, ribbons, decorations, badges, etc., low prices. Sorry – no lists.

Attleborough Accessories
White House, Morley St Peter, Wymondham, Norfolk NR18 9TZ

Contact: C E Pearce
Tel: 01953 454932
Fax: 01953 456744

Militaria and memorabilia, including: Sheffield-made Fairbairn Sykes 3rd pattern and

commemorative knives; K-Bar USMC knives; knife-making supplies, etc. Large SAE for full lists.

Aux Armes d'Antan
1 Avenue Paul Deroulede, 75015 Paris, France

Contact: Maryse Raso
Tel: (0033) 1 47837142
Fax: (0033) 1 47344099

Expert and dealer in sabres, swords and ancient pistols and other weapons from the XIIth to the XIXth century, for 20 years. Publishes a catalogue every 3 months, available by post from the office. All pieces are sold along with a guarantee.

Aviation Antiques
Rivendell, Kings Road, Biggin Hill, Kent TN16 3XU

Contact: Rosemary Sutton
Tel/fax: 01959 576424

Buy and sell aviation and military collectables; books, china, RAF and airline items, travel agent display models, Action Man dolls; all plastic kits. Collections purchased and collected.

Aviation Unlimited
120 Heathbank Road, Cheadle Hulme, Cheshire SK8 6HX

Contact: Kevin Wilson

Specialists in British, American, German WWI/WWII aviation combat clothing and equipment. Lists £1.50. Mail order only.

The Badgesmen
PO Box 1227, Sialkot 51310, Pakistan

Contact: S H Deura
Tel: (0092) 432 86921
Fax: (0092) 432 588417

Specialists in hand embroidered wire blazer badges, emblems, military insignias, uniform regalia, rank and regimental badges, heraldry coats-of-arms, Scottish clan crests, club and association badges, ties, medals, ribbons, leather belts, honore caps, logos, shoulderboards, cap visors, laces, braids, cords, chevrons, pennants, epaulettes, wings – WWI, WWII, American Civil Wars, German Nazi cloth and metal insignias. Also machine embroidered patches available in washable quality.

Bill Gent
173 Westborough Road, Westcliff-on-Sea, Essex SSO 9JD

Tel/fax: 01702 34150

Dealer in a wide range of military and aviation collectables. Occasional lists available – send long SAE. Manufacturer of replica weapons to order – .50 Browning, Vickers K, mortars, etc. – if the genuine item is difficult or expensive to get, ask us.

Black Pig Trading
36 Dundrine Road, Castlewellan, County Down BT31 9EX, N Ireland

Contact: Mark C Myles
Tel: 013967 78618
Fax: 012385 42151
Email: max@mba.dnet.co.uk

All types of military surplus and militaria from all over the world, from 1900 right up to 1998. All stock is guaranteed original.

Blunderbuss Antiques
29 Thayer Street, London W1M 5LJ

Contact: Chris Greenaway
Tel: 0171 486 2444
Fax: 0171 935 1127

Specialists in original antique, World Wars, modern uniforms, equipment, weapons etc.

Boscombe Militaria
86 Palmerston Road, Boscombe, Bournemouth, Dorset BH1 4HU

Contact: E A Browne
Tel: 01202 304250
Fax: 01202 733696

We stock all types of 20th century militaria, e.g. uniforms, head-gear, badges cloth and metal, medals, etc. Closed Wednesday and Sunday. Established 1982.

Bostock Militaria
Pinewood, 15 Waller Close, Leek Wootton, Warwickshire CV35 7QG

Contact: Andrew Bostock
Tel/fax: 01926 56381

Brandenburg Historica
342 A Winchester Street, Ste. 21, Keene, New Hampshire NH 03431, USA

Contact: Diane M Schreiber
Tel: (001) 603 352 1961
Fax: (001) 603 357 5364
Email: preussen@top.monad.net

Brandenburg Historica mail order catalogue of books, military music and militaria; Imperial, Third Reich, DDR. Uniforms, insignia and medals. Bi-monthly listings. We ship worldwide. US$4.00.

Bric-a-Brac
16A Walsingham Place, Truro, Cornwall TR1 2AG

Contact: Richard Bonehill
Tel/fax: 01872 225200

One of the few shops left in Cornwall dealing in medals, badges, helmets, swords and a range of militaria. We specialise in the weird and wonderful (when we can get it!). Original 3rd Reich and DCLI items wanted.

Bryant & Gwynn Antiques
8 Drayton Lane, Drayton Bassett, Staffordshire B78 3TZ

Contact: David Bryant
Tel/fax: 0121 378 4745

All types of military uniforms, medals, swords and militaria supplied. Also deactivated military arms and muskets. Items not stocked can be ordered and traced.

Buckshee Kit
9 St Mary's Row, Moseley, Birmingham, Midlands (W.) B13 8HW

Contact: Paul Mould
Tel/fax: 0121 449 9934

World uniforms, medals, badges and general militaria from the Boer War to the Gulf. Specialists in Soviet Russian uniforms, medals and insignia. New gift range of encased battlefield relics, Victorian toy soldiers, and military prints. The 'one-stop military shop'.

The Bunker
5 Knightrider Street, Maidstone, Kent ME15 6LP

Contact: Bob Ryan
Tel/fax: 01622 686515

Buttons
9 Green Lane, Stopsley, Luton, Bedfordshire LU2 8AS

Contact: Frank Sharp

Mainly military buttons. Mail order only, send SAE for list.

C & D Enterprises
PO Box 7201, Arlington, Virginia 22207-7201, USA

Military elite insignia and parachute badges, MAC-SOG and aviation material. Publisher Elite Insignia Guides. List @ £2.00.

C F Seidler
Stand G12, Grays Antique Mkt,
1-7 Davies Mews, Davies Street,
London W1V 1AR

Contact: Christopher Seidler
Tel/fax: 0171 629 2851

American, British, European,
Oriental edged weapons, antique
firearms, orders and decorations,
uniform items; watercolours,
prints; regimental histories, army
lists; horse furniture; etc. We
purchase at competitive prices
and will sell on a consignment or
commission basis. Valuations for
probate and insurance. Does not
issue a catalogue but will gladly
receive clients' wants lists. Open
Mon-Fri, 11.00am-6.00pm.
Nearest tube station Bond Street
(Central line).

C L Heys
PO Box 615, Middleton,
Tamworth, Staffordshire
B78 2AZ

Contact: Cedric Heys
Tel/fax: 01827 874856

For over 20 years, specialists in the
sale and purchase of genuine
world military and police badges,
titles, collars, buttons and cloth
items. 50p stamp for sample list.

Cairncross and Sons
31 Bellevue Street, Filey,
Yorkshire (N.) YO14 9HU

Contact: George Cairncross
Tel/fax: 01723 513287

Regimental ties, blazer badges,
cap badges, medals, miniature
medals, insignia, blazer buttons,
caps, helmets, tunics, postcards,
cigarette cards, wall plaques,
cufflinks, books, etc.

Caister Surplus
18 High Street, Caister On Sea,
Yarmouth, Norfolk

Tel/fax: 01493 377587

Casque & Gauntlet Militaria
55-59 Badshot Lea Road,
Badshot Lea, Farnham, Surrey
GU9 9LP

Contact: Mrs A Colt
Tel/fax: 01252 20745

We have been dealing in a broad
range of militaria from this same
store for 26 years and offer a
complete service to our customers.
Guaranteed items. Money back if
not satisfied. Item sourcing.
Restoration service.

Castle Armoury
London Road,
Stretton-on-Dunsmore, Rugby,
Warwickshire CV23 9HX

Cathay
Shop 26, Sanlam Centre,
Umhlanga, Natal 4320, South
Africa

Tel: 0027 31 561 5783
Fax: 0027 31 561 4199

Natal's leading militaria shop.
Wide range of medals, badges,
edged weapons and general
collectables. British and South
African. Zulu and Boer Wars a
speciality.

**Central Antique Arms &
Militaria**
7 Smith Street, Warwick,
Warwickshire CV34 4JA

Contact: Chris James
Tel/fax: 01926 400554

Buys and sells antique guns,
swords, bayonets, helmets,
badges, British and German
medals and decorations, original
Third Reich militaria and
documents. Send 10 x 1st class
stamps for illustrated catalogue.
Shop open Mon-Sat,
10.00am-5.00pm. Exhibits at most
major fairs.

Central Militaria Fairs
PO Box 104, Warwick,
Warwickshire CV34 5ZG

Contact: Chris James
Tel/fax: 01926 497340

Specialists in orginal Third Reich
medals, edged weapons,
documents, uniforms, etc.
Japanese swords, medals and
militaria. Sample list, 10 x 1st class
stamps.

Chester Militaria
32 City Road, Chester, Cheshire
CH1 3AE

Contact: James Law
Tel: 01244 328968
Fax: 01244 341818

We are located at Melody's
antique galleries, opening from
9am-5.30pm, Mon-Sat. We buy
and sell all militaria especially
badges and medals.

Chris James Medals & Militaria
Smith Street Antique Centre, 7
Smith Street, Warwick, West
Midlands CV34 4JA

Contact: Chris James
Tel: 01926 400554
Fax: 01926 497340

Dealers in antique guns, swords
and bayonets. British and USSR
medals and decorations. WW2
RAF gallantry, log books and
equipment. Sample list, 10 x 1st
class stamps.

Coburgs Militaria Ltd
142 Stokes Road, Eastham,
London E6 3SE

Contact: Martin Credgington
Tel/fax: 0171 511 2988

Our company deals in military
surplus and US & German
insignia; top quality
reproductions; re-enactment
uniforms and headdress of WWII
period. Annual list; strictly mail
order only.

Coldstream Military Antiques
55A High Street, Marlow,
Buckinghamshire SL7 1BA

Contact: Steven Bosley
Tel/fax: 01628 822503

Fine 19th & 20th century
head-dress and badges of the
British Army. All items are
original and carry a money-back
guarantee. Regret no general
catalogue available, but 'wants'
lists most welcome. Valuations
undertaken for insurance, probate
or private treaty. Postal service
only. Members of LAPADA,
OMRS, MHS, Crown Imperial.

Collectors Corner
51 The Old High Street,
Folkestone, Kent CT20 1RN

Contact: Clive Jennings

Collectors shop specialising in
pre-1945 militaria, medals,
insignia, helmets; also surplus,
postcards, military books.

Collectors Corner Militaria
1 North Cross Rd, East Dulwich,
London SE22 9ET

Collectors Market
London Bridge BR Station,
London Bridge, London EC4

Tel/fax: 0181 398 8065

Over 60 stands including medal
dealers, badges and militaria on
the concourse of London Bridge
Main Line Station every Saturday.

Combat Central Trading
Trendle Street, Sherborne,
Dorset

Contact: Peter Nash
Tel/fax: 01935 872133

Surplus and militaria, British,
German and American 1900 –
1993. Postal and export enquiries
welcome. Open Fri and Sat,
10am-5pm. Established 20 years.

The Cracked Pot
PO Box 527, Buffalo, New York
14212, USA

Contact: Robert Sevier
Tel: (001) 716 634 2611
Fax: (001) 716 633 4124

We specialise in fine German
WWII military and political items,
and are recognised authorities in
the collecting field for over 20
years. We deal in Third Reich
medals and badges, uniforms,
helmets, insignia, headgear and
equipment. All items guaranteed
pre-May 1945 manufacture. We
publish one or two mail listings
per year; air mail catalogue
available for $4.00 US or
equivalent.

Curiosity Shop
127 Old Street, Ludlow,
Shropshire SY8 1NU

Contact: J D Luffman
Tel/fax: 01584 875927

Selection of muskets, pistols,
swords and helmets.

D M S (UK)
Unit 7, Pickwick Workshops,
Park Lane, Corsham, Wiltshire
SN13 0HN

Tel/fax: 01249 715574

David H Brock
46 Manor Road, Akeley,
Buckingham, Buckinghamshire
MK18 5HQ

Third Reich Collector. Awards,
badges and insignia. Full size,
miniature, stickpin. Mail order
only.

Diethard Feibig
Von-Ketteler-Strasse 26, 53229
Bonn, Germany
Tel/fax: (0049) 228484829

East Bloc Militaria
18 Northumberland Avenue,
Reading, Berkshire RG2 7PW

Contact: Michael Passmore
Tel/fax: 0118 9863899

Specialises in East German and
Soviet Forces insignia, medals and
uniforms. Unusual and rare items
available. Mail order only. Search
service provided. Send SAE for
free list. Phone evenings and
weekends only.

East London Armoury Co
128 Gordon Road, Ilford, Essex

Ernst Blass Militaria
Schauenburgerstrasse 59 II,
20095 Hamburg, Germany

Tel/fax: (0049) 1725105216

Uniforms, insignia order and
medals. Armour, arms, books for
collectors. General militaria.

Etzel Militaria
Bossingerweg 31, 73630
Remshalden, Germany

Tel/fax: (0049) 7151 72144

**Flying Spirit Nose Art
Company**
81 Charles Street, Epping, Essex
CM16 7AX

Contact: Simon Boultwood
Tel/fax: 01992 574554

Specialises in reproduction WWII
U.S.A.A.F. 'A2' and U.S.N. 'G1'
flying jackets, with nose-art and
insignia. Jacket painting available
and insignia sold separately. Mail
order only. Trade and film
company orders accepted.
Discounts offered to re-enactment
groups and ex-services members.
Catalogue available £3.50 or SAE
for free list.

Fordeversand Kiel
Rutkamp 4, 24111 Kiel, Germany
Tel/fax: (0049) 431 698419

Fred Wallard
17 Gyllingdime Gardens, Seven
Kings, Essex IG3 9HH

Les Frères d'Armes
16 Rue Lakanal, 38000
Grenobles, France

**Galerie d'Histoire André
Husken**
Dammtorstrasse 12, 20354
Hamburg, Germany

Tel/fax: (0049) 40 343131

Garth Vincent
The Old Manor House, Allington,
Grantham, Lincolnshire NG32
2DH

Contact: Garth Vincent
Tel: 01400 281358
Fax: 01400 282658
*Email: 100653.677
@compuserve.com*

Dealer in antique arms and
armour. Manufacturers/suppliers
of reproduction suits of armour,
banners and weapons for interior
decoration. Registered firearms
dealer.

Gryphon Trading
PO Box 5, Ripon, Yorkshire (N.)
HG4 3YR

Contact: Paul Cordle
Tel: 01765 690216
Fax: 01765 690702

Suppliers of finest English pewter
tankards, flasks, wine flagons,
stainless steel British army knife,
hand painted enamel pill boxes.
All with regimental badges. Send
for catalogue.

Helmut Weitze
Neuer Wall 18, 1 Stock, 20354
Hamburg, Germany

Tel/fax: (0049) 40 352761

Hermann Historica OHG
Sandstrasse 33, Munich 80335,
Germany

Tel: (0049) 89 5237296
Fax: (0049) 89 5237103

Specialist international
auctioneers of arms, armour,
historic militaria, orders, medals,
uniforms, etc.

**Historic Flying Clothing
Company**
192 Broadway, Derby,
Derbyshire DE22 1BP

Contact: David Farnsworth
Tel/fax: 01332 345729
Email: david.farnsworth@virgin.net

Specialist dealer/collector of
WW2 aviation memorabilia.
Flying clothing/equipment
(helmets, goggles, boots, jackets,
etc.) & home front items. Send 2x
1st class stamps for catalogue.

Historical Uniforms Research
Unit 3, Enterprise Centre,
Emmet Road, Ballymote,
Co.Sligo, Ireland

Contact: John Durant
Tel: (00353) 71 83930
Fax: (00353) 71 82404

Manufactures and supplies
world's best reproduction
Waffen-SS reversible camouflage
items to museums, collectors,
re-enactors and film industry.
Smocks, helmet covers,
non-standard field tailored items.
All camouflages and construction
exact to originals. Examples have

been mistakenly judged genuine
by collectors and curators. Can
also supply genuine WSS items
ex-SS 'Skanderbeg' Division; also
1941-1960 Russian camouflages
and other kit used by Albanian
forces. Specialist sourcing, to
order, of Russian WWII weapons,
personal kit, communications
equipment, documents, manuals
etc. Will print to order any of over
200 camouflage patterns;
minimum 1000 metres. Dealer
enquiries welcome. Can also
arrange military interest tours to
Albania. Accept VISA, ACCESS,
MASTERCARD, EUROCARD.

Holland Military Antiques
Bevrijdingsstraat 11, 6701 AA
Wageningen, Netherlands

Contact: Gerard Boon
Tel/fax: (0031) 837021327

HQ84 (The Curiosity Shop)
Southgate St, Gloucester,
Gloucestershire GL1 2DX

Contact: John & Barbara Williams
Tel/fax: 01452 527716

Militaria, badges, medals, genuine
and repro: government surplus.
Mail order lists £1 inland, £2
overseas (stamps/IRCs). Visitors
welcome.

Ian G Kelly (Militaria)
PO Box 18, South District Office,
Manchester, Manchester, Gt.
M14 6BB

Contact: Ian Kelly
*Email: ian_g_kelly_militaria
@compuserve.com*

Original British and
Commonwealth military and police
badges bought and sold with
money-back guarantee. Send large
SAE (or 2 IRCs) for catalogue
including cap, collar, trade and
proficiency badges, shoulder titles,
buttons, formation signs, parachute
wings, Air Force badges and
propaganda leaflets. Specialist in
rare "Staybrite" badges,
N.Ireland-issue cloth beret badges
and DZ flashes. Postal service only.
Web site: http://ourworld.
compuserve.com:80/homepages/ian
_g_kelly_militaria.

III Arm Militaria
Hillbury, Kirkby-cum-Osgodby,
Lincolnshire LN8 3PE

Contact: Duncan Lamb
Tel/fax: 01673 828081

British Victorian medals and
Soviet documented awards a
speciality. A list is published
periodically throughout the year;
send SAE for free sample copy.

International Movie Services
3050 Mountain Highway, North
Vancouver, British Columbia V7J
2P1, Canada

Contact: Major Ian Newby
Tel: (001) 604 985 4020
Fax: (001) 604 985 9642

Collections purchased. Rentals of
military/police etc. uniforms,
vehicles and firearms to the
theatrical and motion picture
industry. Technical advice and
research services available.

Iron Shirt (UK) Ltd
Kenton Hill Cottage, Kenton,
Exeter, Devon EX6 8JD

Over 100 designs – military,
patriotic, firearms – printed on
quality Tee and sweat shirts.
Large SAE for full details and free
catalogue.

J & J Malo
Mail Order Badge Specialists, 20
Brook Drive, North Harrow,
Middlesex HA1 4RT

Tel: 0181 861 0728
Fax: 0181 861 3357

Badge and emblem suppliers.
Hand embroidered in silk, gold
and silver bullion wires, for
blazers, crests, flags, emblems,
sashes, insignia, regalia, banners
etc. Army, navy, air force, schools,
clubs etc. Made to order; first
sample free of charge; send logo,
sketch, photograph or existing
sample. Quality MOD approved.
Machine embroidered badges also
supplied.

J C Cano
Pedro Pérez Fernández, 29, 2°G,
41011-Sevilla, Spain

Contact: Julián Cano
Tel/fax: (0034) 954450130

Guaranteed original items –
Spanish Civil War, Franco era,
Spanish Foreign Legion, Falange
(political movement), books,
uniforms, badges and medals by
mail order.

J T Militaria
40 Severn Road, Pontllanfraith,
Blackwood, Gwent NP2 2GA,
Wales

Contact: Jon Heyworth
Tel/fax: 01850 442899

Always wide range of original
military uniform, equipment, etc.
items in stock; expert specialist in
WWII British.

Jeremy Tenniswood
PO Box 73, Aldershot,
Hampshire GU11 1UJ

Contact: Jeremy Tenniswood
Tel: 01252 319791
Fax: 01252 342339
Email: 100307,1735@compuserve

Established 1966, dealing in
collectable firearms civil and
military, de-activated and for
shooters; also swords, bayonets,
medals, badges, insignia, buttons,
headdress, ethnographica; and
books. Regular lists of Firearms
and Accessories; Medals; Edged
Weapons; Headdress, Headdress
Badges and Insignia;
comprehensive lists Specialist and
Technical Books. Office open
9am-5pm, closed all day Sunday.
Medal mounting service.

Just Military
701 Abbeydale Road, Sheffield
S7 2BE

Tel/fax: 0114 255 0536

Dealers in all types of military
memorabilia and collectables. Full
medal mounting and framing
service including the supply of
miniature and replacement
medals.

Képi Blanc
Conlon Antiques, 22 Lr
Clanbrassil St, Dublin 8, Ireland

Contact: Paul O'Brien

Worldwide militaria stockist.
Postal and export enquiries
welcome. Specialist in Irish Army
insignia. Open Sat 10.30am –
5.30pm.

King & Country
5617 Denny Ave, N Hollywood,
California CA 91601, USA

Contact: Harlan Glenn
Tel: (001) 818 752 0630
Fax: (001) 818 752 8033
Email: kngcntry@westworld.com

Reproduction WWII Denison
smocks, collarless shirts, khaki
drills and maroon para berets (WII
specifics). Also orginal WWII
British collectables.
http://www.westworld.com/~kng
cntry/index.html.

Laurence Corner
62/64 Hampstead Road, London NW1 2NV

Tel: 0171 813 1010
Fax: 0171 813 1413

Established over 30 years, Laurence Corner are famous as a unique source of Government surplus. Militaria and theatrical clothing for sale and hire.

Lawrence & Son
12 Union Street, Andover, Hampshire SP10 1PA

Tel: 01264 355973
Fax: 01264 352616

Liverpool Militaria
48 Manchester Street, Liverpool L1 6ER

Contact: Bill Tagg
Tel/fax: 0151 236 4404

Buying and selling general militaria medals, badges, bayonets, swords, antique guns and helmets. Especially medals to King's Liverpool Regiment 1st and 2nd War gallantry and casualties.

London Militaria Market
Angel Arcade, Camden Passage, London N1

Tel/fax: 01628 822503

Saturdays only, 8am-2pm. About 35 stands dealing in badges, medals, arms, helmets, uniforms, regimental brooches, and many other military collectables.

M & T Militaria (Postal Sales)
The Banks, Banks Lane, Victoria Road, Carlisle, Cumbria CA1 2UD

Contact: Malcolm Bowers
Tel/fax: 01228 31988

Specialists in original Third Reich militaria. Regular mail order catalogue with over 1,000 items. List subscriptions £8 (overseas £10) per year for 4 guaranteed copies.

Marché Serpette
Allee 1, Stand 8, 110 Rue des Rosiers, 93400 Saint Ouen, France

Contact: Jean-Pierre Maury

Markus Eifler Militaria
Engelbertstrasse 10, 40233 Dusseldorf, Germany

Tel/fax: (0049) 211 7332032

Martin Duchemin Display Cabinets
6 Knights Close, Eaton Socon, Cambridgeshire PE19 3DP

Tel/fax: 01480 212843

Manufacturers and stockists of regimental cufflinks, tie grips, lapel badges, blazer buttons, medal display cases, figurines; medals mounted; Royal British Legion approved.

Milan Armouries
21 Cedars Road, Colchester, Essex CO5 7BS

Contact: Christopher Dobson
Tel: 01206 577363
Fax: 01206 761716

Arms and armour of fine quality commissioned. Restoration and conservation for museums and private collectors. No catalogue produced. Visitors strictly by appointment only.

Militaria
c/Bailen 120, 08009 Barcelona, Spain

Contact: Xavier Andrew
Tel/fax: (0034) 3 2075385

Buy, sell and exchange militaria, medals, edged weapons, insignia, headgear etc. Also large selection of military books in stock.

Military Antiques
Shop 3, Phelps Cottage, 357 Upper Street, London N1 0PD

Contact: R Tredwin
Tel/fax: 0171 359 2224

In the heart of Camden Passage, long-established dealers in WWI & WWII uniforms, equipment, headgear, awards, edged weapons. All items original and carry a full money-back cover. Illustrated catalogue; wants lists welcome. Open Tues.-Fri., 11am-5pm; Sat, 10am-5pm; if travelling far, advisable to phone first.

Military Insignia
24 Elton Road, Bristol, Avon BS7 8DD

Military Insignia
7 Peak Road, Clanfield, Hampshire PO8 0QT

Contact: Lew Shotton

Original military insignia, badges, shoulder titles, collar badges, divisional signs. Victorian to date. Postal service only – comprehensive listings.

Millais Antiques
Horsham Antique Centre, Marsh House, Park Place, Horsham, Sussex (W.) RH12 1DF

Contact: Geoffrey Dexter
Tel/fax: 01403 259181

18/19th century US and British service weapons, and British/German scientific instruments, especially telescopes, binoculars, including First/Second World War, compasses, surveying.

Moby Dick Fine Arts
Sint-Jorispoort 13, B-2000 Antwerp, Belgium

Tel/fax: (0032) 3 227 1157

Specialists in maritime antiques and 19th century militaria.

Mons Military Antiques
221 Rainham Road, Rainham, Essex RM13 7SD

Contact: Richard Archer
Tel: 01277 810558
Fax: 01277 811004

Premier buyers and sellers of WW1, interwar and WW2 British and German militaria including headdress, uniforms, equipment, badges and medals. Mail order and militaria fairs only.

Nicholas Morigi
14 Seacroft Road, Broadstairs, Kent CT10 1TL

Contact: Nicholas Morigi
Tel: 01843 602243
Fax: 01843 603940

Specialist in cloth and metal insignia of all countries from 1900 to present day. Specialising in WWII, Vietnam and current. Countries represented include Great Britain, United States, Third Reich Germany, Soviet Union and France. Mail order only. Wholesale enquiries welcome. Full colour catalogue £4.

Norman D Landing Militaria
23 Meadowsweet Road, Creekmore, Poole, Dorset BH17 7XU

Contact: Kenneth Lewis

US Army uniforms and equipment 1900–1945. Mail order catalogues available. Write for list.

Norman Litchfield Militaria Dealer
3 Cedar Drive, Ockbrook, Derby, Derbyshire DE72 3SJ

Tel/fax: 01332 662044

British and Commonwealth uniform badges and insignia 1800 to present day. Single items or collections bought. Part exchanges always considered.

The Old Brigade
10A Harborough Road, Kingsthorpe, Northampton, Northamptonshire NN2 7AZ

Contact: Stewart Wilson
Tel: 01604 719389
Fax: 01684 712489

One of England's leading dealers of original Third Reich militaria. Catalogue £4 + SAE. Shop open 10.30am-5pm, closed Wed. Phone for appointment.

The Old Guard
PO Box 25, Minehead, Somerset TA24 8YX

Contact: Adrian Forman
Tel/fax: 01643 862511

Fine Napoleonic military antiques and books. 'Sales' gazettes (colour illustrated). Write for membership. All items unconditionally guaranteed orginal by Adrain Forman, Esq. Also prints, models and documents, and free book catalogues.

Overlord Militaria
140 Tonge Moor Road, Bolton, Lancashire BL2 4DP

Tel/fax: 01204 398717

Weapons, uniforms and equipment, World War I to present day, also comprehensive stocks of current British Army uniforms and associated equipment. Open 10am-5.30pm.

P G Wing
The Warehouse, Peggy's Walk, Little Bury, Saffron Walden, Essex CB11 4TG

Contact: Peter Wing
Tel: 01799 522196/521801
Fax: 01799 513374

We are the largest Government Surplus Wholesaler in the U.K. We stock uniforms, boots, socks, combat and camouflage trousers, jackets and parkas; French, German, U.S.A., Spanish, British; steel helmets, gasmasks, webbing, bandoliers, leather belts and helmets, bayonets, knives, machetes, ammo boxes, jerricans, rucksacks, kitbags, flags, badges, and many unique and hard-to-find articles. We have a 104 page catalogue, but we are wholesale only. We are five miles from Duxford Museum.

Pastimes
22 Lower Park Row, Bristol, Avon BS16 5BN

Contact: Andy Stevens
Tel/fax: 0117 9299330

We buy and sell medals, badges, swords, bayonets, Nazi and Italian pre-1945 items, second-hand military books etc.

Pete Holder
King of Clubs, 12-14 Quay Street, Gloucester, Gloucestershire GL1 2JS

Tel: 01452 522446
Fax: 01452 310869

Antique American firearms; specialist dealer in Colt percussion, early cartridge and single-action weapons; also Remington, Winchester, Smith & Wesson. I buy, sell and exchange.

Peter Green
4 Barley Croft, Hemel Hempstead, Hertfordshire HP2 4UY

Contact: Peter Green
Tel/fax: 01442 825722

Reproduction British badges from Victorian shakos and glengarrys right up to modern cap badges. Also Nazi, Gurkha, mail order, £1.00 for list.

Pierre De Hugo
80 Rue des Pres aux Bois, 78220 Viroflay, France

The Plumery
16 Deans Close, Whitehall Gardens, Chiswick, London W4 3LX

Tel/fax: 0181 995 7099

Military plume makers to the Household Cavalry, Foot Guards; British and foreign armies, TV, theatre, associations, collectors – all periods and nationalities. Reproduction Napoleonic shakos – £85 + VAT British, £140 + VAT French. Phone enquiries preferred.

Poussières d'Empires SARL
33 Rue Brezin, 75014 Paris, France

Poussières d'Empires sells badges, orders, decorations and military items, particularly French Indochina and other Colonial. Open 11am-7pm; catalogue available for mail order service; Visa and Access accepted.

Preston Military Surplus
142 New Hall Lane, Preston, Lancashire

Tel/fax: 01772 654124

Genuine military surplus and collectables, de-activated weapons, air guns, uniforms, bayonets, swords, Para and SAS smocks, all modern British kit, etc. Open 10am-5pm.

R G Antiques
Main Street, Dunham on Trent, Nottinghamshire NG22 0TY

Contact: R G Barnett
Tel/fax: 01777 228312

R J Brannagan
115 Bonaccord Street, Aberdeen, Strathclyde, Scotland

Tel/fax: 01224 572826

Raven Armoury
Handleys Farm, Dunmow Road, Thaxted, Essex CM6 2NX

Contact: Simon Fearnham
Tel/fax: 01371 870486

Handcrafted swords, chainmail and armour for collectors and re-enactors. Sword range includes historical reproductions on display at The Royal Armouries Shop in The Tower of London. Fully guaranteed.

Regimentals
70 Essex Road, Islington, London N1 8LT

Tel: 0171 359 8579
Fax: 0171 704 0879

Dealers in all aspects of militaria from mid-1850s to 1945, specialising in German WWI and WWII items; 16 years in our present shop address. Postal catalogue sales worldwide are our major priority. Two 1,000-item and colour catalogues are published yearly: subscription rate, overseas air mail £16 for four copies i.e. two years, or £14 UK inland for four copies i.e. two years. Shop opening hours 10am-5pm, Monday to Saturday.

Richard Dunk (Armourer)
23 Overhill Road, Burntwood, Walsall, Midlands (W.) WS7 8SU

Tel/fax: 01543 684726

Manufacturer and restorer of arms, armour and period ironwork.

Rod Akeroyd & Son
101-103 New Hall Lane, Preston, Lancashire PR1 5PB

Tel: 01772 794947
Fax: 01772 654535

Extensive stock of British, Continental, and Japanese guns, swords, helmets and armour. Callers welcome.

Roy King Historical Reproductions
Sussex Farm Museum, Manor Farm, Horam, nr. Heathfield, Sussex (E.) TN21 0JB

Tel/fax: 01435 33733

Reproduction armour, helmets, edged weapons and some associated goods. Theatrical and film props, cannons, siege engines etc. Pyrotechnic service for simulated battles available. Mail order, limited catalogue available – most items made to order. Visitors by appointment only. Surrounding fields and Farm Museum available for location hire.

Ryton Arms
PO Box 7, Retford, Nottinghamshire DN22 7XH

Tel: 01777 860222
Fax: 01777 711297

Wide range of de-activated military weapons, including machine guns etc., complete with inert ammunition, accessories, etc. for impressive displays.

Service Commemoratives Ltd
PO Box 173, Dromana, Victoria 3936, Australia

Tel/fax: (0061) 359 810201

Military service commemorative medals and a range of over 100 different area and specialist clasps/bars available for: combatant service; sea service; foreign service; army service; aviation service; volunteers; national defence; Antarctica service; and POW liberation 1945.

Le Shako Francais
BP 3, 59310 Landas, France

Contact: Michel Deprest

Reproduction and repair of military headdress and equipment.

Soldier of Fortune
Unit 18, Tyn y Llidiart Ind Estate, Corwen, Clwyd LL21 9RR, Wales

Contact: Peter Kabluczenko
Tel: 01490 412225
Fax: 01490 412205

Specialising in reproduction 2nd WW clothing and equipment. British, German and the USA. Catalogue available.

Spitfire Militaria
509 Commissioners Road W, Suite 123, London ON N61 1Y5, Canada

Contact: Ron Taylor
Tel/fax: (001) 519 474 0099
Email: spitfire@wwdc.com

Purveyors of fine aeronautica including headgear, uniforms and accroutrements. Specializing in WW2 RCAF, RAF, RAAF and postwar pieces. Mail and phone orders only please.

Squadron Prints
Terrane House, Whisby Way Ind Estate, Lincoln, Lincolnshire LN6 3LQ

Contact: Jane Rumble
Tel: 01522 697000
Fax: 01522 697154

Embroidery manufacturers and screen printers. Official suppliers to HM Armed Forces and RAF Aerobatic Team, Red Arrows. Please send SAE for lists of mail order goods.

Stephen Gane
Stonegarth, Church Street, Broughton in Furness, Cumbria LA20 6HJ

Tel/fax: 01229 716880

Buy, sell and exchange antique arms, armour and militaria; specialists in rare and interesting bayonets; send large SAE for current Bayonet catalogue.

Stringtown Supplies
29 High Street, Polegate, Sussex (E.) BN26 5AB

Contact: Clive Hodgson
Tel: 01323 488844
Fax: 01323 487799

Military, outdoor and survival clothing & equipment, militaria, swords, antique firearms, replicas, flags etc. New shop now open Mon-Sat 9.30am-5.30pm, Polegate High Street.

Syntown Militaria International
Westway, Rectory Lane, Winchelsea, Sussex (E.) TN36 4EY

Contact: Christopher Coxon
Tel: 01797 223388
Fax: 01797 224834

Stockists of high quality reproduction and original uniforms, equipment and soldiers' personal small kit from WWII, exact repro German camo garments a speciality, Axis and Allied lists quarterly £2.00 each. Access, Visa and Diners Club by post or phone.

T L O Militaria

Longclose House, Common
Road, Eton Wick, nr Windsor,
Berkshire SL4 6QY

Contact: Tony Oliver
Tel: 01753 862637
Fax: 01753 841998

Specialists in DDR collectables at all
levels of scarcity and price. Mail
order; callers welcome seven days a
week, BY PRIOR APPOINTMENT.

Tercio

27 Feversham Crescent, York,
Yorkshire (N.) YO3 7HQ

Contact: G R Henderson
Tel/fax: 01904 63908

Dealer in general militaria
(specialising in Spanish) combat
uniforms and some military books.
Callers welcome evenings &
weekends, but please phone first.

Tiger Collectibles

104 Victoria Avenue, Swanage,
Dorset BH19 1AS

Contact: Mark Bentley
Tel: 01929 423411
Fax: 01929 423410

All militaria and collectible items,
land, sea and air, from both World
Wars, mail order catalogue
available (see advert). Visitors only
by appointment. Militaria fairs
attended.

The Treasure Bunker

Unit 16 Virginia Galleries, Virginia
Street, Glasgow, Strathclyde
G1 1TU, Scotland

Contact: Kenneth J Andrew
Tel: 0141 552 8164
Fax: 0141 552 4651

Specialist dealer in WWII and WWI
militaria, particularly German and
British. Open Mon-Sat
12pm-4.30pm. Pre-WWII militaria
always wanted. Price list £3.

Trenchart

PO Box 3887, Bromsgrove,
Hereford & Worcester B61 0NL

Tel/fax: 0121 453 6329

Suppliers to museums, gift and
surplus shops. Inert ammunition,
all calibres; bullet keyrings,
dogtags, souvenirs, replica
grenades, de-activated weapons,
MG belts and clips, large cartridge
cases, etc. Trade only supplied.

Trident Arms

96-98 Derby Road, Nottingham,
Nottinghamshire NG1 5FB

Contact: Michael Long
Tel: 0115 9413307
Fax: 0115 9414199

The largest variety of antique and
modern arms and militaria in the
UK. Open Monday to Friday
9.30am-5pm, Saturday 10am-4pm.
We buy, sell and exchange.

Ulric of England

6 The Glade, Stoneleigh, Epsom,
Surrey KT17 2HB

Tel: 0181 393 1434
Fax: 0181 393 9555

Ulric of England produces a fully
illustrated catalogue of Third Reich
items. Send £4 (UK/Europe), £5
(Overseas). Personal callers by
appointment only.

Weed Collections

PO Box 4643, Stockton,
California 95204, USA

Contact: Ben K Weed
Email: b.k.weed@worldnet.att.net

Largest private collector of world
wide military flags. Buys flags,
parts and photos. Periodic general
militaria collection disposal lists;

and special request research
service available.
http://members.tripod.com/
noldflagswanted/usa.html or
http://www.collectoronline.com/
collect/wb-old-flags.html

WARGAMING CLUBS & SOCIETIES

A G E M A
6 Studley Road, Linthorpe, Middlesbrough, Cleveland

Contact: I Moran

Bath Wargames Club
7 Heathfield Close, Upper Weston, Bath, Avon BA1 4NW

Contact: S Chasey

Established 1970. Meets every Sunday afternoon at Scout HQ, Grove Street, Bath.

Birmingham Wargames Society
5c Calthorpe Mansions, Calthorpe Road, Birmingham, Midlands (W.) B15 1QS

Contact: Martin Healey
Tel/fax: 0121 242 6718

Meet alternate Sunday afternoons from 2 to 9 pm at Ladywood Community Centre, Ladywood Middleway, Birmingham. All periods & all scales played, new members welcome.

The British Historical Games Society
The White Cottage, 8 Westhill Avenue, Epsom, Surrey KT19 8LE

Tel/fax: 01372 812132
Email: 100523.547
@compuserve.com

Our aim is to organise the UK National Tournament and to collate results from other competitions. Visit our web at http://members.aol.com/BritishHGS/index.htm.

Central London Wargamers
23 Rothay, Albany Street, London NW1 4DH

Contact: S Sola

Chichester Wargames Society
7 Bellemeade Close, Woodgate, Chichester, Sussex (W.) PO20 6YD

Contact: Andy Gilbert
Tel/fax: 01243 544339

Meets Tuesday evenings and once a month for a whole day. All periods covered, but specialise in 15mm Ancients, naval wargaming, 5mm Napoleonic and ECW. Regular entrants in national competitions and World Wargaming Championships.

Colchester Wargames Association
10 Golden Noble Hill, Colchester, Essex

Contact: Pete Anglish
Tel/fax: 01206 563190

Most periods and scales covered. Meets weekly.

Cornwall Miniature Wargamers Assoc
17 Mitchell Hill Terrace, Truro, Cornwall TR1 1HY

Contact: Chris Kent
Tel/fax: 01872 75171

All periods, Ancient to Sci-Fi, catered for. Meetings held fortnightly. Trips throughout year to conventions and museums.

Cornwall Wargames Society
Women's Institute Hall, Cranstock Street, Newquay, Cornwall

Tel/fax: 01209 212357

All periods catered for including fantasy. Membership fees per annum, £10.00 seniors, £5.00 juniors, £5.00 associated. £1.00 table fee per person per game. Meetings bi-monthly.

Devizes & District Wargamers
15 Chiltern Close, Warminster, Wiltshire BA12 8QU

Duke of York's Wargames Club
Duke of York's Military School, Dover, Kent CT15 5EQ

Contact: Michael Carson

Durham Wargames Group
c/o Gilesgate Community Assoc, Vane Tempest Hall, Durham City, Co.Durham

Tel/fax: 0191 384 7659

One of the longest established groups in the area now in its 23rd year. Annual conventions. All periods and scales including some SF and Fantasy. Weekly meetings.

East Leeds Militaria Society
Crossgates Bowling Club, Well Garth, off Station Road, Leeds, Yorkshire (W.)

Eastbourne Men at Arms Wargames Club
17 Rotunda Road, Eastbourne, Sussex (E.) BN23 6LE

Contact: Trevor Body

Essex Warriors
67 Fourth Avenue, Chelmsford, Essex CM1 4EZ

Contact: Peter Grimwood
Tel/fax: 01245 265274

We play historical wargames, fantasy, Sci-fi and role-playing games, and meet at Writtle, near Chelmsford, alternate Sundays 9.30am-5pm. Contact for further details.

Exeter & East Devon Wargames Group
17 Lovelace Crescent, Exmouth, Devon EX8 3PP

Contact: Jonathon Jones

Meeting at Exeter Community Centre, 17 St.Davids Hill, Exeter, first Saturday every month. Membership 16 and above. Periods: Napoleonic, WWII, ACW, Ancients, AWI, WWI Air.

Forlorn Hope Wargames Club
9 Brocklehurst Street, New Cross, London SE14 5QG

Contact: C Barrass
Tel/fax: 0171 639 3240

The French Horn
Oxford Road, Gerrards Cross, Buckinghamshire SL9 7DP

Contact: G Fordham

French-Indian Wargamers
55 Prestop Drive, Westfields, Ashby de la Zouch, Leicestershire LE65 2NA

Contact: Brian Betteridge
Tel/fax: 01530 560164

French-Indian Wars wargamers exchange views, ideas, figures and information. Attend wargame shows with display games, or meet for private battles. Seek like-minded players, collectors; no fees; contact us for details.

Halifax Wargames Society
7 Brookeville Avenue, Hipperholme, Halifax, Yorkshire (W.) HX3 8DZ

Contact: Roger Greenwood
Tel/fax: 01422 203169

Weekly meetings Wednesdays 7.30pm at Halifax Cricket Club, Thrum Hall, adjacent to Halifax Rugby League ground.

Harrogate Wargamers Club
32 Nydd Vale Terrace, Harrogate, Yorkshire (N.) HG1 5HA

Contact: Andrew Baxter
Tel/fax: 01423 541423
Email: ANDY@suc5.demom.co.uk

Organises the annual Sabre Wargames Show. Weekly meetings. Ring above number for further details. All are welcome.

Holts' Tours (Battlefields & History)
15 Market Street, Sandwich, Kent CT13 9DA

Contact: John Hughes-Wilson
Tel: 01304 612248
Fax: 01304 614930
Email: www.battletours.co.uk

Europe's leading military historical tour operator, offering annual world-wide programme spanning history from the Romans to the Falklands War. Holts' provides tours for both the Royal Armouries Leeds and the IWM. Every tour accompanied by specialist guide-lecturer. Send for free brochure.

I N D
9 Camp View Road, St Albans, Hertfordshire

Contact: M Scales

Kingston Games Group
The Swan, 22 High Street, Hampton Wick, Kingston, Surrey KT1 4DB

Contact: Ian Barnett
Tel/fax: 0181 296 0369

The group meets every Tuesday, 6pm-11pm, and on Sundays 12pm-6pm at Wimbledon Community Centre, George Street. All games catered for.

Kirriemuir & District Wargames Society
7 Slade Road, Kirriemuir, Angus DD8 5HN, Scotland

Contact: Dale Smith
Tel/fax: 01575 74128

All periods and scales covered; meet in Glengate Hall, Kirriemuir on alternate Saturday nights from 7.00pm.

Mailed Fist Wargames Group
Hyde Festival Theatre, Corporation Street, Hyde, Cheshire

Maltby & District Wargamers Society
Church Hall, Church Lane, Maltby, Rotherham, Yorkshire (S.) S66 8JB

Contact: Adrian McWalter

We meet every Sunday and are interested in all periods. New members welcome: ring Adrian McWalter on 01542 477977.

Malvern Wargames Club
37 Court Road, Malvern, Hereford & Worcester

Contact: Dave Bolton
Tel/fax: 01684 560383

Manchester Board Wargamers
Sea Scouts Hut, Stockport Road, Romiley, Manchester, Gt.

Contact: Norman Lane
Tel/fax: 0161 494 2604

Meetings first and third Saturdays monthly, 12am-6pm. All board wargames played. The club is active in all areas of the hobby and offers many services.

Manx Wargames Group
4 Wesley Terrace, Douglas, Isle of Man IM1 3HF

Contact: David Sharpe
Tel/fax: 01624 626642

Small, friendly group, regularly playing Ancients, Napoleonics, World Wars, Skirmish and Naval games, plus mainly strategic boardgames. Anything considered. New members, any age, welcome.

Methusalens
5 Fairfax Avenue, Oxford, Oxfordshire OX3 0RP

Contact: M Goddard

Milton Keynes Wargame Society
26 Swift Close, Newport Pagnell, Buckinghamshire MK16 8PP

Contact: B Mills
Tel/fax: 01908 613558

Ragnarock Wargaming Club
216 Powder Mill Lane, Twickenham, Middlesex TW2 6EJ

Contact: Christopher Poore
Tel/fax: 0181 898 2950

Small wargaming club concentrating on Fantasy, Sci-Fi and role-playing games. We meet every Thursday and most weekends.

Rawdons Routers
5 Shiplate Road, Bleadon Village, Weston Super Mare, Avon BS24 0NG

Contact: D Aynsley

Red October
19 Plants Brook Road, Walmley, Sutton Coldfield, Midlands (W.) B76 8EX

Contact: R Boyles
Tel/fax: 0121 313 1053

Reigate Wargames Group
The White Cottage, 8 Westhill Avenue, Epsom, Surrey KT19 8LE

Contact: J D McNeil
Tel/fax: 01372 812132
Email: 100523.547
@compuserve.com

Main interests are ancient, Napoleonics, other periods and boardgames. Meetings on Thursday evenings and alternate Sundays; contact club secretary above for further details.

S E Essex Military Society
202 Westcliff Park Drive, Westcliff on Sea, Essex SS0 9LR

Contact: John Francis
Tel/fax: 01702 431878

Historical wargaming with all periods covered. Produces major demonstration game each year and runs annual wargames show 'Present Arms'. Meets weekly.

S E London Wargames Group
16 West Hallowes, Eltham, London SE9 4EX

Contact: Paul Greenwood
Tel/fax: 0181 857 6107

The South East London Wargames Group (known as SELWG) was formed in 1971 and has a membership of over 100. A wide range of interests are catered for in the wargaming field. The club also runs an annual open day.

S E Scotland Wargames Club
182 Easter Road, Edinburgh, Lothian EH7 5QQ, Scotland

Solo Wargamers Association
120 Great Stone Road, Firswood, Manchester, Gt. M16 0HD

Contact: Steve Moore

Founded 1976; quarterly magazine 'Lone Warrior', by wargamers for wargamers, on solo aspects of wargaming.

South Dorset Military Society
23 Monks Way, Bearwood, Bournemouth, Dorset BH11 9HT

Contact: B Thorburn
Tel/fax: 01202 576423

Stockton Wargamers
c/o Elmwood Community Centre, Green Lane, Hartburn, Stockton on Tees, Cleveland

Contact: Garry Harbottle-Johnson
Tel/fax: 01642 580019

All periods including Ancient, plus Science Fiction and Fantasy. Played in all scales and formats: tabletop, board games, role playing. Frequent 'in club' campaigns, and day trips to conventions, paintball, LRP etc. Annual weekend camping trip; annual show. Miniumum age 12; annual membership £2.00.

Stormin Normans
26 South Norwood Hill, London SE25 6AB

Contact: N Scott

Stourbridge District Wargamers
72 Severn Road, Halesowen, Midlands (W.) B63 2NL

Contact: Ashley Hewitt
Tel/fax: 01384 561389

Weekly meetings. Membership fee dependent on age. Most wargames periods catered for, with role-playing and board games.

Tunbridge Wells Wargames Society
Holly Grove, Cranbrook Road, Hawkhurst, Kent TN18 4AS

Contact: George Gush
Tel/fax: 01580 753680

Wargame society active in periods from Ancient to SF. Own club magazine. Meets first Sunday monthly and every Thursday evening at St Thomas' Hall, Vale Avenue, Southborough, Kent.

Ulster Freikorps A
32 Kilmakee Park, Gilnahirk, Belfast, Co.Antrim BT5 7QY, N. Ireland

Contact: S Sandford
Tel/fax: 01232 797766

Ulster Freikorps Levy
16 Marmont Park, Belfast, Co.Antrim BT4 2GR, N. Ireland

Contact: D Taylor
Tel/fax: 01232 760581

Ulster Wargames Society
35 Kilmakee Park, Belfast, Co.Antrim BT5 7QY, N. Ireland

Contact: Jeremy Dowd
Email: jdowd@dis.n-i.nhs.uk

The society specialises in wargaming, catering for most periods and scales, running multi-player games, campaigns and competitions. Meets monthly at Lough Moss Centre, Carryduff.

Victorian Military Society
20 Priory Road, Newbury, Berkshire RG14 7QN

Contact: Dan Allen
Tel/fax: 01635 48628

The leading society in this field promotes research into all aspects of military history 1837-1914, wargames, re-enactment, etc; and publishes a quartery journal.

Virgin Soldiers
75 Woodford Green Road, Hall Green, Birmingham, Midlands (W.)

Contact: M McVeigh
Tel/fax: 0121 778 5582

Warwager
37 Grove Road, Ilkley, Yorkshire (W.) LS29 9PF

Whitehall Warlords
2 Marsham Stree, London SW1P 3EB

Contact: Seamus Bradley
Tel/fax: 0171 276 5586

Historical,Sci-Fi and Fantasy wargaming – miniatures, board games, map games – and role-playing. Annual membership £15.00; good facilities, no table charges; meets each Wednesday.

Wild Geese II
27 Kingsdale Croft, Stretton, Burton on Trent, Staffordshire

Contact: D McHugh

WARGAMING MODELS, EQUIPMENT & SERVICES

1st Corps / Spirit of Wargaming
44 Cheverton Avenue, Withernsea,
Humberside (N.) HU19 2HP

Contact: Rob Baker
Tel/fax: 01964 613766

Progressive design in 25mm,
featuring the American Civil War
with the Mexican American War.
With exciting new figures and
ranges planned for 1995/6.

A B Figures/Wargames South
24 Cricketers Close, Ockley,
Surrey RH5 5BA

Contact: Mike Hickling
Tel/fax: 01306 627796

Manufacturers of the 15mm AB
Figure range for wargamers and
collectors; plus 10mm WWII and
19th century wargames figures.
Mail order only; send SAE for lists.

A J Dumelow
53 Stanton Road, Stapenhill,
Burton-on-Trent, Staffordshire
DE15 9RP

Tel/fax: 01283 30556

Dealer and painter of wargames
figures in all scales. Comprehensive
stock to supply complete armies or
single figures, painted and
unpainted. Mail order welcome.

Adler Miniatures
129 Bonchurch Road, Brighton,
Sussex (E.) BN2 3PJ

Tel/fax: 01273 85279

Quality 1/300th scale figures for
the collector and wargamer,
perfect for dioramas. Napoleonics,
American Civil War and Seven
Years War ranges.

Adventure Worlds
13 Gillingham Street, London SW1

Alternative Armies
Unit 6, Parkway Court, Glaisdale
Parkway, Nottingham,
Nottinghamshire NG8 4GN

Tel: 0115 9287809
Fax: 0115 9287480

Manufacturers of the finest quality
metal Fantasy and Science Fiction
miniatures, 15mm and 25mm.
Available in all good hobby shops.
SAE for mail order information.

Avalon Hill Games
650 High Road, North Finchley,
London N12 0NL

Awful Dragon Management
3 Ransome's Dock, 35-37 Parkgate
Road, London SW11 4NP

Battlements
Sextons, Wymondham Road,
Bunwell, Norfolk NR16 1NB

Contact: James Main
Tel/fax: 01953 789245

We make quality hand made
buildings, fortifications and
terrain with an excellent tradition
of craftsmanship and attention to
historic detail for wargamers,
museums and collectors.

Bicorne Miniatures
40 Church Road, Uppermill,
Saddleworth, Oldham,
Manchester, Gt. OL3 6EL

Contact: Brian Holland
Tel/fax: 01457 870646

Manufacturers of 25mm cast
white metal soldiers and
equipment pieces – Napoleonic
and American Civil War. List and
sample £1.00; mail order service.

Boutique Jeux Descartes
6 Rue Meissonier, 75017 Paris,
France

Britannia Miniatures
33 St. Mary's Road, Halton Village,
Runcorn, Cheshire WA7 2BJ

Contact: David Howitt
Tel/fax: 01928 564906

25mm wargame figures, Ancients,
Colonial, Napoleonic, Crimea,
Pony Wars etc. 20mm Great War,
early period Germans, British and
equipment. Expanding range of
wargame accessories cast in resin.
Mail order only. Callers by
appointment. Non-illustrated
listings available. Large SAE
required.

Chart Hobby Distributors Ltd
Chart House, Station Road, East
Preston, Littlehampton, Sussex
(W.) BN16 3AG

Tel: 01903 773170
Fax: 01903 782152

UK importers and distributors of
Avalon Hill and Victory Games
Inc. Publishers of military
simulation games distributed in
hobby and games stores
nationally.

Collectair
32 West Hemming Street,
Letham, Angus DD8 2PU,
Scotland

Contact: Peter Fergusson
Tel/fax: 01307 818494

Makers of cast pewter scale model
aircraft (140+ models), 1/300 and
1/200 scales, for collectors and
wargamers. List available; trade
enquiries welcome. Available in
USA from Simtac Inc, 15G Colton
Road, East Lyme, CT 06333.

Computer & Games Centre
34 St Nicholas Cliff,
Scarborough, Yorkshire (N.)
YO11 2ES

Conflict Miniatures
27 Leighton Road, Hartley Vale,
Plymouth, Devon PL3 5RT

Contact: Tim Reader
Tel/fax: 01752 770761

Manufacturer of wargaming
figures/models. Worldwide mail
order. Stockist of rules,
Kreigsspiel, paints, brushes etc.
Painting service available. Trade
enquiries welcome.

Connoisseur Figures
27 Sandycombe Road, Kew,
Richmond, Surrey TW9 2EP

Contact: Chris Cornwell
Tel/fax: 0181 940 8156

Specialists in 25mm figures. Full
mail order catalogue available
with all ranges – Napoleonic,
Colonial, Pony Wars, American
Civil War, Renaissance, etc. 20%
discount on all figures bought
from their shop.

Dean Forest Figures
62 Grove Road, Berry Hill,
Coleford, Gloucestershire
GL16 8QX

Contact: Philip & Mark Beveridge
Tel/fax: 01594 836130

Wargame figure painting,
scratch-built trees, buildings and
terrain features. Large scale figure
painting and scratch-building to
any scale. SAE for full lists.

Dixon Miniatures (Aldo Ltd)
Spring Grove Mills, Linthwaite,
Huddersfield, Yorkshire (W.)
HD7 5QG

Contact: T A Dixon
Tel/fax: 01484 846162

Established 1976, Dixon specialize
in design and production of
25mm historical figures. Main
ranges: American & English Civil
Wars, Wild West, Samurai and
Wolfe's Army.

Donnington Miniatures
15 Cromwell Road, Shaw,
Newbury, Berkshire RG14 2HP

Contact: Graham & Maggie Hyland
Tel/fax: 01635 46627

Manufacturer of 15mm
wargaming figures / models.
Worldwide mail order. Ancient,
Medieval, pike & shot and ACW
ranges covered. Figures sold
singly, not in packets.

Drew's Militia
1 Mosslea Road, Bromley, Kent
BR2 9PS

Contact: Andrew Steven
Tel/fax: 0181 460 0728

20mm 1944 British & US Airborne
figures, their equipment and their
German opponents, inc.light
vehicles, artillery etc. Also
exclusive supplier of Helmet
Miniatures 1/200th scale aircraft
for wargames. SAE for catalogue
and sample.

The Drum
107 Watling Street West,
Towcester, Northamptonshire
NN12 7AG

Contact: Michael Green

Stockists of wargames figures and
boardgames, fantasy & sci-fi role
playing games and mail order
service. Manufacturers of resin
buildings and accessories for the
wargaming and role playing
hobbies. Open Mon-Thur
9.30am-5.30pm, Fri & Sat 9.30am –
6.00pm. Send SAE for resin
castings catalogue.

E M A Model Supplies Ltd
58-60 The Centre, Feltham, Middlesex TW13 4B4

Tel: 0181 890 8404

Fax: 0181 890 5321

Europe's largest range of trees including etched brass palms, cacti, deciduous, firs, etc. Also the home of 'Plastruct' plastic shapes – see our latest glowing rods, ideal for fantasy and Sci-Fi projects. Plus moulding materials: Gelflex, Silicone Rubbers, Latex, powders etc. Closed half day Wednesday; mail order catalogue (100 pages) available for £3.00.

Eagle Miniatures
Wild Acre, Minchinhampton, Gloucestershire GL6 9AJ

Contact: David Atkins

Tel/fax: 01453 835782

Design, manufacture and distribute 15mm, 25mm and 54mm figures from Medieval, Seven Years War, Napoleonic and ACW periods. Design and cast service also available. Callers by appointment.

Elite Miniatures
26 Bowlease Gardens, Bessacarr, Doncaster, Yorkshire (S.) DN4 6AP

Contact: Peter Morbey

Tel/fax: 01302 530038

Quality 25mm metal figures for wargamer and collector; Napoleonic, A.C.W., Seven Years and Punic Wars covered. Send £1 and SAE for catalogue. Credit cards accepted.

Ellerburn Armies
Boxtree, Thornton Dale, Pickering, Yorkshire (N.) YO18 7SD

Tel: 01751 474248

Fax: 01751 477298

Manufacturer of 25mm Hinchcliffe figures and equipments. Catalogue available of all 2,000 figures, from Ancient Assyrians to Colonial Zulu Wars. Mail order and painting service.

Empire Francais
22 Swallow Street, Oldham, Lancashire OL8 4LD

Contact: Christopher Durking

We are very interested in the French First Empire. Our main interest is the French 21eme de Ligne 1812-1815. For more information and talks or displays get in touch with us.

English Computer Wargames
Waterside Cottage, Burton in Lonsdale, Carnforth, Yorkshire (N.) LA6 3NA

Contact: David Millward

Tel: 015242 63087

Fax: 015242 61416

Computer moderated wargames rules: 'Over the Hills' (1660-1789); 'Hard Pounding' (1789-1848); 'Blood and Iron' (1830-1904).

Essex Miniatures
Unit 1, Shannon Centre, Shannon Square, Canvey Island, Essex SS8 9UD

Tel: 01268 682309

Fax: 01268 510151

Manufacturer of 15mm & 25mm metal figures. Send SAE for catalogue. Figures supplied by mail order. All major credit cards accepted.

F J Associates
60 Frederica Road, Winton, Bournemouth, Dorset BH9 2NA

Tel/fax: 01202 511495

Company dealing in wargame insurance.

Figures-Armour-Artillery
17 Oakfield Drive, Upton Heath, Chester, Cheshire CH2 1LG

Tel/fax: 01244 379 399

Highest quality, detailed research, accurate animation – all go to make our 20mm figures amongst the best available for wargamers and modellers. Commissions undertaken. Visitors by appointment.

Frei-Korps 15
25 Princetown Road, Bangor, Co.Down BT20 3TA, N. Ireland

Contact: Antonnio Matassa

Tel/fax: 01247 472860

Large range of 15mm scale wargame figures. Ancient, Thirty Years War, Seven Years War, American Wars, British in India, Europe 1850-1870, 1914-18.

Front Rank Figurines
The Granary, Banbury Road, L.Boddington, Daventry, Northamptonshire NN11 6XY

Contact: Alec Brown

Tel: 01327 262720

Fax: 01327 260569

Manufacturers of 25mm figures.

G J M Figurines
74 Crofton Road, Orpington, Kent BR6 8HY

Contact: Gerard Cronin

Tel/fax: 01689 828474

Wargame figures painted to collectors' standard, 10mm to 30mm. For 15mm/25mm samples send £2.95 or £3.95 respectively. Wargame armies bought/sold. Worldwide mail guaranteed.

Gallia UK / Total System Scenic
PO Box 51E, Worcester Park, Surrey KT4 8NQ

Tel/fax: 0181 330 0239

Manufacturers and suppliers of wargaming scenics, polystyrene terrain modules and scale model trees. Send SAE for free catalogue.

Gamers in Exile
283 Pentonville Road, London N1 9NP

Games
63 Allerton Road, Liverpool, Merseyside L18

Games Corner
76 High Street, Watton, Norfolk IP25 6AH

Contact: Eamon Bloomfield

Tel/fax: 01953 883007

Wargames, fantasy games, general board games. Second-hand listing available by subscription. Trading card games specialists including huge single card inventory.

The Games Room
29A Elm Hill, Norwich, Norfolk

Tel/fax: 01603 628140

The Guardroom
38 West Street, Dunstable, Bedfordshire LU6 1TA

Tel/fax: 01582 606041

Wargames shop – wide range of figures and rules. Open Monday to Friday 9am-5.30pm, Saturday 9am-5pm. Two miles from Junction 11 on the M1.

Hovels Ltd
18 Glebe Road, Scartho, Grimsby, Humberside (S.) DN33 2HL

Contact: Dennis Coleman

Tel/fax: 01472 750552

Ian Weekley Models
The Old Anchor of Hope, Lammas, Norwich, Norfolk NR10 5AF

Contact: Ian Weekley

Tel: 01603 279708

Fax: 01603 278264

Designers of superb quality, well researched architectural features for wargamers, in the usual scales. Model buildings are cast in high definition, light weight urethane resin.

The Iron Duke
Edgehill Cottage, Ropeyard, Wotton Bassett, Wiltshire SN4 7BW

Contact: Ian Barstow

Tel/fax: 01793 850805

Professional wargames figure painter and dealer in second-hand painted figures and armies, all scales. Guaranteed no sub-contracting. All major credit cards accepted.

J G Diffusion
6 Rue Meissonier, 75017 Paris, France

Jeux d'Histoire REK
123 Rue des Mesanges, BP 21, 62155 Merlimont, France

John Mitchell (Peter Laing Figures)
19 Hook Bank Park, Hanley Castle, Hereford & Worcester WR8 0BQ

Tel/fax: 01684 593980

Keep Wargaming
The Keep, Le Marchant Barracks, London Road, Devizes, Wiltshire SN10 2ER

Contact: Paul & Teresa Bailey

Tel/fax: 01380 724558

Shop and mail order service; stockists of wargames figures, books and equipment; some military models and plastic kits. Shop open Tues-Sat 10am to 6pm.

L K M
75 High Road, North Weald, Essex CM16 6HW

Contact: Leslie Mills

Tel/fax: 01992 523130

We provide wargames terrain for any scale of wargames figurers. We are also able to manufacture figures and moulds. Send SAE (A5) for list.

Lancashire Games
20 Platting Road, Lydgate,
Oldham, Manchester, Gt. OL4
4DL

Contact: Allan Lumley
Tel/fax: 01457 872212

15mm castings from Ancient to
Colonial, 25mm Fantasy and
20mm WWII/I, available in
standard and bargain packs.
Painting service available for
figures from 5mm-25mm.

The Land of Gondal
76 Main Street, Haworth,
Yorkshire (W.)

Tel/fax: 01535 44924

Leisure Games
91 Ballard's Lane, Finchley,
London N3 1XY

Tel: 0181 346 2327
Fax: 0181 343 3888

We stock the most comprehensive
range of board war games in
London. Plus 'Minifigs' and a
wide range of rule books, lists and
Osprey publications.

London Wargames Depot
56 Beaumont Place, Mogden
Lane, Isleworth, Middlesex
TW7 7LH

Maltby Hobby Centre
19 Morrell Street, Maltby,
Rotherham, Yorkshire (S.)
S66 7LL

Contact: Richard Moss
Tel/fax: 01709 798287

A wide range of plastic kits and
accessories always in stock. Also
stocked, Fantasy and Sci-Fi games,
model railways and Scalextric.
Please phone for details.

Matchlock Miniatures
816-818 London Road,
Leigh-on-Sea, Essex SS9 3NH

Contact: David Ryan
Tel/fax: 01702 73986

Wargames figures, all periods
from Ancient to Modern. Shop
open six days a week. Send five
1st Class stamps for sample.

Maxart
90 Maple Avenue, Gillingham,
Kent ME7 2NT

Contact: David Mackerracher
Tel/fax: 01634 851939

Medway Games Centre
294-6 High Street, Chatham,
Kent

Tel/fax: 01634 847809

Military Motors
16 Coolhurst Lane, Horsham,
Sussex (W.) RH13 6DH

British World War II vehicles in
white metal, 1/7600 scale, hand
made and painted for display or
wargaming; SAE for catalogue.

Mini-Militaria
12 Ascott Road, Aylesbury
HP20 1HX

Contact: Graham Harrison
Tel/fax: 01296 23118

Second-hand painted and
unpainted wargames figures. All
scales and periods. Regular mail
order lists; send SAE for copy.

Mitregames
77 Burntwood Grange,
Wandsworth Common, London
SW18

Model Figures & Hobbies
4 Lower Balloo Road,
Groomsport, Co.Down BT19
2LU, N. Ireland

Contact: Norman Robinson
Tel: 01247 883187
Fax: 01247 472860

Manufacturer of Platoon 20 WWII
and modern wargame figures.
Also Ensign 1/1200 scale model
ships and aircraft and the
Plastiform scenics range of
Vacforms.

Museum Miniatures
17 Hilderthorpe Road,
Bridlington, Yorkshire (E.) YO15
3AY

Contact: Dave Hoyles
Tel/fax: 01262 670421
Email: museumminiatures
@btinternet

15mm wargaming figures cast in
English pewter, for maximum
definition and environmental
safety (no lead). Ranges include
Ancient, Dark Ages, Medieval,
Renaissance, ACW, Napoleonic.
Equipment includes wagons,
carts, cannons, bombards,
catapults. Mail order service a
speciality with a 7-day
turn-round. Callers welcome but
are advised to ring first.

N B R Games Studios
(Wargaming Specialists), 62
Dickens Avenue, Corsham,
Wiltshire SN13 0AQ

Tel/fax: 01249 715925

Navwar Productions Ltd
11 Electric Parade, Seven Kings,
Ilford, Essex IG3 8BY

Tel/fax: 0181 590 6731

Manufacturers of 1/3000 and
1/1200 scale ships, Roundway
and Naismith 15mm figures;
1/300 figures, tanks & aircraft.
Closed Thursdays.

Newbury Rules
Church Cottage, Church Lane,
Beenham, Reading, Berkshire
RG7 5NN

Contact: Trevor Halsall
Tel/fax: 01189 714473

Oxford Games
6 Harper Road, Summertown,
Oxford, Oxfordshire OX2 7LQ

The Paint Box
Southford Engineering Services,
31 Ely Close, Southminster,
Essex CM0 7AQ

Peter Pig
Maebee, 36 Knightsdale Road,
Weymouth, Dorset DT4 0HS

Tel/fax: 01305 760384

Makers of 15mm metal figures; 17
ranges including WWII, ACW,
SCW, Ancients, Renaissance,
ACW Naval, Sci-Fi, Vietnam.
Send SAE for list and sample.

Picturesque
25/27 Tufton Street, Ashford,
Kent TM23 1QN

Tel: 01233 641682
Fax: 01233 630293

Display cases made to order;
enquiries to address above.

Pireme Publishing Ltd
34 Chatsworth Road,
Bournemouth, Dorset BH8 8SW

Contact: Iain Dickie
Tel: 01202 773490
Fax: 01202 512355

Publishers of 'Miniature
Wargames' magazine – covers all
periods of history and all theatres
of conflict. Inexpensive military
prints, booklets and 15mm
Waterloo card model buildings,
also available.

Raventhorpe Miniatures
2 Bygot Lane, Cherry Burton,
Yorkshire (E.) HU17 7RN

Tel/fax: 01964 551027

20mm, 1/76th scale, figures and
vehicles. Many unique types:
Zulu, Boer, Mexican Revolution,
WWI, WWII, etc. SAE and 2x1st
class stamps for lists.

Richard Newth-Gibbs
Painting Services
59 Victor Close, Hornchurch,
Essex RM12 4XH

Tel/fax: 01708 448785

Any scale from 15mm up, single
figures, groups, dioramas,
artillery pieces, mounted gun
teams; British and Indian Army
specialist. Factual military work
only. Callers by appointment.

Schematica Software
62 Bankbottom, Hadfield, via
Hyde, Cheshire SK15 8BY

Contact: B Pick
Tel/fax: 01457 869452

Wargames rules on disc, for all
periods from Ancients to Future
Wars. SAE for details; campaign
discs available, most discs under
£20.00.

Scotia Micro Models
32 West Hemming Street,
Letham, Angus DD8 2PU,
Scotland

Contact: Robert Fergusson
Tel: 01307 818707
Fax: 01307 818494

Makers of white metal castings for
wargaming and collectors.
Figures, AFVs, landing craft etc. in
1/300, 10mm, 15mm, 20mm,
25mm. Also Sci-Fi figures in 1/300
and 25mm (Acropolis series) and
Kryomek fantasy figures. Contract
casting and modelling
undertaken. Trade enquiries
welcome. Catalogues available.
Also produced in USA by Simtac
Inc., 15G Colton Road, East Lyme,
CT 06333.

Second Chance Games
62 Earlston Road, Wallasey,
Merseyside

Societé Guerre & Plomb
BP 4031, 69615 Villeurbanne
Cedex, France

Spirit Games
98 Station Street, Burton on
Trent, Staffordshire DE14 1BT

Tel/fax: 01283 511293

5mm, 15mm and 25mm figures,
buildings and rules stocked. Open
Tues-Fri 10am-6pm, Sat
10am-5pm. Mail order welcome.

Spot On Models & Hobbies
43 Havelock Street, Swindon,
Wiltshire SN1 1SD

Tel/fax: 01793 617302

Just off the Brunel Centre –
stockists of fantasy role-play
games, dice, paint, brushes,
modelling accessories etc. Open
Mon-Sat, 9.15am-5.30pm.

Tabletop Games
29 Beresford Avenue, Skegness,
Lincolnshire PE25 3JF

Contact: Robert Connor
Tel/fax: 01754 767779

Established in 1976, we are a
specialist mail order company
serving the wargamer throughout
the world. Please send an SAE (2x
IRCs) for our free catalogue.

Terence Wise
Pantiles, Garth lane, Knighton,
Powys LD7 1HH, Wales

Wargames figures: mint castings
and painted figures, all at
competitive prices. Wide range of
manufacturers, mostly 25mm,
some 54mm, 20mm, 15mm. Free
list from Terry Wise at above
address.

Two Dragons Productions
18 Lipscomb Street, Milnsbridge,
Huddersfield, Yorkshire (W.)
HD3 4PF

Contact: Tim Hallam
Tel/fax: 01484 643374

15mm figure manufacturer;
ranges include Samurai, Vikings,
Saxons, Rus, Normans, British
Napoleonic,and Colonial. Stockist
of paints, brushes, range of related
books etc. Mail order service.

U N I T S Wargames Services
40 Cranbrook Street, Barnsley,
Yorkshire (S.) S70 6LP

Tel/fax: 01226 295180

Vandrad
7 Marpool Hill, Exmouth, Devon
EX8 2LJ

Contact: Rick Lawrence
Tel/fax: 01395 278664

20mm figures from 20:20
miniatures, Bataillonfeuer
miniatures & RSM miniatures.
Painting service available. Various
periods available.

Wargames Figures
9 Wargrave Road, Twyford,
Berkshire

Wargames Foundry
4 Victoria Avenue, Norton,
Stockton on Tees, Cleveland
TS20 2QB

Tel/fax: 01642 553787

Designers and manufacturers of
high quality, 25mm white metal
historically accurate miniatures
for collecting, diorama building
and wargaming. Catalogue £2.50
post paid.

Wargames Holiday Centre
Enchanted Cottage, Folkton,
Scarborough, Yorkshire (N.)
YO11 3UH

Contact: Mike Ingham
Tel/fax: 01723 890580

Warlord Games Shop
818 London Road, Leigh on Sea,
Southend, Essex SS9 1NQ

Contact: David Ryan
Tel/fax: 01702 73986

Retail outlet holding wide range
of wargames figures, board
games, rules, books etc. Open 7
days a week, one hour's
drive/train from London.

Warpaint Figure Painting
20 Swaledale Crescent, Barnwell,
Houghton le Spring, Tyne &
Wear DH4 7NT

Tel/fax: 0191 385 7070

Warrior Miniatures
14 Tiverton Avenue, Glasgow,
Strathclyde G32 9NX, Scotland

Contact: John Holt
Tel/fax: 0141 778 3426

Over 1,600 wargaming figures,
many periods, 10mm, 15mm,
20mm, 25mm. Catalogue £1.50, or
overseas $5.00.

Weston Scorpions
5 Shiplate Road, Bleadon Village,
Weston super Mare, Avon
BS24 0NG

Contact: R Aynsley

WEAPONS CLUBS & SOCIETIES

WEAPONS SUPPLIES

Historical Breechloading Smallarms Assoc
P.O. Box 12778, London
SE1 6XB

Contact: David Penn
Tel: 0171 416 5270
Fax: 0171 416 5374

For serious students or collectors. Publishes an annual journal and approx. four newsletters a year. Regular monthly meetings in London. Range practices April-October. Active in monitoring legislation affecting smallarms and ammunition. Occasional national or international symposia. Corresponding membership available for non-UK residents.

National Museum of Military History
10 Bamertal, PO Box 104,
L-9209 Diekirch, Luxembourg

Contact: Roland Gaul
Tel: (00352) 808908
Fax: (00352) 804719
Email: mnhmdiek@pt.lu

Important collections from Battle of the Bulge 1944-45, life-size dioramas, uniforms, vehicles, weapons, equipment. Open from January 1 – March 31: daily 14–18h hrs; April 1 – November 1: daily 10–18 hrs; November 2 – December 31: daily 14–18 hrs; last ticket sold 17.15 hrs.

National Pistol Association (NPA)
21 Letchworth Gate Centre, Protea Way, Pixmore Avenue, Letchworth, Hertfordshire
SG6 1JT

Tel: 01462 679887
Fax: 01462 481183

The NPA promotes the sport of pistol shooting in the UK. It organises shooting meetings, sells merchandise, advises members on shooting problems, etc.

National Smallbore Rifle Assoc (NSRA)
Lord Roberts House, Bisley Camp, Brookwood, Woking, Surrey GU24 ONP

Contact: J D Hoare
Tel: 01483 476969
Fax: 01483 476392
Email: NSRA@dial.pipex.com

National governing body for small-bore, airgun and cross-bow shooting. For details of local clubs please write enclosing SAE.

Outdoor People
PO Box 41, Hastings, Sussex (E.)
TN34 3UN

R R T
26 Lomas Drive, Northfield, Birmingham, Midlands (W.)
B31 5LR

Contact: Ray Thorne
Tel/fax: 0121 476 9078

Specialised metallic/shotgun cartridge display cases. Collections made up. Trade supplied. Send SAE for details. Personal callers by appointment only.

Shooters Rights Association
PO Box 3, Cardigan, Dyfed
SA43 1BN, Wales

Contact: Richard Law
Tel: 01239 698607
Fax: 01239 698614

Membership organisation for collectors and shooters specialising in dealing with legal and licensing problems. Members are insured for public liability and legal costs for £24.

Verband f. Waffentechnik u. Geschichte
Klever Strasse 80, 40477 Dusseldorf, Germany
Tel/fax: (0049) 211 464844

A C S
92 Paddock Road, Kirkburton, Huddersfield, Yorkshire (W.)
HD8 OTW

Tel/fax: 01484 606195

A S Bottomley
The Coach House, Huddersfield Road, Holmfirth, Yorkshire (W.)
HD7 2TT

Contact: Andrew Bottomley
Tel: 01484 685234
Fax: 01484 681551

Established 30 years with clients overseas and in the UK. A fully illustrated mail order catalogue containing a large range of antique weapons and military items despatched world wide. Every item is guaranteed original. Full money back if not satisfied. Deactivated weapons available. Valuations for insurance and probate. Interested in buying weapons or taking items in part exchange. Business hours Mon-Fri 9am – 5pm. Mail order only. All major credit cards welcome. Catalogue UK £5, Euorpe £7, rest of world £10.

Air Gunning Today International
2 Gleave Street, Standish Street, St Helens, Merseyside
WA10 1AU

Tel/fax: 01744 50446

Al-Arms
(RFD Gwent 57), 42 Somerton Road, Newport, Gwent
NP9 7XY, Wales
Tel/fax: 01633 279679

Alan Beadle Antique Arms
320 Upper Street, Islington, London N1

Alanbray Guns & Tackle
5 Waterloo Road, Hinckley, Leicestershire LE10 OQJ

Contact: I M & K R Stevens
Tel/fax: 01455 634317

Albion Small Arms
Unit 4, AML Industrial Estate, Rugeley Road, Hednesford, Staffordshire WS12 5QW

Contact: Ron Curley
Tel/fax: 01543 426113

Alexandra Guns Ltd
54 Alexandra Street, Southend-on-Sea, Essex SS1 1BJ
Tel/fax: 01702 339349

Anne Ford
Maytree House, Woodrow Lane, Bromsgrove, Hereford & Worcester B61 0PL

Contact: Alan Ford
Tel/fax: 0121 453 6329

Suppliers to museums, gift and surplus shops – inert ammunition, all calibres. Bullet keyrings, souvenirs, replica grenades, de-activated weapons, MG belts, etc. Trade and retail lists available.

Antique Vintage Arms
Charmouth Antique Centre, The Street, Charmouth, Dorset DT6 6QH
Tel/fax: 01297 60122/442875

Antiques
Main Street, Durham-on-Trent, Nottinghamshire NG22 0TY

Contact: R G Barnett
Tel/fax: 01777 228312

Antique flintlock and percussion pistols and long arms, swords, daggers and armour our speciality. Also buy and sell general antiques, furniture, etc.

Apollo Firearms Company Ltd
PO Box 404, St Albans, Hertfordshire AH4 9YR

Tel: 0171 739 1616
Fax: 0171 739 6463

Arian Trading
1 The Monkery, Church Road, Great Milton, Oxfordshire OX44 7PB

Tel: 01844 278139
Fax: 01844 278790

Fine quality classic sporting and military firearms of all kinds.

Armstrong's Gunsmiths
360 Carlton Hill, Carlton, Nottinghamshire NG4 1JB
Tel/fax: 0115 9873313

Asgard Armoury
Asgard, 46 Haverhill Road, Stapleford, Cambridgeshire CB2 5BX

Tel/fax: 01223 842926
Email: 101636,3702
@compuserve.com

Certified hand-crafted replica swords and daggers of carbon steel, hardened and tempered in accordance with tests of the Department of Metallurgy, Cambridge University.

Atkin Grant & Lang & Co
6 Lincoln's Inn Fields, London WC2

Tel/fax: 01707 42622

Aux Armes d'Antan
1 Avenue Paul Deroulede, 75015 Paris, France

Contact: Maryse Raso
Tel: (0033) 1 47837142
Fax: (0033) 1 47344099

Expert and dealer in sabres, swords and ancient pistols and other weapons from the XIIth to the XIXth century, for 20 years. Publishes a catalogue every 3 months, available by post from the office. All pieces are sold along with a guarantee.

Barnett International
Dept GM, Dock Meadow Drive, Lanesfield, Wolverhampton, Midlands (W.) WV4 6UD

Tel/fax: 01902 405121

Bartrop's Shooting and Fishing
206 Knutsford Road, Warrington, Cheshire WA4 1AU

Contact: Peter H Bartrop
Tel/fax: 01925 572509

Battle Orders Ltd
71A Eastbourne Road, Lower Willingdon, Eastbourne, Sussex (E.) BN20 9NR

Contact: Graham Barton
Tel: 01323 485182
Fax: 01323 487309

We specialise in Airsoft pistols, metal replica knives, Japanese and European historic swords. Find us on the A22 just two miles north of Eastbourne.

Battle Orders Scotland Ltd
76 Coburg Street, Leith, Edinburgh, Lothian EH6 6HJ, Scotland

Contact: Angus Neilson
Tel: 0131 538 8383
Fax: 0131 555 2071

Suppliers of replica/re-enactment arms and armour; medieval, Scottish Highlander and 20th century weaponry a speciality for stage/re-enactment; instructional videos.

Benjamin Wild & Son
Price Street Works, Price Street, Birmingham, Midlands (W.) B4 6JZ

Tel/fax: 0121 359 2303

Bill Gent
173 Westborough Road, Westcliff-on-Sea, Essex SSO 9JD

Tel/fax: 01702 34150

Dealer in a wide range of military and aviation collectables. Occasional lists available – send long SAE. Manufacturer of replica weapons to order – .50 Browning, Vickers K, mortars, etc. – if the genuine item is difficult or expensive to get, ask us.

Blade & Bayonet
884 Christchurch Rd, Boscombe, Bournemouth, Dorset

Tel/fax: 01202 429891

Bond & Bywater
42 Fylde Street, Preston, Lancashire

Tel/fax: 01772 58980

Bryant & Gwynn Antiques
8 Drayton Lane, Drayton Bassett, Staffordshire B78 3TZ

Contact: David Bryant
Tel/fax: 0121 378 4745

All types of military uniforms, medals, swords and militaria supplied. Also deactivated military arms and muskets. Items not stocked can be ordered and traced.

C F Seidler
Stand G12, Grays Antique Mkt, 1-7 Davies Mews, Davies Street, London W1V 1AR

Contact: Christopher Seidler
Tel/fax: 0171 629 2851

American, British, European, Oriental edged weapons, antique firearms, orders and decorations, uniform items; watercolours, prints; regimental histories, army lists; horse furniture; etc. We purchase at competitive prices and will sell on a consignment or commission basis. Valuations for probate and insurance. Does not issue a catalogue but will gladly receive clients' wants lists. Open Mon-Fri, 11.00am-6.00pm. Nearest tube station Bond Street (Central line).

C Pell Antiques & Militaria
PO Box 3584, London N3 1RP

Tel: 0181 343 1225
Fax: 0181 343 3883

Carry Arms
39 St Davids Road, Allhallows, Rochester, Kent ME3 9PW

Tel/fax: 01634 271340

Suppliers of Black Powder arms and accroutrements to Re-enactment societies and black powder users. Lock tuning, frizzen hardening service, gun bags, flints, cleaning kits.

Castle Armoury
London Road, Stretton-on-Dunsmore, Rugby, Warwickshire CV23 9HX

Castle Keep (Weaponsmith)
Viginish Lodge, Dunvegan, Isle of Skye IV55 8ZR, Scotland

Contact: Rob Miller
Tel/fax: 01478 612114

Hand forged swords, knives, daggers and dirks. Damascus steels, traditional Scottish weaponry. Catalogue available, £1.

Central Antique Arms & Militaria
7 Smith Street, Warwick, Warwickshire CV34 4JA

Contact: Chris James
Tel/fax: 01926 400554

Buys and sells antique guns, swords, bayonets, helmets, badges, British and German medals and decorations, original Third Reich militaria and documents. Send 10 x 1st class stamps for illustrated catalogue. Shop open Mon-Sat, 10.00am-5.00pm. Exhibits at most major fairs.

Centrevine Ltd
76 Stevensons Way, Formby Business Park, Formby, Merseyside

Tel: 017048 70598
Fax: 017048 71613

Chris Blythman
The Flat, Brook House Farm, Middleton, Ludlow, Shropshire SY8 2DZ

Tel/fax: 01584 878591

Quality hand forged military and domestic ironwork, for museums, re-enactment, TV and film.

Coach Harness
Haughley, Stowmarket, Suffolk IP14 3NS

Tel/fax: 01449 673258

Mail order specialists; new books on antique and early guns and Western artifacts. Suppliers of restoration parts for flintlock and percussion guns. Gunsmithing service available.

David Partridge Enterprises
Gable End, Pitchcombe, Stroud, Gloucestershire GL6 6LN

Tel: 01452 812166
Fax: 01452 812719

Antique vintage and modern firearms and accessories.

Derbyshire Shooting Centre
Gella Mills, Bonsall, Matlock,
Derbyshire
Tel/fax: 01629 824580

Dragon Forge
Y Bryn, Llaneilian Road, Amlwch,
Anglesey LL68 9HU, Wales

Contact: T K M Craddock
Tel/fax: 01407 831076

Manufacture of fully researched
weaponry, iron work and
furniture for re-enactment or the
collector. Sword blades, spring
steel, predominantly 9th / 10th
century catered for.

Dunmore of Abingdon Ltd
Wootton Road, Abingdon,
Oxfordshire OX13 6BH

Tel: 01235 520168
Fax: 01235 555738

Classic guns, pistols and rifles;
visit the largest gun shop and
shooting centre in the country.

Eagle Classic Archery
41 Spring Walk, Worksop
S80 1XQ

Contact: Roy Simpson
Tel: 01909 478935
Fax: 01909 488115

Longbows & replica Mongol
bows, any poundage; arrows and
replica Medieval arrow heads.
Telephone (01909) 478935, Fax
(01909) 488115 anytime.

East Coast Services
169 High Road West, Felixstowe,
Suffolk IP11 9BD

Tel/fax: 01394 271685

Edgar Brothers
Catherine Street, Macclesfield,
Cheshire SK11 6SG

Tel: 01625 613177
Fax: 01625 615276

Federal Arms
PO Box 191, Southport,
Merseyside PR8 5AD

Period and classic Bowie knives
for the collector, dealer and
enthusiast. Send £5 and large SAE
for colour catalogue.

Fieldsport Equipe
20A Elwy Street, Rhyl, Clwyd,
Wales

Tel/fax: 01745 353476

Les Frères d'Armes
16 Rue Lakanal, 38000
Grenobles, France

Gaunt d'Or (Weaponsmith)
58 Springfield Road,
Wolverhampton, Midlands (W.)
WV10 0LJ

Contact: Brian Gunter
Tel/fax: 01902 683875
Email: 106135.2771
@compuserve.com

For authentic reproduction
swords, daggers and other
weapons from the Dark Ages
through to Renaissance. Suitable
for re-enactment, display or wall
hangings.

**Hartmut Burger Antique
Firearms GmbH**
Moerser Strasse 106, 47803
Krefeld, Germany

Henry Krank
100/102 Lowton, Pudsey,
Yorkshire (W.) LS28 9AY

Tel: 0113 2569163
Fax: 0113 2574962

Historical Uniforms Research
Unit 3, Enterprise Centre,
Emmet Road, Ballymote,
Co.Sligo, Ireland

Contact: John Durant
Tel: (00353) 71 83930
Fax: (00353) 71 82404

Manufactures and supplies
world's best reproduction
Waffen-SS reversible camouflage
items to museums, collectors,
re-enactors and film industry.
Smocks, helmet covers,
non-standard field tailored items.
All camouflages and construction
exact to originals. Examples have
been mistakenly judged genuine
by collectors and curators. Can
also supply genuine WSS items
ex-SS 'Skanderbeg' Division; also
1941-1960 Russian camouflages
and other kit used by Albanian
forces. Specialist sourcing, to
order, of Russian WWII weapons,
personal kit, communications
equipment, documents, manuals
etc. Will print to order any of over
200 camouflage patterns;
minimum 1000 metres. Dealer
enquiries welcome. Can also
arrange military interest tours to
Albania. Accept VISA, ACCESS,
MASTERCARD, EUROCARD.

Le Hussard
BP 69, 38353 La Tour du Pin
Cedex, France

Contact: J Jacques Buigné

Old weapons. Firearms from 18th
and 19th centuries, edged
weapons: sabres, bayonets,
halberds, swords, from all ages.
Illustrated free catalog on request
(in French).

Imperial Arms Company
PO Box 883, Selly Oak,
Birmingham, Midlands (W.) B29
5TT

Tel/fax: 0121 478 3518

Purveyors of fine classic pistols,
bullets and accessories for
sporting gentlemen and their
ladies.

Jack Greene Longbows
Oldwood Pits, Tanhouse Lane,
Yate, Bristol BS17 5PZ

Contact: Jack Greene
Tel/fax: 01454 227164

Utility English longbows in
degame/hickory or yew. Varied
for different periods, e.g. plain or
horn nocks etc. Linen strings.
Visitors welcome by appointment.

Jeremy Tenniswood
PO Box 73, Aldershot,
Hampshire GU11 1UJ

Contact: Jeremy Tenniswood
Tel: 01252 319791
Fax: 01252 342339
Email: 100307,1735@compuserve

Established 1966, dealing in
collectable firearms civil and
military, de-activated and for
shooters; also swords, bayonets,
medals, badges, insignia, buttons,
headdress, ethnographica; and
books. Regular lists of Firearms
and Accessories; Medals; Edged
Weapons; Headdress, Headdress
Badges and Insignia;
comprehensive lists Specialist and
Technical Books. Office open
9am-5pm, closed all day Sunday.
Medal mounting service.

Liverpool Militaria
48 Manchester Street, Liverpool
L1 6ER

Contact: Bill Tagg
Tel/fax: 0151 236 4404

Buying and selling general militaria
medals, badges, bayonets, swords,
antique guns and helmets.
Especially medals to King's
Liverpool Regiment 1st and 2nd
War gallantry and casualties.

Lock Stock & Barrel
19 Stamford Street, Mossley,
Lancashire OL5 0LL

Tel/fax: 01457 834564

Manton International Arms
140 Bromsgrove Street,
Birmingham, Midlands (W.) B5
6RG

Tel: 0121 666 6066
Fax: 0121 622 5002

We sell a wide range of
de-activated weapons, swords
and bayonets. Export and retail
enquiries welcomed. Shop open
Mon-Fri 9am- 5pm; Sat 10am-4pm.

Marché Serpette
Allee 1, Stand 8, 110 Rue des Rosiers, 93400 Saint Ouen, France

Contact: Jean-Pierre Maury

Mars
5 Rue Jean-Marie Michel, 69410 Champagne au Mont d'Or, France

McAvoy Guns
3 High Street, Standish, Wigan, Lancashire

Tel: 01257 426129
Fax: 01257 472248

Sales and repairs, reloading equipment and supplies. Rifle scopes, binoculars and night vision. Good range of clothing, wellingtons and dog training aids.

Military Antiques
Shop 3, Phelps Cottage, 357 Upper Street, London N1 0PD

Contact: R Tredwin
Tel/fax: 0171 359 2224

In the heart of Camden Passage, long-established dealers in WWI & WWII uniforms, equipment, headgear, awards, edged weapons. All items original and carry a full money-back cover. Illustrated catalogue; wants lists welcome. Open Tues.-Fri., 11am-5pm; Sat, 10am-5pm; if travelling far, advisable to phone first.

Musket, Fife & Drum
27 Church Road Business Centre, Sittingbourne, Kent ME10 3RS

Contact: Phil Bleazey
Tel: 01795 470149
Fax: 01795 474109

Manufacturers of black powder weapons – 17th Century muskets, and earlier pieces (made to order).

Pemberton Sporting Services
23 A J Cook Terrace, Shotton Colliery, Co.Durham DH6 2PR

Contact: P Lambert
Tel/fax: 0191 526 6225

Also HQ of North Eastern Target & Sporting Association (NETSA).

Richard Wells Sporting Guns
The Gunshop, Haslemere, Surrey

Tel/fax: 01428 651913

Rod & Gun Shop
18 Church Street, Woodbridge, Suffolk

Tel/fax: 01394 32377

Rod Akeroyd & Son
101-103 New Hall Lane, Preston, Lancashire PR1 5PB

Tel: 01772 794947
Fax: 01772 654535

Extensive stock of British, Continental, and Japanese guns, swords, helmets and armour. Callers welcome.

Roding Amoury
Silver Street, Market Place, Abridge, nr.Romford, Essex

Tel/fax: 01992 813570/813005

Roy King Historical Reproductions
Sussex Farm Museum, Manor Farm, Horam, nr. Heathfield, Sussex (E.) TN21 0JB

Tel/fax: 01435 33733

Reproduction armour, helmets, edged weapons and some associated goods. Theatrical and film props, cannons, siege engines etc. Pyrotechnic service for simulated battles available. Mail order, limited catalogue available – most items made to order. Visitors by appointment only. Surrounding fields and Farm Museum available for location hire.

Ryton Arms
PO Box 7, Retford, Nottinghamshire DN22 7XH

Tel: 01777 860222
Fax: 01777 711297

Wide range of de-activated military weapons, including machine guns etc., complete with inert ammunition, accessories, etc. for impressive displays.

Sabre Sales
85-87 Castle Road, Southsea, Portsmouth, Hampshire PO5 3AY

Contact: Nick Hall
Tel: 01705 833394
Fax: 01705 837394

Open six days a week near D-Day Museum; we have an extensive stock at our shops and warehouse, specialising in WWII memorabilia, and all collectables.

Service Arms (UK)
PO Box 21, Llandeilo, Carmarthenshire SA19 6XE, Wales

Contact: Brian C Knapp
Tel/fax: 01558 823610

Leading specialists in 18th and 19th century military firearms. Send SAE for free catalogue.

Sherwood Armoury
34 Lambs Walk, Whitstable, Kent CT5 4PJ

Contact: David Lee
Tel/fax: 01227 262217

Sidelock Guns
125 Long Street, Atherstone, Warwickshire

Tel/fax: 01827 712903

Soldier of Fortune
Unit 18, Tyn y Llidiart Ind Estate, Corwen, Clwyd LL21 9RR, Wales

Contact: Peter Kabluczenko
Tel: 01490 412225
Fax: 01490 412205

Specialising in reproduction 2nd WW clothing and equipment. British, German and the USA. Catalogue available.

Stephen Gane
Stonegarth, Church Street, Broughton in Furness, Cumbria LA20 6HJ

Tel/fax: 01229 716880

Buy, sell and exchange antique arms, armour and militaria; specialists in rare and interesting bayonets; send large SAE for current Bayonet catalogue.

Tasco Sporting Goods
2 Tewin Court, Tewin Road, Welwyn Garden City, Hertfordshire AL7 1AU

Tel: 01707 376660
Fax: 01707 373013

Terry Abrams
3 Ongar Road, Margaret Roding, Dunmow, Essex CM6 1QP

Tel/fax: 0124 531753

Dealer in de-activated firearms including vintage military handguns. Parts available for modern and vintage weapons including magazines and grips. Callers by appointment only.

Trenchart
PO Box 3887, Bromsgrove, Hereford & Worcester B61 0NL

Tel/fax: 0121 453 6329

Suppliers to museums, gift and surplus shops. Inert ammunition, all calibres; bullet keyrings, dogtags, souvenirs, replica grenades, de-activated weapons, MG belts and clips, large cartridge cases, etc. Trade only supplied.

Wyte Stone Metalcrafts
2 Laggotts Close, Hinton Waldrist, Oxfordshire SN7 8RY

Contact: Paul Ellison

Bronze Age to ECW: jewellery, costume accessories, esoterica and other properties to your requirements. Film, stage and fantasy. Illustrated catalogue available.

Yakusa
48 Manchester Street, Liverpool L1 6ER

Contact: Bill Tag
Tel/fax: 0151 236 4404

Buys and sells antique and WWII Japanese swords. Also armour and Tsuba repair and re-polishing service. Part exchange and deals considered. Closed Wednesday–Sunday.

INDEX BY ORGANISATION NAME

INDEX BY COUNTY/COUNTRY

Entries are broken down by county or other area (UK) or country (overseas).

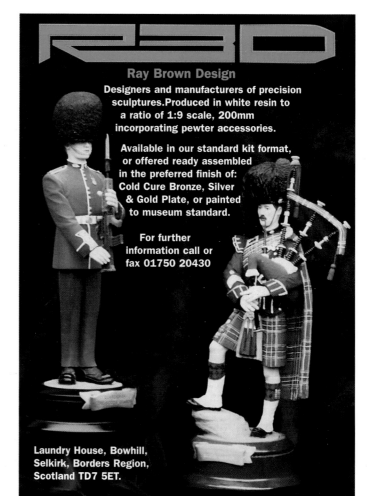